Numerical Methods
for Physics

Numerical Methods
for Physics

ALEJANDRO L. GARCIA

San Jose State University

 Prentice Hall, Englewood Cliffs, New Jersey 07632

Library of Congress Cataloging-in-Publication Data

Garcia, Alejandro L., (date)
    Numerical methods for physics / Alejandro L. Garcia.
        p.   cm.
    Includes index.
    ISBN 0-13-151986-7
    1. Mathematical physics.   2. Differential equations, Partial—
Numerical solutions.   3. Physics.   4. Numerical calculations.
I. Title.
    QC20.G37   1994   93-46708
    530.1'5—dc20   CIP

Acquisitions editor: *Ray Henderson*
Editorial/production supervision
  and interior design: *Kathleen M. Lafferty*
Proofreader: *Bruce D. Colegrove*
Manufacturing buyer: *Trudy Pisciotti*

 © 1994 by Prentice-Hall, Inc.
A Paramount Communications Company
Englewood Cliffs, New Jersey 07632

MATLAB is a registered trademark of The MathWorks, Inc.

Printed in the United States of America

10 9 8 7 6 5 4 3 2 1

ISBN 0-13-151986-7

Prentice-Hall International (UK) Limited, *London*
Prentice-Hall of Australia Pty. Limited, *Sydney*
Prentice-Hall Canada Inc., *Toronto*
Prentice-Hall Hispanoamericana, S.A., *Mexico*
Prentice-Hall of India Private Limited, *New Delhi*
Prentice-Hall of Japan, Inc., *Toyko*
Simon & Schuster Asia Pte. Ltd., *Singapore*
Editora Prentice-Hall do Brasil, Ltda., *Rio de Janeiro*

To
Josefina Ovies García
and
Miriam González López

Contents

Chapter 3

Ordinary Differential Equations II: Advanced Methods

57

Chapter 4

Solving Systems of Equations

90

Chapter 5

Analysis of Data

119

* Optional section.

Chapter 6

Partial Differential Equations I:
Explicit Methods   **161**

Chapter 7

Partial Differential Equations II:
Relaxation and Spectral Methods   **204**

Chapter 8

Partial Differential Equations III:
Stability and Implicit Methods   **235**

Chapter 9

Chapter 10

Preface

When I was an undergraduate, computers were just beginning to be introduced into the university curriculum. Physics majors were expected to take a single semester of FORTRAN. Because the course was taught by the computer science department, we wrote programs to sort lists, process a payroll, and so forth. Mastering the tools of scientific computing was something we were expected to pick up as we went along. Most of us wasted many human and computer hours learning them by trial and error.

In recent years, many departments have added a computational physics course—taught by physicists—to their curricula. However, there is still considerable debate as to how this course should be organized. My philosophy is to use the upper division/graduate mathematical physics course as a model. Consider the following parallels between this text and a typical math physics book: A variety of numerical and analytical techniques used in physics are covered. Topics include ordinary and partial differential equations, linear algebra, Fourier transforms, integration, and probability. Because the text is written for physicists, these techniques are applied to solving realistic problems, many of which the students have encountered in other courses.

Numerical Methods for Physics is organized to cover what I believe are the most important, basic computational methods for physicists. The structure of the book differs considerably from the generic numerical analysis text. For example, nearly a third of the book is devoted to partial differential equations. This emphasis is natural considering the fundamental importance of Maxwell's equations, the

Schrödinger equation, the Boltzmann equation, and so forth. Chapter 6 introduces some methods in computational fluid dynamics, an increasingly important topic in the fields of nonlinear physics, environmental physics, and astrophysics.

Numerical techniques may be classified as basic, advanced, and cutting edge. On the whole, this text covers only fundamental techniques; to work effectively with advanced numerical methods requires that the user first understand the basic algorithms. The discussion in the ''Beyond This Chapter'' section at the end of each chapter guides the reader to advanced algorithms and indicates when it is appropriate to use them. Unfortunately, the cutting edge moves so quickly that any attempt to summarize the latest algorithms would quickly be out of date.

The material in this text may be arranged in various ways to suit anything from a 10-week, upper-division class to a full-semester, graduate course. Most chapters include optional sections that may be omitted without loss of continuity. Furthermore, entire chapters may be skipped; chapter interdependencies are indicated in the flow chart.

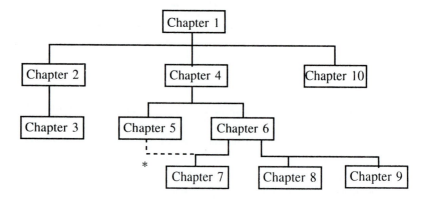

The over 200 exercises should be regarded as an essential part of the text. The length of time needed to complete most of the exercises ranges from 30 minutes to 2 days; in my classes, I assign about five exercises per week. While some texts emphasize month-long projects, I find that shorter exercises allow the class to move at a brisker pace, covering a wider variety of topics. Some instructors may wish to assign a single, large, end-of-semester project, and many of the exercises may be expanded into such projects.

I have tried to present the algorithms in a clear, universal form that would allow the reader to use them comfortably in *any* language. MATLAB® is used as the primary computer language in the text; FORTRAN versions of the programs appear in the appendices. My reasons for selecting MATLAB are discussed at length in Section 1.1. Briefly, MATLAB runs on IBM PC compatibles, Apple Macintoshes, and UNIX workstations. MATLAB is easy to learn since most of its

---

* Chapter 5 is required for optional Section 7.2.

syntax is similar to FORTRAN and C. In my classes, students are allowed to use any language, yet I find that over 90% use MATLAB.

The reader may take the MATLAB programs in this text (most of which are under 50 lines), run them on almost any platform, and obtain results, *including graphics*. Copies of the MATLAB and FORTRAN programs, plus additional materials, may be obtained by mailing in the reply card or by contacting The MathWorks, Inc., at (508) 653-1415 (e-mail address: info@mathworks.com).

An additional benefit is the availability of a powerful but inexpensive student edition of MATLAB for the Macintosh and PC platforms. The student edition has sold nearly 50,000 copies at this writing; I can only imagine that this early success will broaden its appeal as a classroom tool. Prentice Hall is the publisher for the student editions of MATLAB (and also this text) and will make a bundled version of the two products available for a considerable savings. Please contact your local representative for more information about this arrangement.

I wish to thank the people in my department, especially D. Strandburg, P. Hamill, and A. Tucker, for their strong support; my students and teaching assistants (J. Stroh, S. Moon, and D. Olson), who braved the rough waters of the early drafts; the National Science Foundation for its support of the computational physics program at San Jose State University; my editors at Prentice Hall and the technical staff of The MathWorks, Inc., for their assistance; and the National Oceanic and Atmospheric Administration Climate Monitoring and Diagnostics Laboratory for the CO_2 data used in Chapter 5. In addition, I appreciate the comments of the following reviewers: Wolfgang Christian, Davidson College; David M. Cook, Lawrence University; Harvey Gould, Clark University; Cleve Moler, The MathWorks, Inc.; and Cecile Penland, University of Colorado, Boulder. Finally, I owe a special dept of gratitude to my entire family for their moral support as I wrote this book.

Alejandro L. Garcia

Numerical Methods
for Physics

Chapter 1
Preliminaries

This chapter has no physics; the first half includes an introduction to the text and a synopsis of the MATLAB® programming language. In the second half we discuss an important concept: numerical differentiation. Before you studied physics, you had to learn calculus, so it should not be surprising that we start with this topic.

1.1 PROGRAMMING

General Thoughts

Before we get started, let me warn you that this book does not teach programming. Presumably you have already learned some programming language (it doesn't really matter which one) and have had some practice in writing programs. This book covers numerical algorithms, specifically those that are most useful in physics. The style of presentation is informal. Instead of rigorously deriving all the details of all possible algorithms, I'll only cover the essential points and stress the practical methods.

If you've had a math course in numerical analysis, you may see some old friends (such as Romberg integration). This book emphasizes the application of such methods to physics problems. You will also learn some specialized techniques generally not presented in a mathematics course. If you have not had numerical analysis, don't worry. The book is organized assuming no prior knowledge of numerical methods.

In your earlier programming course I hope you learned about good programming style. I try to use what I consider good style in the programs, but everyone has personal preferences. The point of good style is to make your life easier by organizing and structuring your program development. Many programs in this book sacrifice efficiency for the sake of clarity. After you understand how a program works, you should make it a regular exercise to improve it. However, always be sure to check your improved version against the original.

Most of the exercises in this book involve programming projects. In the first few chapters many exercises only require that you modify an existing program. In the later chapters you are asked to write more and more of your own code. The exercises are purposely organized in this fashion to allow you to come up to speed on whatever computer system you choose to use. Unfortunately, computational physics is often like experimental work in the following regard: Debugging and testing is a slow, tedious, but necessary task similar to aligning optics or fixing vacuum leaks.

Programming Languages

In writing this book, one of the most difficult decisions I had to make was choosing a language. I adopted the following criteria:

1. Powerful
2. Clean
3. Good graphics
4. Standard/portable

The obvious choices were Basic, FORTRAN, Pascal, and C. I also considered symbolic manipulators such as Maple® and Mathematica®. These five choices are rated on a grade scale (A–F) in Table 1.1.

You see that I am a pretty tough grader. Let me explain some of these marks. A powerful language should be compiled, handle different data types (such

TABLE 1.1 My opinions on the relative merits of standard programming languages

Language	Powerful	Clean	Good Graphics	Standard/ Portable
Basic	C−	B+	B−	B−
FORTRAN	B	C+	C−	C+
Pascal	C+	B	C	B−
C	B+	B−	C	B−
Symbolic manipulators	C+	B−	A−	B+

as complex numbers), and have good standard libraries available. A clean language should be easy to read, easy to use, and easy to debug. For a language to have good graphics, it should have not just the graphics primitives, but high-level routines (e.g., contour plots) as well. Portability is important because I want you to be able to use IBM PCs (or clones) or Macintoshes or UNIX workstations.

Each language has its faults. Basic is not really powerful enough; it is a nice language, but not really designed for scientists. FORTRAN is perhaps the "obvious" choice, but it is not well implemented on micros. Graphics in FORTRAN is problematic, and no one would accuse it of being a clean language. Pascal shares the problems of Basic and FORTRAN and has some of its own. In my research work I often use C, but it has a steep learning curve. C is a powerful language but it is very unforgiving and there is no standard graphics package.

One problem with symbolic manipulators is performance; I've found them to be significantly slower than even interpreted Basic. This is not surprising since they are designed to do symbolic rather than numeric calculations. A more serious difficulty is that their logical structure (which is usually based on LISP) is radically different from that used in the standard languages. As such, translating a program written for a symbolic manipulator into C or FORTRAN usually requires completely reworking its structure.

Finally, some books in numerical analysis present their programs in metacode. This makes the programs very readable since one is not tied to a precise syntax. The disadvantage is that the programs are not machine readable, so they may only be tested manually. Ultimately, the user still has to translate the programs into a real computer language.

MATLAB

The main language used in this book is MATLAB. Those of you who have never used it may be apprehensive. Since you may prefer to use a different language, I have made every effort to ensure that the MATLAB programs are easily translatable. There are also FORTRAN versions of the programs in the appendices. However, let me first try to convince you that you would enjoy using MATLAB.

Although you may not be familiar with MATLAB, it is widely used in both academia and industry. It is especially popular in the engineering community and with applied mathematicians. MATLAB is very portable; it runs on a wide range of machines from AT clones and Macintoshes to workstations and supercomputers.

MATLAB's basic data structure is the matrix (its name is an acronym for MATrix LABoratory). In most languages, an array is just a set of sequential memory locations; array operations are done element by element. In MATLAB, if A and B are matrices, then A*B is the matrix product. You will see that this feature greatly simplifies most programs by eliminating many loops. Physicists are comfortable using vectors and matrices, and writing programs is more natural when we use them. In this sense, MATLAB borrows from object-oriented programming by "overloading" the basic operators, such as multiplication. MATLAB also handles complex numbers gracefully and automatically without making a big deal out of them like most languages do.

Another strength of MATLAB is that it is an interpreted language with excellent compiled libraries. These libraries are especially designed for scientific computing. Being an interpreted language it is easy to use interactively, while the compiled libraries improve its performance. Being interpreted also makes MATLAB very clean. Many details (such as dimensioning matrices) are handled automatically. MATLAB has very good graphics facilities, including high-level routines.

MATLAB also has its weaknesses. It would be much improved if it were possible to compile MATLAB programs fully. It could also use better editing and debugging facilities. You might wonder what "grade" I would give to MATLAB: I think it gets a B+ or A- in each of the four categories.

1.2 BASIC ELEMENTS OF MATLAB

Variables

Even if you plan to use a different language, you should learn at least the basic elements of MATLAB because it is used throughout the text. MATLAB is very similar to other structured languages, so I think you can pick up the basics with no trouble. Advanced features will be introduced in later chapters as we need them.

MATLAB has only one data type: the matrix. A scalar is a 1×1 matrix, a row vector is a $1 \times N$ matrix, and a column vector is an $N \times 1$ matrix. Variables are not declared explicitly; MATLAB just dimensions them as they are used. In the examples below I will use the scalars x and y, the vectors $\mathbf{a}$ and $\mathbf{b}$, and the matrices $\mathbf{C}$, $\mathbf{D}$, and $\mathbf{E}$. For the examples, let's give them the values

$$x = 3; \quad y = -2; \quad \mathbf{a} = \begin{bmatrix} 1 \\ 2 \\ 3 \end{bmatrix}; \quad \mathbf{b} = [0 \quad 3 \quad -4];$$

$$\mathbf{C} = \begin{bmatrix} 1 & 0 & 1 \\ 0 & 1 & -1 \\ 1 & 2 & 0 \end{bmatrix}; \quad \mathbf{D} = \begin{bmatrix} 0 & 1 & 1 \\ 2 & 3 & -1 \\ 0 & 0 & 1 \end{bmatrix}; \quad \mathbf{E} = \begin{bmatrix} 1 & \pi \\ 0 & -1 \\ x & \sqrt{-1} \end{bmatrix}$$

In MATLAB these variables would be set by the assignment statements

```
x = 3        % These are some simple assignments
y = -2
a = [1; 2; 3]
```

```
b = [0 3 -4]
C = [1 0 1; 0 1 -1; 1 2 0]
D = [0 1 1; 2 3 -1; 0 0 1]
E = [1 pi; 0 -1; x sqrt(-1)]
```

Notice that rows are separated by semicolons. Anything following a percent sign is considered a comment in MATLAB. The variable pi is preassigned to 3.14159 . . . in MATLAB.

Arithmetic

The basic arithmetic operations are defined in the natural manner. For example, the MATLAB statements

```
z = x - y        % Some basic operations
t = b * a
F = C + D
G = C * E
```

assign the values

$$z = 5 \qquad t = -6 \qquad \mathbf{F} = \begin{bmatrix} 1 & 1 & 2 \\ 2 & 4 & -2 \\ 1 & 2 & 1 \end{bmatrix} \qquad \mathbf{G} = \begin{bmatrix} 4 & \pi + i \\ -3 & -1 - i \\ 1 & \pi - 2 \end{bmatrix}$$

Notice that b*a is just the dot product of these vectors. For matrices, division is implemented by using Gaussian elimination (discussed in Chapter 4). The power operator is ^, so 2^3 is $2^3 = 8$. MATLAB will balk if you try to do a matrix operation when the dimensions don't match (e.g., it will *not* compute C+E).

Sometimes we want to perform operations element by element, so MATLAB defines the operators .* ./ and .^ . Here are some examples of these array operations:

```
H = C .* D     % These operations are performed
J = E .^ x     % element by element
```

give the values

$$\mathbf{H} = \begin{bmatrix} 0 & 0 & 1 \\ 0 & 3 & 1 \\ 0 & 0 & 0 \end{bmatrix}; \qquad \mathbf{J} = \begin{bmatrix} 1 & \pi^3 \\ 0 & -1 \\ 27 & -i \end{bmatrix}$$

Individual elements of a matrix may be addressed by using their indices. For example, J $(1, 2)$ equals π^3 and J $(3, 1)$ equals 27. Similarly, for vectors, b (3) (or b $(1, 3)$) equals -4. Notice that matrix indices start at one and not zero.

Matrix **B** is the transpose of matrix **A** if $A_{ij} = B_{ji}$, that is, the rows and columns are exchanged. The Hermitian conjugate of a matrix is the transpose of its complex conjugate. In MATLAB

```
K = E'      % Hermitian conjugate
L = E.'     % Transpose
```

give the values

$$\mathbf{K} = \begin{bmatrix} 1 & 0 & 3 \\ \pi & -1 & -i \end{bmatrix}; \qquad \mathbf{L} = \begin{bmatrix} 1 & 0 & 3 \\ \pi & -1 & i \end{bmatrix}$$

The Hermitian conjugate of E is K, while L is the transpose of E.

Loops and Conditionals

Repeated operations are performed by using loops. Here is an example of a for loop in MATLAB:

```
for i=1:5        % Your basic loop; i goes from 1 to 5
   p(i) = i^2
end              % This is the end of the loop
```

This loop assigns the value p = [1 4 9 16 25]. Notice that **p** is created as a row vector. If we wanted it to be a column vector, we could build it as

```
for i=1:5        % Your basic loop; i goes from 1 to 5
   p(i,1) = i^2  % p is a column vector
end              % This is the end of the loop
```

or

```
for i=1:5        % Your basic loop; i goes from 1 to 5
   p(i) = i^2
end              % This is the end of the loop
p = p.'          % Transpose p into a column vector
```

using the transpose operator.

In a for loop, the default step is +1, but it is possible to use a different increment. For example, the loop

```
for i=1:2:5       % Loop over odd values of i
   q(i) = i
   q(i+1) = -i
end
```

assigns the values q = [1 -1 3 -3 5 -5].

M<small>ATLAB</small> also has while loops; here is a simple example:

```
while( x > 1 )
   x = x/2
end
```

A while command executes the statements in the body of the loop while the loop condition is true. If $x = 5$ before the loop, then x will equal 5/8 when the loop completes. The break statement can be used to terminate for loops or while loops.

Here are some examples of how conditionals are implemented; you see that it is quite standard.

```
if( x > 5 )        % A simple conditional
   z = z-1
end

if( x >= x_min & x <= x_max )
   status = 1
else                        % Another conditional using else
   status = 0
end

if( x == 0 | x == 1)      % Another conditional using elseif
   flag = 1
elseif( x < 0 & x ~= -1)    % Notice that elseif is ONE WORD
   flag = -1
else
   flag = 0
end
```

Notice that equals and not equals are == and ~=, respectively. Logical "and" is & (ampersand) and logical "or" is | (vertical bar). The end command terminates both loops and conditionals.

Input and Output

MATLAB has various types of input and output facilities. The `input` command prints a prompt to the screen and accepts input from the keyboard. Here is a simple example of its use:

```
x = input('Enter the value of x - ')
```

In this example, you can enter a scalar, a matrix, or even any valid MATLAB expression.

The `disp` command may be used to display the value of a variable or to print text:

```
disp('The value of z is ')
disp(z)
```

If the variable z is a matrix, `disp(z)` will display it in tabular form.

Formatted output is also available with the `fprintf` command,

```
fprintf('The values of x and y are %g and %g meters \n',x,y)
```

The \n in the line above indicates a carriage return (new line). C language programmers will immediately recognize an old friend; others should refer to the manual for details on the use of `fprintf`. It is also possible to read data from and write data to files using `load` and `save`.

Standard Functions and Graphics

One of MATLAB's strengths is its collection of built-in functions. Below is a short list of some of the basic functions.

`abs(x)`	Absolute value or complex magnitude
`sqrt(x)`	Square root
`sin(x)` `cos(x)`	Sine and cosine
`exp(x)` `log(x)`	Exponential and natural logarithm
`rem(i,n)`	Remainder (also known as modulo function)
`floor(x)` `ceil(x)`	Round down (floor) or round up (`ceil`) to nearest integer (e.g., `floor(3.2)=3`, `ceil(3.2)=4`)
`rand(N)`	Returns an N × N matrix filled with random numbers from the interval [0, 1).

inv (X)   Returns the inverse of the matrix X.

fft (x)   Returns the Fourier transform of the vector x.

There are many more functions that will be introduced as we need them.

MATLAB has various graphics commands for creating xy plots, contour plots, and three-dimensional (3-D) wire-mesh plots. Below is a list of a few of the basic graphics commands; they and others are discussed as we use them in the programs.

plot (x, y)	Plot vector x versus vector y
loglog (x, y)	Plot vector x versus vector y using log or semilog scales
semilogx (x, y)	
semilogy (x, y)	
polar (theta, rho)	Polar plot
contour (z)	Contour plot of matrix z
mesh (z)	3-D wire-mesh plot of matrix z
title (' *text* ')	Write title and labels on plots
xlabel (' *text* ')	
ylabel (' *text* ')	

On the Macintosh you can copy a plot from the graphics window into the Scrapbook for later use. For IBM PCs, the Print Screen key sends the plot to your printer. On some platforms, the meta command allows you to save graphics to a file. The graph may later be redisplayed or printed using gpp (graphics postprocessor).

MATLAB Session

When you first enter the MATLAB environment, you are at the command level as indicated by the >> prompt. From the command line you can enter individual MATLAB commands (see Figure 1.1). To end your MATLAB session, type quit or exit.

For programming, it is more convenient to enter a set of commands to be executed in a file. In the MATLAB terminology such a script of commands is called an M-file. Our programs and functions will all be M-files. You run an M-file by invoking its name (the name of the file) on the command line. After MATLAB executes the commands in the M-file, it returns control to the command line. You can then enter individual commands (for example, to display the values of various variables).

A few more points: MATLAB has an interactive help that may be used from the command line. For example,

>> help bessel

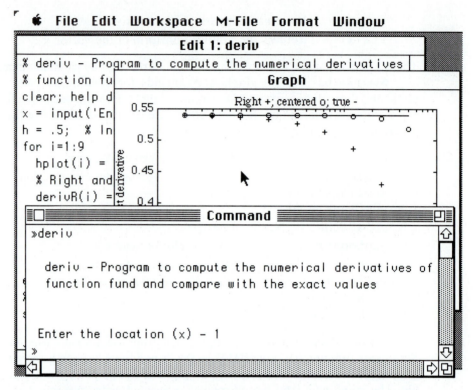

Figure 1.1   MATLAB on the Macintosh. The command window is in the foreground. Behind it is the Graph window and the Edit window. Program deriv is discussed in Section 1.5.

tells you about MATLAB's Bessel function routines. You can also get help on special characters (e.g., try help %). As an M-file is executed, each line is echoed to the screen. This echoing is sometimes useful in debugging, but is usually just annoying. If you put a semicolon at the end of a line, MATLAB executes it without echoing the output (see also the echo command).

Sometimes you want to keep a record of your MATLAB session, as part of a homework assignment, for example. The command diary *filename* saves the subsequent output (except for graphics) to the file *filename*. The diary may be toggled on and off using diary on and diary off.

After this introduction to MATLAB, some of you may still want to use another language for your projects. FORTRAN versions of all the programs are listed in the appendices. Using either the MATLAB or the FORTRAN listings, it should not be difficult to translate the programs to any other language. If you use the FORTRAN versions, you'll need a way to produce plots. The programs write data files that may be read by a graphics application or a spreadsheet. You may even use MATLAB as your graphics application; see the load and save commands in the MATLAB manual.

EXERCISES[1]

1. For the matrix A = [1 2;3 4], compute (a) A*A; (b) A.*A; (c) A^2; (d) A.^2; (e) A/A; (f) A./A.

2. Given the vectors **x** = [1 2 3 ... 10] and **y** = [1 4 9 ... 100], plot them in MATLAB using: (a) plot(x,y); (b) plot(x,y,'+'); (c) plot(x,y,'-', x,y,'+'); (d) plot(x,y,'-',x(1:2:10),y(1:2:10),'+'); (e) semilogy (x,y); (f) loglog(x,y,'+').

3. The MATLAB function inv(A) returns the inverse of matrix A. Find the inverse of the matrices (a) [1 2 3;0 4 5;0 0 6]; (b) [1 0 0;0 2 0;0 0 3]; (c) [1 2 0; 3 4 5; 0 6 7]; (d) [1 2 3; 4 5 6; 7 8 9]. Check that a matrix times its inverse equals the identity matrix.

4. The exponential of a matrix is defined by the expansion

$$e^{\mathbf{A}} = \mathbf{I} + \mathbf{A} + \frac{1}{2!}\mathbf{A}^2 + \frac{1}{3!}\mathbf{A}^3 + \ldots$$

For the various matrices, estimate $e^{\mathbf{A}}$ using the first six terms of the expansion. Compare with the exact result given by the MATLAB function expm(A). When does the estimate work and when does it fail?

5. Reproduce the plots shown below. Try to be as accurate as possible in your reconstruction.

(a)
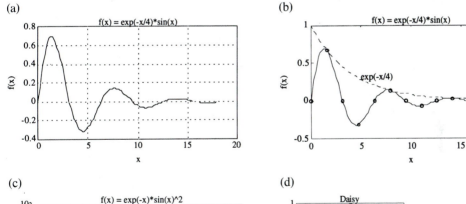

(b)

(c)
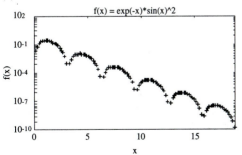

(d)
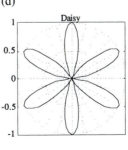

---

[1] Recommended exercises are indicated with boldface numbers.

6. It is always good to know your limits. For your computer system find: (a) the largest possible floating point number; (b) the largest integer I such that $(I + 1) - 1$ equals I; (c) the smallest possible positive floating point number; (d) the smallest positive floating point number x such that $(1 + x) - 1$ does not equals zero; (e) the largest $N \times N$ matrix allowed by memory; (f) the longest row vector allowed by memory; (g) the maximum number of array dimensions allowed (e.g., $a(i, j, k)$ is a three-dimensional array).

7. For your computer system, write a program to estimate the number of floating point operations (e.g., multiplications) that can be performed in 1 s.

1.3 SIMPLE MATLAB PROGRAMS

Orthogonality Program

In this section we write some simple programs using MATLAB; our first example is given in Listing 1.1. This program, called orthog, tests whether two vectors are orthogonal by computing their dot product. This sample program is very simple and can be significantly improved (see the exercises).

LISTING 1.1 Program orthog. Determines if a pair of vectors is orthogonal by computing their dot product.

```
1   % orthog - Program to test if a pair of vectors
2   % is orthogonal.  Assumes vectors are in 3D space
3   clear;  help orthog;  % Clear the memory and print header
4   va = input('Enter the first vector - ');
5   vb = input('Enter the second vector - ');
6   % Compute the dot product of va and vb
7   adotb = 0;
8   for i=1:3
9     adotb = adotb + va(i)*vb(i);
10  end
11  % Vectors are orthogonal if dot product equals zero
12  if ( adotb == 0 )
13    disp('Vectors are orthogonal')
14  else
15    disp('Vectors are NOT orthogonal')
16    fprintf('Dot product = %g \n',adotb)
17  end
```

Here is what a typical run would look like:

```
>>orthog

orthog - Program to test if a pair of vectors
is orthogonal. Assumes vectors are in 3D space
```

```
Enter the first vector - [1 1 1]

Enter the second vector - [1 -2 1]
Vectors are orthogonal
```

A few notes about the program: The line numbers in the listing are not in the script file. I will put line numbers on listings to indicate specific points in the programs easily. Notice that I put semicolons at the ends of most lines so that they will be executed without echoing.

The first two lines of orthog are comments; if you type help orthog from the command line, MATLAB displays these lines. Notice that unlike most languages, there is no "PROGRAM" statement on the first line. It is the file's name that gives the program its name (this program is in a file called orthog or orthog.m). Each MATLAB program or function needs to be in a separate file.

The clear command on the third line clears the memory. The help statement on this line serves to display the first two lines each time you run orthog. The vectors are entered using the input command on lines 4 and 5. The body of the program is in lines 7–10, where the dot product is computed. A slicker way to do it would be to replace lines 7–10 with

```
adotb = va * vb';
```

Finally, according to the value of adotb, the program displays one of the two possible responses.

Interpolation Program

It is well known that given three (x, y) pairs, one may find a quadratic that fits the points. There are various ways to find this polynomial and various ways to write it. The Lagrange form of the polynomial is

$$p(x) = \frac{(x - x_2)(x - x_3)}{(x_1 - x_2)(x_1 - x_3)} y_1 + \frac{(x - x_1)(x - x_3)}{(x_2 - x_1)(x_2 - x_3)} y_2 + \frac{(x - x_1)(x - x_2)}{(x_3 - x_1)(x_3 - x_2)} y_3$$

$$(1\text{-}1)$$

where (x_1, y_1), (x_2, y_2), (x_3, y_3) are the three data points to be fit. Commonly, such polynomials are used to interpolate between data points.

A simple interpolation program, called interp, is given in Listing 1.2. Almost all this program is input/output and comments. First, the three (x, y) pairs (lines 5–9) and the range of values for which the data is to be interpolated (line 10) are read in. The interpolated value $y = p(x)$ is computed by the function intrpf for a range of values from xr(1) to xr(2).

LISTING 1.2   Program `interp`. Uses the `intrpf` (Listing 1.3) function to interpolate between data points.

```
1    % interp - Program to interpolate data using Lagrange
2    % polynomial to fit quadratic to three data points
3    clear; help interp;   % Clear memory and print header
4    disp('Enter data points as x,y pairs (e.g. [1 2])');
5    for i=1:3
6      temp = input('Enter data point - ');
7      x(i) = temp(1);
8      y(i) = temp(2);
9    end
10   xr = input('Enter range of x values as [x_min x_max] - ');
11   nplot = 100;      % Number of points for interpolation curve
12   for i=1:nplot
13     xi(i) = xr(1) + (xr(2)-xr(1))*(i-1)/(nplot-1);
14     yi(i) = intrpf(xi(i),x,y); % Use intrpf function
15   end
16   plot(x,y,'*',xi,yi,'-');
17   xlabel('x');
18   ylabel('y');
19   title('Three-point interpolation');
```

Finally, the original data and the interpolated points are plotted (lines 16–19). The data are marked by asterisks and the interpolated data with a solid line (see Figure 1.2).

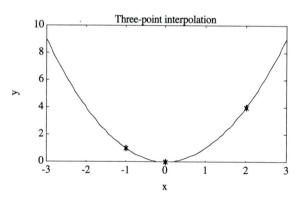

Figure 1.2   Graphical output from the `interp` program. Input data is $[-1 \quad 1]$, $[0 \quad 0]$, and $[2 \quad 4]$; interpolation range is $[-3 \quad 3]$.

LISTING 1.3   Function `intrpf`. Uses a quadratic to interpolate between three data points.

```
1    function yi = intrpf(xi,x,y)
2    % Function to interpolate between data points
```

```
3    % using Lagrange polynomial (quadratic)
4    % Input parameters - x,y are the data vectors
5    %                    xi is the x value for interpolation
6    % Output value - y is the interpolated value
7    yi = (xi-x(2))*(xi-x(3))/((x(1)-x(2))*(x(1)-x(3)))*y(1) ...
8       + (xi-x(1))*(xi-x(3))/((x(2)-x(1))*(x(2)-x(3)))*y(2) ...
9       + (xi-x(1))*(xi-x(2))/((x(3)-x(1))*(x(3)-x(2)))*y(3);
10   return;
```

The real work is done by the function `intrpf` (see Listing 1.3). This function is fairly straightforward since it does nothing but evaluate equation (1-1). Notice that lines 7–9 are all one MATLAB command. The ellipsis[2] (. . .) ending the first two lines signals that they are continued on the next line.

Functions in MATLAB are implemented as in most languages, except each function must be in a separate file. It is the file's name that gives the function its name (notice that this function is in a file called `intrpf` or `intrpf.m`). The first line of a function is always the keyword `function`.

As in most languages, variables in a function are local to the function. For example, if we were to modify the value of `xi` inside the function `intrpf`, the variable `xi` in the main program (`interp`) would be unaffected. Variables in the calling sequence are passed by value (as in C) and not by reference (as in FORTRAN). MATLAB does not have subroutines (such as in FORTRAN) since they may always be implemented as functions (C also works this way).

Colon Operator

The colon operator, : , is one of MATLAB's handiest tools. Let's consider a few examples of its use. First, the `for` loop,

```
tau = 0.1
for i=1:100
   time(i) = tau * i
end
```

could be replaced with

```
tau = 0.1
i=1:100
time = tau * i
```

---

[2] An ellipsis is the punctuation mark given by three points that indicates an incomplete thought. An ellipse is a geometric object whose name indicates that it is an imperfect circle.

In the latter, a vector i = [1 2 . . . 100] is created. We may further abbreviate this to

```
tau = 0.1
time = tau * (1:100)
```

In all three cases, we create the vector time = [0.1 0.2 . . . 10.0].

Not only does using the colon operator abbreviate our code, but it also makes our code run faster. The program becomes more efficient because we are explicitly executing a vector operation instead of performing an element-by-element calculation. In the interp program we could eliminate line 13 by inserting

```
xi = xr(1) + (xr(2)-xr(1))*(0:(nplot-1))/(nplot-1)
```

just above the for loop on line 12.

The colon operator is also useful for selecting parts of a matrix. For example,

```
z = A(:,3)
```

assigns the third column of matrix A to the vector z. As another example, suppose that x and y are vectors of length N. The statement

```
plot(x, y)
```

graphs the values. On the other hand,

```
plot(x(1:5:N), y(1:5:N))
```

only graphs every fifth data point. This can be useful if N is excessively large.

At first the colon operator can be tricky to use. However, once you've mastered it you won't know how you ever programmed without it. Some modern languages, such as FORTRAN 90, have similar colon operators.

Finally, two handy MATLAB functions for creating matrices are zeros and ones. The statement A=zeros(N) creates an N × N matrix with all elements set to zero. Similarly, the statement A=zeros(M, N) creates an M × N matrix with all elements set to zero. The ones function works in the same fashion but creates matrices filled with ones.

EXERCISES

8. Modify the orthog program so that it can handle vectors of any length (see the MATLAB function length). Your program should detect and gracefully handle erroneous inputs such as vectors of unequal length.

9. Modify the orthog program so that it accepts a pair of three-dimensional vectors and outputs the unit vector that is orthogonal to the input vectors.

10. Modify orthog so that if the second vector is not orthogonal to the first the program computes a new vector that is orthogonal to the first vector, has the same length as the second vector, and is in the same plane as the two input vectors. This orthogonalization is often used with eigenvectors and is commonly performed using the *Gram–Schmidt procedure*.

11. From tables, we find the following values for the zeroth-order Bessel function: $J_0(0) = 1.0$; $J_0(0.5) = 0.9385$; $J_0(1.0) = 0.7652$. Using intrp, find the estimated values of $J_0(x)$ for the range $x = 0$ to $x = 2.0$. For the values of $x > 1$, we are actually extrapolating instead of interpolating. Compare your results with the known values (see the MATLAB command bessel (n, x)).

12. Write a function that is similar to intrpf but that returns the estimated derivative at the interpolation point. The function will accept three (x, y) pairs, fit a quadratic to the data, then return the value of the derivative of the quadratic at the desired point.

13. Modify interp so that it can handle any number of data points by using higher-order polynomials. After testing your program, give it the following values of the Bessel function: $J_0(0) = 1.0$; $J_0(0.2) = 0.9900$; $J_0(0.4) = 0.9604$; $J_0(0.6) = 0.9120$; $J_0(0.8) = 0.8463$; $J_0(1.0) = 0.7652$, and repeat Exercise 1.11. Do your estimates improve?

1.4 NUMERICAL ERRORS

Round-off Error

In most languages, floating point numbers are represented with only finite precision. Typically, single precision allocates 4 bytes (32 bits) for the representation of a number, while double precision uses 8 bytes (64 bits). Exactly how a computer handles the representation is not as important as knowing the number of significant digits and the maximum range.

For single precision, the number of significant digits is typically six or seven decimal digits. By default, MATLAB uses double precision, giving about 16 significant digits. Thus, in double precision, the operation $3 + 10^{-20}$ returns an answer of 3 because of round-off. Round-off error is especially troublesome when we subtract two numbers of nearly the same magnitude. Consider the operation

$$\frac{10^{-20}}{(3 + 10^{-20}) - 3}$$

Round-off in the denominator could be catastrophic to the entire calculation.

MATLAB defines a permanent variable, eps, whose value is the distance between 1.0 and the next largest floating point number. This value is useful in testing for round-off tolerance. In other languages it may be computed as

```
eps = 1.
while( (1+eps) > 1 )
  eps = eps/2
end
eps = 2*eps
```

On my computer, eps $\cong 2.22 \times 10^{-16}$.

The program `rndoff` demonstrates the effect of round-off error in a simple calculation (see Listing 1.4). On line 5, the variable `temp = (10+h) - 10` is computed; in the absence of round-off it should always equal h. Figure 1.3 shows the absolute fractional error as a function h. Notice that the error is 100% when h < eps since `temp = 0`.

LISTING 1.4   Program `rndoff`. Illustrates round-off error in a simple calculation.

```
1   % rndoff - Demo program illustrating round-off error
2   clear; help rndoff;  % Clear memory and print header
3   h=1;   % Initial value for h
4   for i=1:21
5     temp = (10+h)-10;   % Without round-off, temp=h
6     hplot(i) = h;
7     eplot(i) = abs(temp-h)/h;   % Absolute fractional error
8     h = h/10;         % Decrement magnitude of h
9   end
10  loglog(hplot,eplot,'*');   % Plot data using log-log scale
11  xlabel('h');
12  ylabel('Fractional error');
13  title('Round-off Error');
```

Sometimes round-off problems may be cured by using double precision. The disadvantages of double precision are that it requires more storage and that is

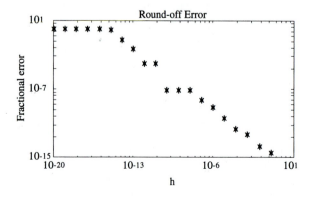

Figure 1.3   Absolute fractional error due to round-off for the calculation of (10+h) − 10 as a function of h (see Listing 1.4).

sometimes (but not always) more computationally costly. Modern math processors are designed to work in double precision, so it can actually be slower to work in single precision.

Using double precision sometimes only forestalls the difficulties of round-off. For example, a matrix inverse calculation may work fine in single precision for small matrices of 50×50 elements but fail because of round-off for larger matrices. Double precision may allow us to work with matrices of 100×100, but if we need to solve even larger systems we will have to use a different algorithm. The best approach is to work with algorithms that are robust against round-off error.

Range Error

The second consideration in the representation of a number is the maximum range, or how large an exponent we may use. For single precision, a typical value is $10^{\pm 37}$; for double precision it is typically $10^{\pm 308}$. Exceeding the single precision range is not difficult. Consider, for example, something as common as the Bohr radius,

$$a_0 = \frac{4\pi \varepsilon_0 \hbar^2}{m_e e^2} \cong 5.3 \times 10^{-11} \text{ m} \qquad (1\text{-}2)$$

in MKS units. While its value lies within the range of a single precision number, the range is exceeded in the calculation of the numerator ($4\pi \varepsilon_0 \hbar^2 \cong 1.24 \times 10^{-78}$ kg $\cdot$ C^2 $\cdot$ m) and the denominator ($m_e e^2 \cong 2.34 \times 10^{-68}$ kg $\cdot$ C^2). The best solution for dealing with this type of range difficulty is to work in a set of units natural to a problem (e.g., for atomic problems work in units of angstroms, the charge of an electron).

Sometimes range problems arise not from the choice of units but because the numbers in the problem are inherently large. Let's consider an important example, the factorial function. Using the definition

$$n! = n \times (n - 1) \times (n - 2) \times \ldots \times 3 \times 2 \times 1 \qquad (1\text{-}3)$$

it is easy to write a naive program to evaluate $n!$ (see Listing 1.5). Unfortunately, due to range problems, we cannot evaluate $n!$ for $n > 170$ using double precision.

LISTING 1.5 Program `factn`. Computes factorial using the naive approach.

```
1   % factn - Naive program for computing factorial
2   clear;  help factn;  % Clear memory and print header
3   n = input('Enter value of n - ');
4   nf = 1;
5   for i=1:n
6      nf = nf*i;
```

```
7   end
8   fprintf(' n! = %g \n',nf);
```

A common solution to working with very large numbers is to use their logarithm. For the factorial,

$$\log(n!) = \log(n) + \log(n-1) + \ldots + \log(3) + \log(2) + \log(1) \qquad (1\text{-}4)$$

However, this scheme is computationally expensive if n is large. An even better approach is to combine the use of logarithms with Stirling's formula,[3]

$$n! = \sqrt{2n\pi}\, n^n e^{-n} \left(1 + \frac{1}{12n} + \frac{1}{288n^2} + \ldots \right) \qquad (1\text{-}5)$$

The program facts (Listing 1.6) uses equation (1-3) for small values of n (< 30) and Stirling's formula to compute $\log_{10}(n!)$ for larger values. In the latter case, the value of $n!$ is reconstituted as

$$n! = (\text{mantissa}) \times 10^{(\text{exponent})} \qquad (1\text{-}6)$$

where the exponent is the integer part of $\log_{10}(n!)$ and the mantissa is $10^{\wedge}$(fractional part of $\log_{10}(n!)$).

LISTING 1.6   Program facts. Computes factorial using Stirling's approximation.

```
1    % facts - Program for computing factorial
2    % using Stirling's approximation
3    clear;  help facts;  % Clear memory and print header
4    n = input('Enter value of n - ');
5    if( n < 30 )   % For small values of n use product
6        nf = 1;
7        for i=1:n
8            nf = nf*i;
9        end
10       fprintf(' n! = %g \n',nf);
11   else           % For larger values use Stirling's apprx.
12       log_nf = .5*log10(2*pi) + (n+.5)*log10(n) ...
13               - n*log10(exp(1)) + log10(1+1/(12*n)+1/(288*n^2));
14       exponnt = floor(log_nf);
15       mantisa = 10^(log_nf - exponnt);
16       fprintf(' n! = %g e %g \n',mantisa,exponnt);
17   end
```

---

[3] M. Abramowitz and I. Stegun, *Handbook of Mathematical Functions* (New York: Dover, 1972), section 6.1.

Languages that implement IEEE arithmetic (MATLAB included) have a special value to represent infinity. Infinity is obtained as a result of operations such as 1.0/0.0. In MATLAB, the permanent variable, Inf, is set to this value. A related quantity is the IEEE arithmetic representation for Not-a-Number; the permanent variable NaN is set to this value. NaN is obtained as a result of undefined operations such as 0.0/0.0.

It is possible to perform calculations of arbitrary precision in symbolic-based languages. For example, by default Mathematica stores numbers as rational quotients (fraction with integer numerator and denominator). While some calculations call for such accuracy (e.g., some orbital mechanics problems), the computations can be very slow.

EXERCISES

14. Suppose that you are standing at 40° latitude and are 2 m tall.
    (a) Find the velocity of your feet, v, due to the rotation of Earth. Assume that Earth is a perfect sphere of radius $R = 6378$ km and that a day is exactly 24 h long.
    (b) Using the result from (a), compute the centripetal acceleration at your feet as $a = v^2/r$.
    (c) Repeat parts (a) and (b) for your head and compute the difference between the acceleration at your head and feet. Show how round-off can corrupt your calculation and how to fix the problem.

15. Run rndoff but change line 8 to read: h=h/2. The results are strikingly different; explain why.

16. Consider the Taylor expansion for the exponential

$$e^x = 1 + x + \frac{x^2}{2!} + \frac{x^3}{3!} + \ldots = \sum_{i=0}^{\infty} \frac{x^i}{i!} = \lim_{N \to \infty} S(x, N)$$

    where $S(x, N)$ is the partial sum with $N + 1$ terms.
    (a) Write a program that plots the absolute fractional error of the sum, $|S(x, N) - e^x|/e^x$, versus N (up to $N = 60$) for a given value of x. Test your program for $x = 10, 2, -2,$ and -10. From the plots, explain why this is not a good way to evaluate e^x when $x < 0$.[4]
    (b) Modify your program so that it uses the identity $e^x = 1/e^{-x} = 1/S(-x, \infty)$ to evaluate the exponential when x is negative. Explain why this approach works better.

17. Suppose that we use electron volts as our unit of energy, the mass of the electron as our unit of mass, and set Planck's constant equal to unity. Convert 1 kg, 1 m, and 1 s into these units.

---

[4] G. E. Forsythe, M. A. Malcolm, and C. B. Moler, *Computer Methods for Mathematical Computation* (Englewood Cliffs, N.J.: Prentice Hall, 1977), section 2.3.

18. Suppose that we use the mean radius of Earth–Sun orbit as the unit of length, the mass of Earth as the unit of mass, and a year as the unit of time. Convert the gravitational constant, G, into these units. What is the force of attraction between Earth and the Moon in these units? Between Earth and the Sun?

19. The probability of flipping N coins and obtaining m "heads" is given by the binomial distribution to be

$$P_N(m) = \frac{N!}{m!(N-m)!} \left(\frac{1}{2}\right)^N$$

What is more probable, flipping 10 coins and getting no heads or flipping 10,000 coins and getting exactly 5000 heads?

1.5 NUMERICAL DIFFERENTIATION

Naive Formula

In calculus you learned the following formula for the derivative,

$$f'(x) \equiv \lim_{h \to 0} \frac{f(x+h) - f(x)}{h} \tag{1-7}$$

We know that we cannot set $h = 0$, but if we use a "small enough" value for h then we get a "good enough" answer. In other words, we could say

$$f'(x) \approx \frac{f(x+h) - f(x)}{h} \quad (h > 0) \tag{1-8}$$

or

$$f'(x) = \frac{f(x+h) - f(x)}{h} + \text{(error term)} \tag{1-9}$$

On the computer, we run into several difficulties. First, the allowed range limits our values of h (e.g., for single precision $h > 10^{-37}$). A more serious problem is round-off; there is round-off in the calculation of $x + h$ and in the subtraction in the numerator. The bottom line is that we cannot use an arbitrarily small value for h. The idea now is to do as well as possible given this constraint.

Our first step is to obtain an explicit expression for the error term. The simplest approach is to use the Taylor expansion. As physicists, we usually see the Taylor series expressed as

$$f(x+h) = f(x) + hf'(x) + \tfrac{1}{2}h^2 f''(x) + \dots \tag{1-10}$$

where the symbol (. . .) means higher-order terms that are usually dropped from the derivation by the next line. An alternative, equivalent form of the Taylor series used in numerical analysis is

$$f(x + h) = f(x) + hf'(x) + \tfrac{1}{2}h^2 f''(\zeta) \qquad (1\text{-}11)$$

where ζ is a value between x and $x + h$. We have not dropped any terms; this expansion has a *finite* number of terms. Taylor's theorem guarantees that there exists *some* value ζ for which (1-11) is true, but it doesn't tell us what that value is.

The previous equation may be rewritten to give

$$f'(x) = \frac{f(x + h) - f(x)}{h} - \tfrac{1}{2}hf''(\zeta) \qquad (1\text{-}12)$$

where $x \le \zeta \le x + h$. This equation is known as the *right derivative* formula. The last term on the right is the *truncation error*; it is the error introduced by the truncation of the Taylor series.

In other words, if we keep the last term in (1-12), our expression for $f'(x)$ is exact. But we can't evaluate that term because we don't know ζ; all we know is that ζ lies somewhere between x and $x + h$. So we drop the $f''(\zeta)$ term (truncate) and say that the error we make by neglecting this term is the truncation error. Do not confuse this with the round-off error discussed in Section 1.4. Round-off error depends on hardware; truncation error depends on the approximations used in an algorithm.

Sometimes you will see equation (1-12) written as

$$f'(x) = \frac{f(x + h) - f(x)}{h} + O(h) \qquad (1\text{-}13)$$

where the truncation error term is now just specified by its order in h. The essential point is that the truncation error is *linear* in h. But if we use a very small value of h to reduce truncation error, then we aggravate the round-off error. Fortunately, there are better ways to compute the derivative.

Centered Formulas

An equivalent definition for the derivative is

$$f'(x) = \lim_{h \to 0} \frac{f(x + h) - f(x - h)}{2h} \qquad (1\text{-}14)$$

This formula is said to be "centered" in x. While this formula looks very similar to (1-12), there is a big difference when h is finite. Again, using the Taylor expansion,

$$f'(x) = \frac{f(x + h) - f(x - h)}{2h} - \tfrac{1}{6}h^2 f^{(3)}(\zeta) \tag{1-15}$$

where $f^{(3)}$ is the third derivative of $f(x)$ and $x - h \leq \zeta \leq x + h$. The key point is that the truncation error is now *quadratic* in h, which is a big improvement.

Using the Taylor expansions for $f(x + h)$ and $f(x - h)$ we can build a centered formula for the second derivative. It has the form

$$f''(x) = \frac{f(x + h) + f(x - h) - 2f(x)}{h^2} - \tfrac{1}{12}h^2 f^{(4)}(\zeta) \tag{1-16}$$

where $x - h \leq \zeta \leq x + h$. Again, the truncation error is quadratic in h. The best way to understand this formula is to think of the second derivative as being composed of a right derivative and left derivative.

You might think that the next step would be to cook up a more involved formula, maybe using $f(x \pm h)$ and $f(x \pm 2h)$, which has an even smaller truncation error. While such formulas exist and are occasionally used, equations (1-12), (1-15), and (1-16) serve as the workhorses for computing first and second derivatives.

Selecting a Value for *h*

A question that is always raised is, "What do you pick for h?" and the answer is often an inane comment such as, "A small number like 10^{-8}." Let's try to do better. First, we define the absolute error in our answer as

$$\varepsilon \equiv |(\text{true value}) - (\text{computed value})| \tag{1-17}$$

If round-off error is negligible, then we only have to worry about truncation error. Computing $f'(x)$ by the central difference formula, equation (1-15), we should choose h as

$$h < \sqrt{\frac{6\varepsilon}{|f^{(3)}(\zeta)|}} \tag{1-18}$$

Generally, we don't know $f^{(3)}(\zeta)$, but often we can set a bound. For example, if $f(x) = \sin(x)$, then $|f^{(3)}(\zeta)| \leq 1$ so if we want an absolute error of $\varepsilon \approx 10^{-6}$, then we should take $h \approx 2 \times 10^{-3}$.

In the real world, we often don't do such a nice analysis, for a variety of reasons (little information on $f(x)$, problems with round-off, laziness, etc.). However, we may often use physical intuition. You should ask yourself, "On what scale is the function almost linear?" For example, consider the flight of a baseball.

While the motion is roughly parabolic, on the scale of a millisecond the velocity is approximately constant. If our function $y(t)$ is the height as a function of time for the trajectory of a baseball, we know that to find the vertical velocity, dy/dt, we should be able to use an increment of a millisecond.

An alternative approach is to pick a value of h, use it, try a smaller value of h, compare the two answers, and if they are close enough, assume everything is fine. Sometimes we automate testing various values of h; the program is then said to be "adaptive" (we will build such a program in Chapter 3). As with any numerical method, blind application of this technique is discouraged, although with just a bit of care it can be used successfully.

Numerical Differentiation Program

Let's test these new formulas by writing a simple MATLAB program to compute the first derivative of the sine function. The program deriv is given in Listing 1.7. The program is very simple; first the user is prompted for the value of x at which $f'(x)$ is to be computed. The step size starts at $h = \frac{1}{2}$ (line 5) but is halved with each pass of the for loop (see line 13). The two different versions (right and centered) of the first derivative are computed on lines 9 and 10.

LISTING 1.7 Program deriv. Uses the right and centered formulas to compute first derivatives. Plots and compares results with the exact answer. Uses fund (Listing 1.8).

```
1    % deriv - Program to compute the numerical derivatives of the
2    % function fund and compare with the exact values
3    clear; help deriv;   % Clear memory and print header
4    x = input('Enter the location (x) - ');
5    h = .5;   % Initial value of the grid spacing
6    for i=1:9
7      hplot(i) = h;   % Record the value of h for plotting
8      % Right and centered first derivatives
9      derivR(i) = (fund(x+h,0) - fund(x,0))/h;
10     derivC(i) = (fund(x+h,0) - fund(x-h,0))/(2*h);
11     % True first derivative
12     derivT(i) = fund(x,1);
13     h = h/2;    % Halve the grid spacing at each iteration
14   end
15   % Plot numerical and true derivative on semi-log scale
16   semilogx(hplot,derivR,'+',hplot,derivC,'o',...
17                              hplot,derivT,'-')
18   xlabel('Interval size (h)')
19   ylabel('First derivative')
20   title('Right +; centered o;  true -')
```

To modularize the deriv program, we define a MATLAB function which returns either $f(x)$ or its derivatives. The function fund (x, n) (see Listing 1.8) returns the nth derivative of the sine function. If called as fund (x, 0), it returns the value of the 0th derivative, which is just the original function.

LISTING 1.8 Function fund. Used by deriv (Listing 1.7) to return $f(x)$ or its derivative.

```
1    function f = fund(x,n)
2    %   Return function value or derivative
3    %   fund(x,0) returns the value of the function
4    %   fund(x,n) returns the value of the nth derivative
5    %                if n=1 or 2 else an error
6    if ( n == 0 )
7       f = sin(x);
8    elseif ( n == 1 )
9       f = cos(x);
10   elseif ( n == 2 )
11      f = -sin(x);
12   else
13      error('Error; bad value for n in fund')
14   end
15   return;
```

Finally, lines 16–20 of deriv plot the data in the most natural format, a semilog graph. The right derivative is plotted as +'s, the centered derivative is plotted as o's, and the true value of the derivative is a solid line. For the value $x = 2.5$, we obtain the result shown in Figure 1.4 (see also Figure 1.1). As expected, the centered formula does a much better job than the right derivative formula.

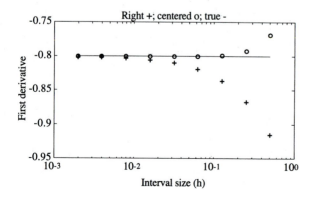

Figure 1.4 Graphical output from the deriv program for $x = 2.5$. Notice that the two numerical estimates for the first derivative improve as h decreases.

EXERCISES

20. Derive the centered formulas, equations (1-15) and (1-16), for $f'(x)$ and $f''(x)$. (Hint: Write out the Taylor series for $f(x + h)$ and $f(x - h)$, combine, and cancel terms.)

21. Check that the centered formulas for $f'(x)$ and $f''(x)$, equations (1-15) and (1-16), are exact for a quadratic (i.e., for $f(x) = ax^2 + bx + c$). Why is this true?

22. Modify deriv so that the absolute error between the approximate and exact derivative is computed, that is, $|g(x) - f'(x)|$, where $g(x)$ is the approximation to $f'(x)$ given by the right or centered formula. Using loglog graph the absolute error versus h for $x = 0, \pi/8, \pi/4, \pi/2, 3\pi/4$, and π. Confirm that the truncation errors for the right and centered first derivative formulas go as h and h^2, respectively.

23. Use the program from the previous exercise but increase the number of times the for loop is performed and notice when round-off error dominates truncation error. Try the values $x = 0, \pi/8, \pi/4, \pi/2, 3\pi/4$, and π. Does the round-off depend on x? Explain why. Compare the value of h where round-off error is significant with the value of eps.

24. Modify deriv so that it computes $f''(x)$ using the centered formula, equation (1-16). Obtain the absolute error between the approximate and exact values, that is, $|g(x) - f''(x)|$, where $g(x)$ is the approximation to $f''(x)$ given by the centered formula. Using loglog graph the absolute error versus h for $x = 0, \pi/8, \pi/4, \pi/2, 3\pi/4$, and π. Confirm that the truncation error goes as h^2.

25. For some values of x, the right derivative is more accurate than the centered derivative. To illustrate this, modify deriv so that for a fixed value of h the program computes the two estimates for $f'(x)$ and the true value for a range of values for x between 0 and 2π. Plot the absolute error for each estimate as a function of x. Explain why the right derivative formula is sometimes more accurate.

26. Modify fund so that instead of returning the sine function and its derivatives, it works with one of the following functions:
    (a) $(x - 1)^4$ (b) $\ln(x)$
    (c) e^{-x}/x^2 (d) $\tanh(x)$
    Repeat Exercise 1.22 using a variety of values for x appropriate for your function. For what values of x is the truncation error the greatest? Can you correlate your findings with the formulas for truncation error?

27. (a) Using Taylor expansion, derive the three-point forward difference formula

$$f'(x) = \frac{-3f(x) + 4f(x + h) - f(x + 2h)}{2h} + O(h^2)$$

and obtain an explicit expression for the error term.

    (b) Modify deriv so that it also computes $f'(x)$ using the above formula. Try a variety of values for x and comment on your results.

28. Using Taylor expansion, derive the centered difference formula for the third derivative

$$f^{(3)}(x) = \frac{f(x + 2h) - 2f(x + h) + 2f(x - h) - f(x - 2h)}{2h^3} + O(h^2)$$

but obtain an explicit expression for the error term.

BEYOND THIS CHAPTER

This chapter gives only a brief introduction to MATLAB. More elements of the language will be discussed as we need them in later chapters. The MATLAB manual includes several tutorial chapters along with a complete reference to the language. MATLAB's demos also cover a variety of interesting applications. Since their source code is included, the demos are also good samples of MATLAB programming. The functions in MATLAB's standard toolbox are also worth opening up and reading.

MATLAB can be upgraded with the addition of supplemental toolboxes. These are collections of advanced routines for specialized topics. Some of the currently available toolboxes include signal processing, spline curve fitting, and nonlinear optimization.

The Lagrange formulation for assembling an interpolating polynomial has only one advantage: It is easy to understand and remember. Algorithmically, it is not the best method for polynomial interpolation for a variety of reasons, including susceptibility to round-off error when fitting higher-order polynomials. A superior approach is to build a divided difference table and use the Newtonian formulation.[5] In general, round-off error in numerical schemes can be quantitatively estimated using backward error analysis.[6]

A few difference formulas for estimating derivatives are presented in this chapter. Higher-order difference formulas are not difficult to construct using the Taylor expansion. One common use of specialized, high-accuracy formulas is for estimating derivatives at boundaries to compute the flux at the boundary. Partial derivatives (e.g., Laplacian operator) may be constructed by applying our standard formulas in each direction; specialized formulas also exist for partial derivatives. Tables of finite difference formulas are found in many numerical analysis references.[7]

Finally, you'll discover that most of the fundamental elements of numerical analysis were introduced long before the invention of the electronic computer. This is clear from the names of the basic algorithms; the litany includes Newton, Euler, Gauss, Jacobi, and many other famous physicists and mathematicians. For a historical account of the development of numerical analysis, see Goldstine.[8]

---

[5] S. Conte and C. de Boor, *Elementary Numerical Analysis* (New York: McGraw-Hill, 1980), chapter 2.

[6] J. Wilkinson, *Rounding Errors in Algebraic Processes* (Englewood Cliffs, N.J.: Prentice Hall, 1963).

[7] For example, M. Abramowitz and I. Stegun, *Handbook of Mathematical Functions* (New York: Dover, 1972), section 25.3; D. Anderson, J. Tannehill, and R. Pletcher, *Computational Fluid Mechanics and Heat Transfer* (New York: Hemisphere, 1984), chapter 3.

[8] H. H. Goldstine, *A History of Numerical Analysis from the 16th through the 19th Century* (New York: Springer-Verlag, 1977).

APPENDIX 1A: FORTRAN LISTINGS

LISTING 1A.1 Program orthog. Determines if a pair of vectors is orthogonal by computing their dot product.

```fortran
      program orthog
! Program to test if a pair of vectors is orthogonal
      real va(3),vb(3)

      write (*,*) 'Enter the first vector as x,y,z'
      read (*,*) va(1),va(2),va(3)
      write (*,*) 'Enter the second vector as x,y,z'
      read (*,*) vb(1),vb(2),vb(3)
! Compute the dot product of va and vb
      adotb = 0
      do i=1,3
        adotb = adotb + va(i)*vb(i)
      end do
! Vectors are orthogonal if dot product is zero
      if( adotb .eq. 0 ) then
        write (*,*) 'Vectors are orthogonal'
      else
        write (*,*) 'Vectors are NOT orthogonal'
        write (*,*) 'Dot product = ',adotb
      end if
      stop
      end
```

LISTING 1A.2 Program interp. Uses the intrpf (Listing 1A.3) function to interpolate between data points.

```fortran
      program interp
! Program to interpolate data using Lagrange polynomial to
! fit quadratic to three data points
      parameter( maxplot = 1000 )
      real x(3),y(3),xr(2),xi(maxplot),yi(maxplot),intrpf

      write (*,*) 'Enter data points as x,y pairs'
      do i=1,3
        write (*,*) 'Enter the data point'
        read (*,*) x(i),y(i)
      end do
      write (*,*) 'Enter range of x values as x_min,x_max'
      read (*,*) xr(1),xr(2)
      nplot = 100        ! Number of points for interpolation curve
```

```
      do i=1,nplot
        xi(i) = xr(1) + (xr(2)-xr(1))*(i-1.)/(nplot-1.)
        yi(i) = intrpf(xi(i),x,y)   ! Find interpolation point
      end do
! Print out the plotting variables -
!    x,y,xi,yi
!
      open(11,file='x.dat')
      open(12,file='y.dat')
      open(13,file='xi.dat')
      open(14,file='yi.dat')
      do i=1,3
        write (11,*) x(i)
        write (12,*) y(i)
      end do
      do i=1,nplot
        write (13,*) xi(i)
        write (14,*) yi(i)
      end do
      stop
      end
```

LISTING 1A.3   Function `intrpf`. Uses a quadratic to interpolate between three data points.

```
      real function intrpf(xi,x,y)
! Function to compute interpolation point using quadratic
! Lagrange polynomial
      real x(3),y(3)

      yi = (xi-x(2))*(xi-x(3))/((x(1)-x(2))*(x(1)-x(3)))*y(1)
    &  + (xi-x(1))*(xi-x(3))/((x(2)-x(1))*(x(2)-x(3)))*y(2)
    &  + (xi-x(1))*(xi-x(2))/((x(3)-x(1))*(x(3)-x(2)))*y(3)
      intrpf = yi
      return
      end
```

LISTING 1A.4   Program `rndoff`. Illustrates round-off error in a simple calculation.

```
      program rndoff
! Demo program illustrating round-off error
      parameter(maxn = 100)
      real hplot(maxn),eplot(maxn)

      h=1.
      do i=1,21
        temp = (10.+h)-10.
```

```
      hplot(i) = h
      eplot(i) = abs(temp-h)/h
      h = h/10.
   end do
! Print out the plotting variables -
!   hplot,eplot
!
      open(11,file='hplot.dat')
      open(12,file='eplot.dat')
      do i=1,21
         write (11,*) hplot(i)
         write (12,*) eplot(i)
      end do
      stop
      end
```

LISTING 1A.5 Program `factn`. Computes factorial using the naive approach.

```
      program factn
! Naive program for computing factorial
      real nf

      write (*,*) 'Enter the value of n'
      read (*,*) n
      nf = 1
      do i=1,n
         nf = nf * i   ! Compute n! = n*(n-1)*...*3*2*1
      end do
      write (*,*) 'n! = ',nf
      stop
      end
```

LISTING 1A.6 Program `facts`. Computes factorial using Stirling's approximation.

```
      program facts
! Program to compute n! using Stirling's approximation
      real nf,log_nf,mantisa
      integer exponnt

      pi = 4.*atan(1.)   ! = 3.14159...
      write (*,*) 'Enter the value of n'
      read (*,*) n
      if ( n .lt. 30 ) then   ! If n < 30, use the naive method
         nf = 1                ! to find as n! = n*(n-1)*...*3*2*1
         do i=1,n
            nf = nf * i
```

```
      end do
      write (*, *)  'n! = ', nf
   else
      ! Use Stirling's approximation
      temp = 1. + 1./(12.*n) + 1./(288.*n**2)
      log_nf = .5*alog10(2*pi) + (n+0.5)*alog10(float(n))
&            - n*alog10(exp(1.)) + alog10(temp)
      exponnt = int(log_nf)             ! Exponent of n!
      mantisa = 10**(log_nf-exponnt)    ! Mantissa of n!
      write (*, *)  'n! = ', mantisa, ' * 10** ', exponnt
   end if
   stop
   end
```

LISTING 1A.7 Program `deriv`. Uses the right and centered formulas to compute first derivatives. Uses `fund` (Listing 1A.8).

```
      program deriv
! Program to compute the numerical derivatives of the function
! fund and compare with exact values
      parameter (maxn=100)
      real derivR(maxn), derivC(maxn), derivT(maxn), hplot(maxn)

      write (*, *) 'Enter the location (x) -'
      read (*, *) x
      h = 0.5    ! Initial grid spacing for numerical derivatives
      do i=1, 9
        hplot(i) = h  ! Record the value of h for plotting
        ! Right and centered first derivatives
        derivR(i) =    (fund(x+h, 0)-fund(x, 0))/h
        derivC(i) =    (fund(x+h, 0)-fund(x-h, 0))/(2*h)
        ! True first derivative
        derivT(i) = fund(x, 1)
        h = h/2.       ! Decrease the grid spacing
      end do
! Print out the plotting variables -
!    hplot, derivR, derivC, derivT
!
      open(11, file='hplot.dat')
      open(12, file='derivR.dat')
      open(13, file='derivC.dat')
      open(14, file='derivT.dat')
      do i=1, 9
        write (11, *)  hplot(i)
        write (12, *)  derivR(i)
        write (13, *)  derivC(i)
        write (14, *)  derivT(i)
```

```
      end do
      stop
      end
```

LISTING 1A.8 Function fund. Used by deriv (Listing 1A.7) to return $f(x)$ or its derivative.

```
          function fund(x,n)
! Return function value or derivative
! fund(x,0) returns the value of the function
! fund(x,n) returns the value of the nth derivative
!             if n=1 or 2 else an error
          if( n .eq. 0 ) then
            fund = sin(x)             ! Original function
          else if( n .eq. 1 ) then
            fund = cos(x)             ! First derivative
          else if( n .eq. 2 ) then
            fund = -sin(x)            ! Second derivative
          else
            write (*,*) 'Error; bad value for n in fund'
            stop
          end if
          return
          end
```

Chapter 2
Ordinary Differential Equations I:
Basic Methods

In this chapter we solve one of the first problems you considered in freshman physics: the flight of a baseball. Without air resistance the problem is easy to solve. However, to include realistic drag, we need to compute the solution numerically. In the process we cover some basic methods for solving ordinary differential equations (ODEs). In the latter half of the chapter we visit another old friend, the simple pendulum, but without the small angle approximation.

2.1 PROJECTILE MOTION

Basic Equations

Consider simple projectile motion, say the flight of a baseball. To describe the motion we must compute the vector position $\mathbf{r}(t)$ and vector velocity $\mathbf{v}(t)$ of the particle. The basic equations of motion are

$$\frac{d\mathbf{v}}{dt} = \frac{1}{m}\,\mathbf{F}_a(\mathbf{v}) - g\hat{\mathbf{y}}; \qquad \frac{d\mathbf{r}}{dt} = \mathbf{v} \qquad (2\text{-}1)$$

where m is the mass of the particle. The force due to air resistance is $\mathbf{F}_a(\mathbf{v})$, the gravitational acceleration is g, and $\hat{\mathbf{y}}$ is the unit vector in the y-direction. The motion is two-dimensional, so we may ignore the z-component and work in the xy plane.

Air resistance increases with the velocity of the particle. The precise form for $\mathbf{F}_a$ depends on the flow around the projectile. Commonly, it is approximated as

$$\mathbf{F}_a(\mathbf{v}) = -\tfrac{1}{2} C_d \rho A |\mathbf{v}| \mathbf{v} \tag{2-2}$$

where C_d is the drag coefficient, ρ is the density of the air, and A is the cross-sectional area of the projectile. The drag coefficient is a dimensionless parameter that depends on the geometry of the projectile. The more streamlined the object, the smaller the coefficient. Notice that the drag force varies as the square of the magnitude of the velocity ($|\mathbf{F}_a| \propto |\mathbf{v}|^2$) and, of course, acts in the direction opposite the velocity.

We know how to solve the equations of motion if air resistance is negligible. For example, say that the particle starts at the origin [$\mathbf{r}(t = 0) = 0$] with an initial velocity $v_o \equiv |\mathbf{v}(t = 0)|$ at an angle θ from the horizontal. The horizontal range is then

$$x_{\text{max}} = \frac{2v_o^2}{g} \sin \theta \cos \theta \tag{2-3}$$

and the maximum height achieved is

$$y_{\text{max}} = \frac{v_o^2}{2g} \sin^2 \theta \tag{2-4}$$

The time of flight is

$$t_{\text{fl}} = \frac{2v_o}{g} \sin \theta \tag{2-5}$$

Again, these expressions only hold when there is no air resistance. It is easy to check that the maximum horizontal range is achieved when the initial velocity makes an angle of $\theta = 45°$ with the horizontal. We want to keep this information in mind when we build our simulation. If one knows the exact solution for a special case, one should always check that the program works correctly for that case.

The projectile we consider is a baseball; from common experience, we know that MKS units are suitable for describing its motion. Some important parameters in the problem are:

Mass of a baseball: 0.145 kg
Radius of a baseball: 3.7 cm
Density of air at 20°C and 1 atm: 1.20 kg/m³
Typical drag coefficient for a baseball: 0.35
Typical distance to center field fence: 125 m
Typical velocity of a batted ball: 35–55 m/s

The drag coefficient is an average value for the typical range of velocities found in baseball.[1]

Euler Method

The equations of motion that we want to solve numerically may be written as

$$\frac{d\mathbf{v}}{dt} = \mathbf{a}(\mathbf{r}, \mathbf{v}); \qquad \frac{d\mathbf{r}}{dt} = \mathbf{v} \tag{2-6}$$

where $\mathbf{a}$ is the acceleration. Notice that this is the more general form of the equations. In our case the acceleration is only a function of $\mathbf{v}$ (because of drag); in later problems (e.g., orbits of comets) the acceleration will depend on position.

After what you learned in Chapter 1, you are probably ready to replace the time derivatives with the "centered" formula. For the moment, however, let's start by using the right derivative formula

$$\frac{df}{dt} = \frac{f(t + \tau) - f(t)}{\tau} + O(\tau) \tag{2-7}$$

where τ is called the time increment or time step. Our equations of motion are

$$\frac{\mathbf{v}(t + \tau) - \mathbf{v}(t)}{\tau} + O(\tau) = \mathbf{a}(\mathbf{r}(t), \mathbf{v}(t)) \tag{2-8a}$$

$$\frac{\mathbf{r}(t + \tau) - \mathbf{r}(t)}{\tau} + O(\tau) = \mathbf{v}(t) \tag{2-8b}$$

or

$$\mathbf{v}(t + \tau) = \mathbf{v}(t) + \tau\mathbf{a}(\mathbf{r}(t), \mathbf{v}(t)) + O(\tau^2) \tag{2-9a}$$

$$\mathbf{r}(t + \tau) = \mathbf{r}(t) + \tau\mathbf{v}(t) + O(\tau^2) \tag{2-9b}$$

Notice that $\tau O(\tau) = O(\tau^2)$. This numerical scheme is called the *Euler method*. Before discussing the relative merits of this approach, let's see how it would be used in practice.

First, we introduce the notation,

$$f_n \equiv f((n - 1)\tau); \qquad n = 1, 2, \ldots \tag{2-10}$$

[1] C. Frohlich, "Aerodynamic drag crisis and its possible effect on the flight of baseballs," *Am. J. Phys.*, **52**, 325–34 (1984).

so $f_1 = f(t = 0)$. Our equations for the Euler method (dropping the error term) now take the form

$$\mathbf{v}_{n+1} = \mathbf{v}_n + \tau \mathbf{a}_n \tag{2-11a}$$

$$\mathbf{r}_{n+1} = \mathbf{r}_n + \tau \mathbf{v}_n \tag{2-11b}$$

The calculation of the trajectory would proceed as follows:

1. Specify the initial conditions, $\mathbf{r}_1$ and $\mathbf{v}_1$.
2. Choose a time step τ.
3. Calculate the acceleration given the current $\mathbf{r}$ and $\mathbf{v}$.
4. Use the Euler method to compute the new $\mathbf{r}$ and $\mathbf{v}$.
5. Go to step 3 until enough trajectory points have been computed.

The method computes a set of values for $\mathbf{r}_n$ and $\mathbf{v}_n$ that gives us the trajectory, at least at a discrete set of points.

Euler–Cromer and Midpoint Methods

A simple (and for now unjustified) modification of the Euler method is to use the following equations:

$$\mathbf{v}_{n+1} = \mathbf{v}_n + \tau \mathbf{a}_n \tag{2-12a}$$

$$\mathbf{r}_{n+1} = \mathbf{r}_n + \tau \mathbf{v}_{n+1} \tag{2-12b}$$

Notice the subtle change: The updated velocity is used in the second equation. This formula is called the *Euler–Cromer method*.[2] The truncation error is still of order τ^2, so it doesn't look like we gain much. Interestingly, we will see that this form is markedly superior to the Euler method in some cases.

Our democratic nature makes us think of using the *midpoint method*,

$$\mathbf{v}_{n+1} = \mathbf{v}_n + \tau \mathbf{a}_n \tag{2-13a}$$

$$\mathbf{r}_{n+1} = \mathbf{r}_n + \tau \frac{\mathbf{v}_{n+1} + \mathbf{v}_n}{2} \tag{2-13b}$$

Notice that we average the two velocities. Plugging the velocity equation into the position equation, we see that

$$\mathbf{r}_{n+1} = \mathbf{r}_n + \tau \mathbf{v}_n + \tfrac{1}{2}\mathbf{a}_n \tau^2 \tag{2-14}$$

---

[2] A. Cromer, "Stable solutions using the Euler approximation," *Am. J. Phys.*, **49**, 455–59 (1981).

which really makes this look appealing. The truncation error is still of order τ^2 in the velocity equation, but for position the truncation error is now τ^3. Indeed, for projectile motion this method works better than the other two. Unfortunately, we will later be disappointed to find that the midpoint method gives relatively poor results for other physical systems.

Local Error versus Global Error

To judge the accuracy of these methods we need to distinguish between local truncation error and global truncation error. So far, the truncation error we have discussed has been the local error, that is, the error made in a single time step. In a typical problem we want to evaluate a trajectory from $t = 0$ to $t = T$. The number of time steps is $N_\tau = T/\tau$; notice that if we reduce τ we have to take more steps. If the local error is $O(\tau^n)$, then we estimate the global error as

$$\text{global error} \propto N_\tau \times (\text{local error})$$
$$= N_\tau O(\tau^n) = (T/\tau)O(\tau^n) = TO(\tau^{n-1}) \tag{2-15}$$

For example, the Euler method has a local truncation error of $O(\tau^2)$ but a global truncation error of $O(\tau)$. Of course, this analysis only gives us an estimate since we don't know if the local errors will accumulate or cancel (i.e., constructively or destructively interfere). The actual global error for a numerical scheme is highly dependent on the problem we are studying.

Baseball Program

Listing 2.1 gives a simple program, called `balle`, that uses the Euler method to compute the trajectory of a baseball. The initial position and velocity are input as vectors in the form [x-component   y-component]. The `for` loop (lines 15–25) computes values of $\mathbf{r}_n$ and $\mathbf{v}_n$ until the y-component of $\mathbf{r}_n$ is less than zero (line 22), that is, the ball hits the ground. On line 18 the acceleration due to air resistance is computed in vector form; the MATLAB function `norm(v)` returns $|\mathbf{v}|$.

LISTING 2.1   Program `balle`. Computes the trajectory of a baseball, including air resistance.

```
1    % balle - Program to compute the trajectory of a baseball
2    % using the Euler method.
3    clear;  help balle;  % Clear memory and print header
4    r = input('Enter initial position r=[x,y] - ');   % (meters)
5    v = input('Enter initial velocity v=[vx,vy] - '); % (m/sec)
6    tau = input('Enter timestep, tau - ');  % (sec)
7    Cd = 0.35;      % Drag coefficient (dimensionless)
8    rho = 1.2;      % Density of air (kg/m^3)
9    area = 4.3e-3;  % Cross-sectional area of projectile (m^2)
```

```
10   grav = 9.81;      % Gravitational acceleration (m/s^2)
11   mass = 0.145;     % Mass of projectile (kg)
12   air_const = -0.5*Cd*rho*area/mass;   % Air resistance constant
13   maxstep = 1000; % Maximum number of steps
14   %%%% MAIN LOOP %%%%
15   for istep=1:maxstep
16     xplot(istep) = r(1);     % Record trajectory for plot
17     yplot(istep) = r(2);
18     accel = air_const*norm(v)*v;    % Air resistance
19     accel(2) = accel(2)-grav;       % Gravity
20     r = r + tau*v;                  % Euler step
21     v = v + tau*accel;
22     if ( r(2) < 0 )   % Break out of loop when ball hits ground
23       break;          % i.e. when y-position is < 0
24     end
25   end
26   xplot(istep+1) = r(1);   yplot(istep+1) = r(2);
27   fprintf('Maximum range is %g meters\n',r(1))
28   fprintf('Time of flight is %g seconds\n',istep*tau)
29   % Mark the location of the ground by a straight line
30   xground = [0 xplot(istep+1)]; yground = [0 0];
31   % Graph the trajectory of the baseball
32   plot(xplot,yplot,'+',xground,yground,'-');
33   xlabel('Range (m)')
34   ylabel('Height (m)')
35   title('Projectile motion')
```

The Euler scheme is used to update the position and velocity on lines 20 and 21. Notice that if we interchange these lines, we have the Euler–Cromer method. The trajectory is recorded in the arrays `xplot` and `yplot`; a graph of the trajectory is drawn on line 32. Finally, the maximum range and time of flight are displayed by lines 27 and 28.

Numerical Results

Before we try the program, let's use our physical intuition to determine some reasonable values to take as inputs. The initial conditions $r_1 = [0 \quad 0]$ m and $v_1 = [10 \quad 10]$ m/s give us a weakly hit ball (about 30 mph) for which air resistance should be small. Here the time of flight, neglecting air resistance, is about 2 s. To get a reasonable sketch of the trajectory we want about 40 to 50 data points, so we take $\tau = 0.05$ s. Here is what the output to the screen looks like when we run the program:

```
>>balle

    balle - Program to compute the trajectory of a baseball
    using the Euler method.
```

```
Enter initial position r=[x,y] - [0 0]

Enter initial velocity v=[vx,vy] - [10 10]

Enter timestep, tau - .05
Maximum range is 19.1440 meters
Time of flight is 2.05000 seconds
```

The range and time of flight are within a few percent of the values predicted by equations (2-3) and (2-5) (neglecting air resistance). The trajectory is shown in Figure 2.1; it looks parabolic, as expected.

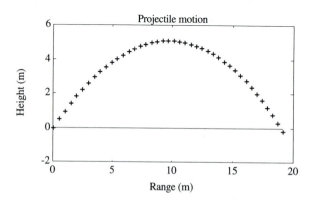

Figure 2.1 Output from balle for $r_1 = [0 \quad 0]$ m, $v_1 = [10 \quad 10]$ m/s, and $\tau = 0.05$ s.

Next, let's use a larger value for the velocity, $v_1 = [40 \quad 40]$ m/s (about 125 mph). Since the time of flight is longer, we try a time step of $\tau = 0.1$ s. Due to the air resistance, we find the range reduced to less than half of its theoretical maximum. The trajectory is shown in Figure 2.2; notice how it changes from a parabola to a sharply dropping curve. This is why a ball driven deep into the outfield always appears to be caught as though it is falling almost straight down.

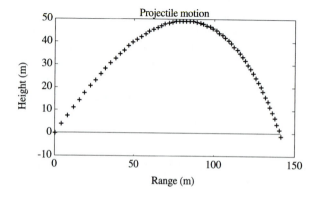

Figure 2.2 Output from balle for $r_1 = [0 \quad 0]$ m, $v_1 = [40 \quad 40]$ m/s, and $\tau = 0.1$ s.

This ordinary differential equation is easily solved to obtain

$$\theta(t) = C_1 \cos(2\pi t/T_s + C_2) \tag{2-18}$$

where the constants C_1 and C_2 are determined by the initial values of θ and $\omega = d\theta/dt$. The small angle period, T_s, is

$$T_s = 2\pi \sqrt{\frac{L}{g}} \tag{2-19}$$

This approximation is reasonably good for initial displacements of about 20° or less.

Without the small angle approximation, the equation of motion is more difficult to solve. However, we know from experience that the motion is still periodic. In fact, it is possible to obtain an expression for the period without explicitly solving for $\theta(t)$. The total energy is

$$E = \tfrac{1}{2} mL^2\omega^2 - mgL \cos \theta \tag{2-20}$$

where m is the mass of the bob. The total energy is conserved and equal to $E = -mgL \cos \theta_m$, where θ_m is the maximum angle. From the above, we have

$$\tfrac{1}{2} mL^2\omega^2 - mgL \cos \theta = -mgL \cos \theta_m \tag{2-21}$$

or

$$\omega^2 = \frac{2g}{L} (\cos \theta - \cos \theta_m) \tag{2-22}$$

or

$$\frac{d\theta}{dt} = \sqrt{\frac{2g}{L} (\cos \theta - \cos \theta_m)} \tag{2-23}$$

so

$$dt = \frac{d\theta}{\sqrt{\frac{2g}{L} (\cos \theta - \cos \theta_m)}} \tag{2-24}$$

In one period, the pendulum swings from $\theta = \theta_m$ to $\theta = -\theta_m$ and back to $\theta = \theta_m$. Thus, in half a period, the pendulum swings from $\theta = \theta_m$ to $\theta = -\theta_m$. Last, by the same argument, in a quarter period the pendulum swings from $\theta = \theta_m$ to $\theta = 0$, so integrating both sides,

$$\frac{T}{4} = \sqrt{\frac{L}{2g}} \int_0^{\theta_m} \frac{d\theta}{\sqrt{\cos\theta - \cos\theta_m}} \tag{2-25}$$

This integral may be rewritten in terms of special functions by using the identity $\cos 2\theta = 1 - 2\sin^2\theta$, so

$$T = 2\sqrt{\frac{L}{g}} \int_0^{\theta_m} \frac{d\theta}{\sqrt{\sin^2(\theta_m/2) - \sin^2(\theta/2)}} \tag{2-26}$$

Introducing $K(x)$, the complete elliptic integral of the first kind,[6]

$$K(x) \equiv \int_0^{\pi/2} \frac{dz}{\sqrt{1 - x^2\sin^2 z}} \tag{2-27}$$

we may write the period as

$$T = 4\sqrt{\frac{L}{g}}\, K(\sin\tfrac{1}{2}\theta_m) \tag{2-28}$$

using the change of variable $\sin z = \sin(\theta/2)/\sin(\theta_m/2)$. For small values of θ_m, we may expand $K(x)$ to obtain

$$T \approx 2\pi\sqrt{\frac{L}{g}}\left(1 + \frac{1}{16}\theta_m^2 + \ldots\right) \tag{2-29}$$

Notice that the leading term is the small angle approximation discussed at the beginning of the section.

Verlet Method

Before programming the pendulum problem, let's look at another method for computing the motion of an object. For the pendulum, the generalized position and velocity are θ and ω, but to maintain the same notation as in Section 2.1 we'll

---

[6] I. S. Gradshteyn and I. M. Ryzhik, *Table of Integrals, Series and Products* (New York: Academic Press, 1965).

work with $\mathbf{r}$ and $\mathbf{v}$. To try a fresh approach, suppose we start with

$$\frac{d\mathbf{r}}{dt} = \mathbf{v} \tag{2-30a}$$

$$\frac{d^2\mathbf{r}}{dt^2} = \mathbf{a} \tag{2-30b}$$

Using the central difference formulas for first and second derivatives from the previous chapter, we have

$$\frac{\mathbf{r}_{n+1} - \mathbf{r}_{n-1}}{2\tau} + O(\tau^2) = \mathbf{v}_n \tag{2-31a}$$

$$\frac{\mathbf{r}_{n+1} + \mathbf{r}_{n-1} - 2\mathbf{r}_n}{\tau^2} + O(\tau^2) = \mathbf{a}_n \tag{2-31b}$$

Again we use the notation $f_n = f(t = (n-1)\tau)$ where τ is the time step. Equations (2-31a, b) may be rewritten as

$$\mathbf{v}_n = \frac{\mathbf{r}_{n+1} - \mathbf{r}_{n-1}}{2\tau} + O(\tau^2) \tag{2-32a}$$

$$\mathbf{r}_{n+1} = 2\mathbf{r}_n - \mathbf{r}_{n-1} + \tau^2 \mathbf{a}_n + O(\tau^4) \tag{2-32b}$$

These equations, known as the *Verlet method*, may look strange at first but they are easy to use. Suppose that we know $\mathbf{r}_1$ and $\mathbf{r}_2$. Using equation (2-32b) we get $\mathbf{r}_3$ and then compute $\mathbf{v}_2$ using (2-32a). Knowing $\mathbf{r}_2$ and $\mathbf{r}_3$ we may now compute $\mathbf{r}_4$ and $\mathbf{v}_3$, and so forth.

The Verlet method has the disadvantage that it is not "self-starting." Usually we have the initial conditions $\mathbf{r}_1$ and $\mathbf{v}_1$ and not $\mathbf{r}_1$ and $\mathbf{r}_2$. What we do is use one of the other methods (e.g., Euler) to compute $\mathbf{r}_2$ (e.g., $\mathbf{r}_2 = \mathbf{r}_1 + \mathbf{v}_1\tau$). After the first time step, the other steps are evaluated using the Verlet method.

The Verlet method has several advantages. First, the position equation has a good truncation error. Second, if the force is only a function of position and if we only care about the trajectory of the particle and not its velocity (as in many celestial mechanics problems), we can completely skip the velocity calculation. The method is popular for computing trajectories in systems with many particles, for example, the study of fluids at the microscopic level.

Simple Pendulum Program

The equations of motion are

$$\frac{d\omega}{dt} = \alpha(\theta); \qquad \frac{d\theta}{dt} = \omega \tag{2-33}$$

where the angular acceleration $\alpha(\theta) = -(g/L)\sin\theta$. The Euler method for solving these ODEs is to iterate the equations

$$\theta_{n+1} = \theta_n + \tau\omega_n$$
$$\omega_{n+1} = \omega_n + \tau\alpha_n$$

(2-34)

For the Verlet method, since we only need the angle and not the velocity, we just use the equation

$$\theta_{n+1} = 2\theta_n - \theta_{n-1} + \tau^2\alpha_n \qquad (2\text{-}35)$$

The Euler scheme may be used to start the iteration by taking a single backward step to compute θ_0 given the initial conditions, θ_1 and ω_1.

Instead of using MKS units, we'll use the dimensionless units natural to the problem. There are only two parameters in the problem, g and L, and they always appear in the ratio g/L. Setting this ratio to unity, the small amplitude period is $T_s = 2\pi$. In other words, we need only one unit in the problem: a time scale. We set our unit of time such that the small amplitude period is 2π.

Listing 2.2 gives a simple program, called pendul, that computes the motion of a simple pendulum. The pendulum is initially at rest (omega = 0) and at an angle theta; the angle is input and output in degrees but used in radians in the calculations. The main loop computes nstep points on the trajectory using the Euler method (lines 19 and 20). The angle theta as a function of time is recorded (lines 15 and 16) and plotted (line 36).

LISTING 2.2 Program pendul. Computes the time evolution of a simple pendulum using the Euler method.

```
1    % pendul - Program to compute the motion of a simple pendulum
2    % using the Euler method
3    clear;  help pendul        % Clear the memory and print header
4    theta = input('Enter initial angle (in degrees) - ');
5    theta = theta*pi/180;      % Convert angle to radians
6    omega = 0;                 % Set the initial velocity
7    tau = input('Enter time step - ');
8    g_over_L = 1;              % The constant g/L
9    time_old = -1;             % Fake value (see below)
10   irev = 0;                  % Used to count number of reversals
11   nstep = 300;               % Number of time steps
12   time = 0;
13   %%%% MAIN LOOP %%%%
14   for istep=1:nstep
15     t_plot(istep) = time;             % Record time and angle
16     th_plot(istep) = theta*180/pi;  % for plotting
```

```
17    accel = -g_over_L*sin(theta);    % Gravitational
                                            acceleration
18    theta_old = theta;
19    theta = theta + tau*omega;       % Euler method
20    omega = omega + tau*accel;
21    time = time + tau;
22    if ( theta*theta_old < 0 )  % Test position for sign change
23       fprintf('Turning point at time t= %f \n',time);
24       if ( time_old < 0 )         % If this is the first change,
25          time_old = time;         % just record the time
26       else
27          irev = irev + 1;         % Increment the number of
                                            reversals
28          period(irev) = 2*(time - time_old);
29          time_old = time;
30       end
31    end
32 end
33 fprintf('Average period = %g +/- %g\n',mean(period),...
34                              std(period)/sqrt(irev));
35 % Graph the oscillations
36 plot(t_plot,th_plot,'+');
37 xlabel('Time');
38 ylabel('Theta (degrees)');
39 title('Simple pendulum');
```

The half period is computed by looking for when the angle changes sign (lines 22–31). Notice that if theta and theta_old are of opposite sign their product is negative. The variable time_old is the time of the last reversal; initially it is set to the dummy value of −1 (line 9). One could improve this estimate by using the intrpf function (see Exercise 2.14).

Each reversal gives us an estimate for the period, presumably accurate to within one time step. The estimated period from each reversal is recorded (line 28) and the average value is output (line 33). The MATLAB function mean(a) returns the sample mean of the data list $a(i)$,

$$\text{mean}(\text{a}) = \langle a \rangle = \frac{1}{N} \sum_{i=1}^{N} a(i) \tag{2-36}$$

The function std(a) returns the sample standard deviation of the data list $a(i)$,

$$\text{std}(\text{a}) = \sqrt{\frac{1}{N-1} \sum_{i=1}^{N} [a(i) - \langle a \rangle]^2} \tag{2-37}$$

Using the standard deviation, we obtain an estimate for the error in the period (line 34).

The pendul program is easy to modify to implement the Verlet scheme. First, replace lines 18–20 with

```
theta_new = 2*theta - theta_old + tau^2*accel;
theta_old = theta;              % Verlet method
theta = theta_new;
```

The Verlet method needs an initial value for theta_old; a simple way to set it is to add the line

```
% Take one backward Euler step to start Verlet
theta_old = theta - omega*tau;
```

just outside the for loop on line 14.

Numerical Results

To check the program, we first try a small value for the initial angle, θ_m, since we know the period in this case should be 2π. Taking $\tau = 0.1$ we have about 60 data points per oscillation; the program takes 300 steps, so we should get about five oscillations. Here is what a typical run using the Euler method looks like:

```
>>pendul

   pendul - Program to compute the motion of a simple
pendulum using the Euler method


   Enter initial angle (in degrees) - 10

   Enter time step - .1
Turning point at time t= 1.600000
Turning point at time t= 4.800000
Turning point at time t= 8.000000
Turning point at time t= 11.100000
Turning point at time t= 14.300000
Turning point at time t= 17.500000
Turning point at time t= 20.700000
Turning point at time t= 23.900000
Turning point at time t= 27.100000
Average period = 6.375 +/- 0.025
```

The average period is about 1.5% larger than the expected $T_s = 2\pi$. Our estimated error for the period is about $\pm \tau/2$ for each measurement. Five oscilla-

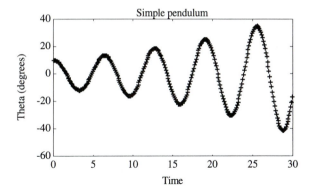

Figure 2.4 Output from pendul program using the Euler method. Initial angle is $\theta_m = 10°$ and the time step $\tau = 0.1$.

tions gives us nine reversals, so our estimated error for the period should be about $(\tau/2)/\sqrt{9} \cong 0.02$. Notice that this estimate is in good agreement with the result obtained using the sample standard deviation. So far, everything looks reasonable.

Unfortunately, the graphical output (Figure 2.4) shows us that the Euler method has a problem. The amplitude of the oscillation is steadily growing in time. Since the energy is proportional to the maximum angle, this means that the total energy is also increasing in time. The global truncation error in the Euler method is, in this system, accumulating. By lowering the time step to $\tau = 0.05$ and increasing the number of steps to 600, we may improve the results, as shown in Figure 2.5. The midpoint method (2-13) also suffers from this same numerical instability.

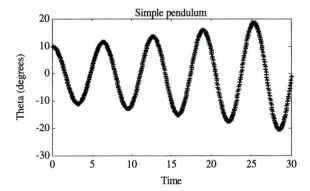

Figure 2.5 Output from pendul program using the Euler method. Initial angle is $\theta_m = 10°$ and the time step $\tau = 0.05$. Compare with Figure 2.4; note difference in axes scales.

Using the Verlet method with $\theta_m = 10°$ and $\tau = 0.1$, we obtain the plot shown in Figure 2.6. These results look much better; the amplitude of the oscillation stays between $\pm 10°$. Fortunately, neither the Verlet nor the Euler–Cromer method suffers from the instability found using the Euler method.

For larger angles, the form of the oscillation changes markedly. For the very large angle of $\theta_m = 170°$, using Verlet we get the trajectory shown in Figure 2.7.

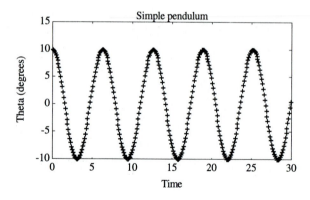

Figure 2.6 Output from pendul
program using the Verlet method.
Initial angle is $\theta_m = 10°$ and the time
step $\tau = 0.1$.

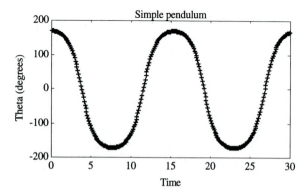

Figure 2.7 Output from pendul
program using the Verlet method.
Initial angle is $\theta_m = 170°$ and the time
step $\tau = 0.1$.

Notice how the curve tends to flatten at the turning points. In this case the period
is about $T = 15.4$, or two and a half times larger than the small angle period.

EXERCISES

8. In the small angle approximation, the total energy of a simple pendulum is

$$E = \tfrac{1}{2} mL^2\omega^2 + \tfrac{1}{2} mgL\theta^2 - mgL$$

Show analytically that E monotonically increases with time when the Euler method is
used to compute the motion.

9. The *leap-frog method* discretizes the equations of motion using centered first derivatives as

$$\frac{\mathbf{r}_{n+2} - \mathbf{r}_n}{2\tau} = \mathbf{v}_{n+1}; \qquad \frac{\mathbf{v}_{n+1} - \mathbf{v}_{n-1}}{2\tau} = \mathbf{a}_n$$

Notice that this scheme, like Verlet, is not self-starting. Also, the positions are only computed on odd steps ($r_1, r_3, \ldots$) while the velocities are found on even steps ($v_2, v_4, \ldots$). Modify pendul to use this method and compare with the other methods. Specifically, show its relation to the Euler–Cromer method.

10. Obtain a plot of the period T as a function of the initial angle θ_m using the Verlet method. Be sure to use a small enough value of τ to get at least 1% accuracy in T. On the same plot sketch the small angle approximation ($T = T_s$) and the approximation given by equation (2-29). Estimate the values of θ_m where each approximation breaks down.

11. Modify the pendul program so that it plots $\omega(t)$ versus $\theta(t)$, that is, a phase space plot. Instead of running for a fixed number of steps, have your program halt the calculation when the pendulum completes one period. Plot the data for initial angles of 10°, 45°, 90°, 120°, and 170°. Notice the difference in the shapes of the phase space orbits as a function of the initial angle. Be sure to use the Verlet method.

12. Consider a particle of mass m moving along the x-axis under influence of the force

$$F = \begin{cases} F_o & x < 0 \\ -F_o & x > 0 \end{cases}$$

where F_o is a constant.
    (a) Using the analysis presented in this section, show that the period is $T = 4\sqrt{2mx_m/F_o}$ where x_m is the maximum value for the particle's position.
    (b) Write a program to compute the trajectory of the particle and confirm the result from part (a).

13. Write a version of the pendul program that uses (a) the Euler–Cromer method or (b) the midpoint method. Run your program for the cases shown in Figures 2.4 to 2.7. Compare your results with those using Euler and Verlet.

14. The pendul program computes the period in a rather crude fashion. Improve it by using interpolation to estimate the time when θ changes sign using its three most recent values. You may want to use the intrpf function from Chapter 1. Comment on the relative improvement in the error.

15. Consider a pendulum with a harmonically driven pivot. The equation of motion is

$$\frac{d^2\theta}{dt^2} = -\frac{g + a_d(t)}{L} \sin\theta$$

where $a_d(t) = A_o \sin(2\pi t/T_d)$ is the time-varying acceleration of the pivot. Write a program that simulates this system; be sure to use a time step appropriate to the driving period, T_d. Show that when the amplitude of the driving acceleration is sufficiently high ($A_o \gg g$), the pendulum is stable in the *inverted* position (i.e., if $\theta(t = 0) \cong \pi$, then the pendulum oscillates about the point $\theta = \pi$).[7]

---

[7] L. D. Landau and E. M. Lifshitz, *Mechanics* (Oxford: Pergamon Press, 1969), section 30; J. A. Blackburn, H. J. T. Smith, and N. Grønbech-Jensen, "Stability and Hopf bifurcations in an inverted pendulum," *Am. J. Phys.*, **60**, 903–8 (1992).

16. The "velocity Verlet" scheme is defined as[8]

$$\mathbf{r}_{n+1} = \mathbf{r}_n + \tau\mathbf{v}_n + \tfrac{1}{2}\tau^2\mathbf{a}_n$$

$$\mathbf{v}_{n+1} = \mathbf{v}_n + \tfrac{1}{2}\tau(\mathbf{a}_n + \mathbf{a}_{n+1})$$

Notice that this scheme *is* self-starting. Prove that the values of $\mathbf{r}_n$ computed by this scheme are the same as those obtained by the standard Verlet algorithm.

17. Show that for ODEs of the form $d^2x/dt^2 = f(t)x(t)$, where f is a known function, the Verlet algorithm may be improved as

$$x_{n+1} = \frac{2x_n - x_{n-1} + \tau^2 a_n + \tfrac{1}{12}\tau^4(a_{n-1} - 2a_n)}{1 - \tfrac{1}{12}\tau^4 f_{n+1}} + O(\tau^6)$$

where $a(t) \equiv f(t)x(t)$. This scheme is known as *Numerov's method*; notice its excellent local truncation error. Numerov's method is often used to solve the time-independent Schrödinger equation in one dimension. [Hint: Apply $\left(1 + \dfrac{\tau^2}{12}\dfrac{d^2}{dt^2}\right)$ to both sides of (2-30b).]

BEYOND THIS CHAPTER

In this chapter we introduced some basic techniques for solving ordinary differential equations and applied them in two fundamental physics problems (projectile motion and the simple pendulum).[9] In the next chapter we cover some advanced techniques for solving ordinary differential equations. However, there are many instances for which you would want to stick with the basic methods. The most common scenario is the simulation of a large system of interacting particles (e.g., stars in a galaxy, electrons and ions in a plasma). In this case we are not interested in high-accuracy trajectories for individual particles. Instead, you would measure the collective, statistical properties in the system such as density and temperature. These statistical quantities are measured as time averages, so computational efficiency is essential. The Verlet method is a popular algorithm for these types of simulations.[10]

[8] W. C. Swope, H. C. Andersen, P. H. Berens, and K. R. Wilson, "A computer simulation method for the calculation of equilibrium constants for the formation of physical clusters of molecules: application to small water clusters," *J. Chem. Phys.*, **76**, 637–49 (1982).

[9] For many more examples, see H. Gould and J. Tobochnik, *An Introduction to Computer Simulation Methods*, part 1 (Reading, Mass.: Addison-Wesley, 1988); or M. L. De Jong, *Introduction to Computational Physics* (Reading, Mass.: Addison-Wesley, 1991).

[10] M. Allen and D. Tildesley, *Computer Simulation of Liquids* (Oxford: Clarendon Press, 1987); J. M. Haile, *Molecular Dynamics Simulation* (New York: Wiley, 1992).

The ODE problems in this chapter and Chapter 3 are all initial value problems. In each case we are given complete information as to the state of the system at time $t = 0$ and from that information computed future values. An alternative type of ODE problem is a *boundary value problem*. Suppose that we are given incomplete information for the state of the system for two points in time, say $t = 0$ and T. An example of such a problem would be if we knew the initial and final positions of a baseball (but not the velocities) and wished to compute the entire trajectory.

A common way of solving boundary value problems is to guess several initial conditions (e.g., initial velocities for the baseball). We then compute the solutions up to time T. We hope that some of the guesses will be close to the specified boundary value at time T. We then refine our guesses and continue until finding an acceptable result. This procedure is called a *shooting method*, and of course there are systematic numerical methods for updating the guesses (e.g., Newton's method). For more insight on how such iterative methods work, see Section 4.3. An alternative way to solve boundary value problems is by *relaxation*. This technique is described for linear partial differential equations in Section 7.1. For a complete discussion of numerical techniques for solving boundary value problems, see Ascher, Mattheij, and Russell.[11]

APPENDIX 2A: FORTRAN LISTINGS

LISTING 2A.1 Program `balle`. Computes the trajectory of a baseball, including air resistance.

```
      program balle
! Program to compute the trajectory of a baseball using Euler
! method
      parameter (maxstep = 1000)   ! Maximum number of steps
      real r(2),v(2),accel(2),xplot(maxstep),yplot(maxstep)
      real mass, norm_v

      write (*,*) 'Enter initial position r= x,y'  ! (meters)
      read (*,*) r(1),r(2)
      write (*,*) 'Enter initial velocity v= vx,vy' ! (m/sec)
      read (*,*) v(1),v(2)
      write (*,*) 'Enter timestep, tau'    ! (sec)
      read (*,*) tau
      Cd = 0.35      ! Drag coefficient (dimensionless)
      rho = 1.2      ! Density of air (kg/m^3)
      area = 4.3e-3  ! Cross-sectional area of projectile (m**2)
      grav = 9.81    ! Gravitational acceleration (m/s**2)
      mass = 0.145   ! Mass of projectile
```

[11] U. M. Ascher, R. M. M. Mattheij, and R. D. Russell, *Numerical Solution of Boundary Value Problems for Ordinary Differential Equations* (Englewood Cliffs, N.J.: Prentice Hall, 1988).

```
          air_const = -0.5*Cd*rho*area/mass ! Air resistance
                                            ! constant
!!!!! MAIN LOOP !!!!!
          do istep=1,maxstep
             xplot(istep) = r(1)     ! Record trajectory for plot
             yplot(istep) = r(2)
             norm_v = sqrt(v(1)**2 + v(2)**2)
             accel(1) = air_const*norm_v*v(1)          ! x-acceleration
             accel(2) = air_const*norm_v*v(2) - grav   ! y-acceleration
             do i=1,2
                r(i) = r(i) + tau*v(i)          ! Euler step
                v(i) = v(i) + tau*accel(i)
             end do
             if( r(2) .lt. 0 ) then
                goto 100          ! Break out of loop when ball
                                  ! hits
             end if                       ! ground, when y-position < 0
          end do
100       continue   ! Jump to here when particle reaches the ground
          xplot(istep+1) = r(1)   ! Be sure to record last position
          yplot(istep+1) = r(2)   ! for plotting
          write (*,*) 'Maximum range is ',r(1),' meters'
          write (*,*) 'Time of flight is ',istep*tau,' seconds'
! Print out the plotting variables -
!    xplot,yplot
!
          open(11,file='xplot.dat')
          open(12,file='yplot.dat')
          do i=1,istep+1
             write (11,*) xplot(i)
             write (12,*) yplot(i)
          end do
          stop
          end
```

LISTING 2A.2   Program pendul. Computes the time evolution of a simple pendulum using the Euler or Verlet methods.

```
      program pendul
! Program to compute the motion of a simple pendulum using
! Euler method or Verlet method
      parameter( nstep = 300, maxrev = 50 )
      real mean,t_plot(nstep),th_plot(nstep),period(maxrev)

      pi = 4*atan(1.) ! = 3.14159...
      write (*,*) 'Enter initial angle (in degrees)'
      read (*,*) theta
```

```fortran
      theta = theta*pi/180.        ! Convert angle to radians
      omega = 0.                   ! Set the initial velocity
      write (*,*) 'Enter time step'
      read (*,*) tau
      g_over_L = 1                 ! The constant g/L
      time_old = -1                ! Fake value (see below)
      irev = 0                     ! Used to count number of
                                   ! reversals
      time = 0
      ! Take one backward Euler step to start Verlet
      theta_old = theta - omega*tau
!!!!! MAIN LOOP !!!!!
      do istep=1,nstep
        t_plot(istep) = time            ! Record time and angle
        th_plot(istep) = theta*180./pi  ! for plotting
        accel = -g_over_L * sin(theta)  ! Gravitational
                                        ! acceleration
!!!! Euler method !!!!!!!!!!!!!!!!!!!!!!!!!!!!!!
!         theta_old = theta             ! Used to find
                                        ! half-period
!         theta = theta + tau*omega     ! Euler method
!         omega = omega + tau*accel
!!!!!!!!!!!!!!!!!!!!!!!!!!!!!!!!!!!!!!!!!!!!!!!!
!!!! Verlet method !!!!!!!!!!!!!!!!!!!!!!!!!!!!!
        theta_new = 2*theta - theta_old + tau**2*accel
        theta_old = theta
        theta = theta_new
!!!!!!!!!!!!!!!!!!!!!!!!!!!!!!!!!!!!!!!!!!!!!!!!
        time = time + tau
        ! Test position for sign change
        if( theta_old*theta .lt. 0 ) then
          write (*,*) 'Turning point at time = ',time
          if( time_old .lt. 0 ) then    ! If this is the first
                                        ! change
            time_old = time             ! just record the time
          else
            irev = irev + 1             ! Increment number of
                                        ! reversals
            period(irev) = 2.0*(time - time_old)
            time_old = time
          end if
        end if
      end do
      sum = 0
      do i=1,irev
        sum = sum + period(i)
      end do
      mean = sum/irev                   ! Mean value for the period
```

```
      sum = 0
      do i=1,irev
        sum = sum + (period(i)-mean)**2
      end do
      stdev = sqrt(sum/(irev-1))   ! Standard deviation of the
                                   ! period
      write (*,*) 'Average period = ',mean,' +/- ',
     &                               stdev/sqrt(float(irev))
! Print out the plotting variables -
!    t_plot,th_plot
!
      open(11,file='t_plot.dat')
      open(12,file='th_plot.dat')
      do i=1,nstep
        write (11,*) t_plot(i)
        write (12,*) th_plot(i)
      end do
      stop
      end
```

Chapter 3
Ordinary Differential Equations II: Advanced Methods

In Chapter 2 we learned how to solve ordinary differential equations (ODEs) using some simple methods. In this chapter we do some basic celestial mechanics beginning with the Kepler problem. Computing the orbit of a small satellite about a large body (e.g., a comet orbiting the Sun) we discover that more sophisticated methods are needed to handle even this simple two-body system. The second problem we consider in this chapter is the Lorenz model. This nonlinear system of ODEs was one of the first in which chaotic dynamics was found.

3.1 KEPLER PROBLEM AND THE ORBITS OF COMETS

Basic Equations

Consider a small satellite, such as a comet, orbiting the Sun. We use a Copernican coordinate system and fix the Sun at the origin. For now, consider only the gravitational force between the comet and the Sun and neglect all other forces (e.g., forces due to the planets, solar wind). The force on the comet is

$$\mathbf{F} = \frac{-GmM}{|\mathbf{r}|^3}\mathbf{r} \tag{3-1}$$

where $\mathbf{r}$ is the position of the comet, m is its mass, M ($= 1.99 \times 10^{30}$ kg) is the mass of the Sun, and G ($= 6.67 \times 10^{-11}$ m³/kg · s²) is the gravitational constant.

The natural units of length and time for this problem are not meters and seconds. As a unit of distance we will use the astronomical unit (AU = 1.496 × 10^{11} m), which equals the mean Earth–Sun distance. The unit of time will be years. In these units, the product $GM = 4\pi^2$ AU³/yr². We take the mass of the comet, m, as unity; in MKS units the typical mass of a comet is $10^{15\pm3}$ kg.

We now have enough to assemble our program, but before doing so let's quickly review what we know about orbits. For a complete treatment, see any of the standard mechanics texts.[1] The total energy of the satellite is

$$E = \tfrac{1}{2}mv^2 - \frac{GMm}{r} \tag{3-2}$$

where $r = |\mathbf{r}|$ and $v = |\mathbf{v}|$. This total energy is conserved as is the angular momentum,

$$\mathbf{L} = \mathbf{r} \times (m\mathbf{v}) \tag{3-3}$$

Since this problem is two-dimensional, we will use only the xy plane. The only nonzero component of the angular momentum is in the z-direction.

When the orbit is circular, the centripetal force equals the gravitational force,

$$\frac{mv^2}{r} = \frac{GMm}{r^2} \tag{3-4}$$

or

$$v = \sqrt{GM/r} \tag{3-5}$$

To put in some values, in a circular orbit at $r = 1$ AU the velocity is $v = 2\pi$ AU/yr (about 30,000 km/h). Using (3-5) in (3-2), the total energy in a circular orbit is

$$E = -\frac{GMm}{2r} \tag{3-6}$$

In an elliptical orbit, the semimajor and semiminor axes, a and b, are unequal. The eccentricity, e, is defined as

$$e = \sqrt{1 - b^2/a^2} \tag{3-7}$$

---

[1] For example, K. Symon, *Mechanics* (Reading Mass.: Addison-Wesley, 1971); L. Landau and E. Lifshitz, *Mechanics* (Oxford: Pergamon, 1976).

Earth's eccentricity is $e = 0.017$, so its orbit is nearly circular. The distance from the Sun at perihelion (closest approach) is $q = (1 - e)a$; the distance from the Sun at aphelion is $Q = (1 + e)a$.

Equation (3-6) also holds for elliptical orbits if we replace the radius with the semimajor axis, that is, the total energy is

$$E = -\frac{GMm}{2a} \tag{3-8}$$

From (3-2) and (3-8), we find that the orbital speed as a function of radial distance is

$$v = \sqrt{GM\left(\frac{2}{r} - \frac{1}{a}\right)} \tag{3-9}$$

Finally, using conservation of angular momentum we may derive Kepler's third law,

$$T^2 = \frac{4\pi^2}{GM} a^3 \tag{3-10}$$

where T is the period of the orbit.

The orbital data for a few well-known comets are given in Table 3.1. The inclination, i, is the angle between the orbital plane of the comet and the ecliptic plane (the plane of the orbit of the planets). When the inclination is less than $90°$ the orbit is said to be direct, when it is greater than $90°$ the orbit is retrograde (i.e., the comet orbits the Sun in the opposite sense from the planets).

TABLE 3.1 Orbital data for selected comets

Comet Name	T (yrs)	e	q (AU)	i	First Pass
Encke	3.30	0.847	0.339	12.4°	1786
Biela	6.62	0.756	0.861	12.6°	1772
Schwassmann–Wachmann 1	16.10	0.132	5.54	9.5°	1925
Halley	76.03	0.967	0.587	162.2°	239 B.C.
Grigg–Mellish	164.3	0.969	0.923	109.8°	1742

Orbit Program

A simple program, called orbe, that computes orbits for the Kepler problem using the Euler method is given in Listing 3.1. A few comments about the program: The main loop is between lines 12–25; the rest of the program is input/

output. All the dynamics is contained in lines 21–24. By interchanging the two lines where position and velocity are updated (lines 22–23), we can change the algorithm to Euler–Cromer. At each iteration we record the radial and angular position (r and θ) of the satellite in the variables rplot and thplot. These are used to create a polar plot (lines 28–31), which is the most natural way to view the orbit. The kinetic, potential, and total energy are recorded at each iteration and plotted as a function of time (lines 33–37). Notice that the subplot command is used to create side-by-side graphs.

LISTING 3.1   Program orbe. Computes the orbit of a comet about the Sun using the Euler method.

```
1    % orbe - Program to compute the orbit of a comet
2    % using the Euler method.
3    clear; help orbe;    % Clear memory and print header
4    r0 = input('Enter initial radial distance - ');   % (AU)
5    r = [r0 0];         % Initial position is on x-axis
6    v0 = input('Enter initial tangential velocity - '); % (AU/yr)
7    v = [0 v0];         % Initial velocity is in the y-direction
8    tau = input('Enter time step, tau - ');   % (yr)
9    GM = 4*pi^2;        % Grav. const. * Mass of Sun (AU^3/yr^2)
10   mass = 1.;          % Mass of projectile
11   %%%% MAIN LOOP %%%%%%
12   time = 0;
13   nstep = 200;        % Number of time steps
14   for istep=1:nstep
15     rplot(istep) = norm(r); % Record orbit for polar plot
16     thplot(istep) = atan2(r(2),r(1));
17     tplot(istep) = time;       % Record time for plot
18     kinetic(istep) = .5*mass*norm(v)^2;  % Record energies
19     potential(istep) = - GM*mass/norm(r);
20     % Calculate new position and velocity
21     accel = -GM*r/norm(r)^3; % Gravity
22     r = r + tau*v;         % Euler step
23     v = v + tau*accel;
24     time = time + tau;
25   end
26   % Graph the trajectory of the comet
27   subplot(121)
28     polar(thplot,rplot,'+')   % Polar plot of trajectory
29     grid                      % Include a grid on the plot
30     ylabel('Distance (AU)')
31     title('Orbital motion')
32   subplot(122)
33     totalE = kinetic + potential;   % Total energy
34     plot(tplot,kinetic,'-.',tplot,potential,'--',tplot,totalE,'-')
35     xlabel('Time (yr)')
```

```
36     ylabel('Energy')
37     title('KE(dot); PE(dash); Total(solid)')
38  subplot(111)
```

The simplest test case is a circular orbit. For an initial radial distance of 1 AU we have a circular orbit when the initial tangential velocity is 2π AU/yr. Fifty data points per orbital revolution should give us a smooth curve, so $\tau = 0.02$ yr (or about 1 week) is a reasonable time step. With these values, the orbe program gives the results shown in Figure 3.1. We immediately see that the orbit is not a circle but an outward spiral. The reason is clear when we look at the energy graph. The total energy (solid line) is not constant but increasing. This type of instability is also observed in the Euler method for the simple pendulum problem (see Section 2.2).

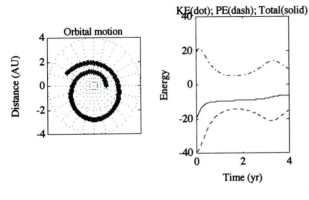

Figure 3.1 Graphical output from the orbe program (Euler method). Initial radial distance is 1 AU and the initial tangential velocity is 2π AU/yr. The time step is $\tau = 0.02$ yr. Results disagree with theory since it predicts a circular orbit and constant total energy.

Fortunately, there is a simple solution to this problem. By interchanging lines 22 and 23 in orbe, we change the algorithm to Euler–Cromer. With this modification, the program will be called orbec. For the same initial conditions and time step, the Euler–Cromer method gives much better results, as shown in Figure 3.2. The orbit is now almost exactly circular and the total energy is con-

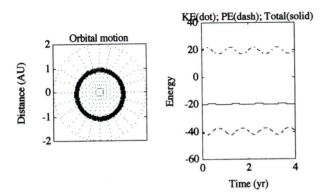

Figure 3.2 Graphical output from the orbec program (Euler–Cromer method). Initial radial distance is 1 AU and the initial tangential velocity is 2π AU/yr. The time step is $\tau = 0.02$ yr. Theory predicts a circular orbit and constant total energy.

served. The kinetic and potential energies are not constant, but this problem may be improved by using a smaller time step.

Although the Euler–Cromer method does a good job for low eccentricity orbits, it has problems with more elliptical orbits. For example, for an initial radial distance of 1 AU, an initial tangential velocity of π AU/yr, and a time step of $\tau = 0.02$ yr, the orbec program produces the results shown in Figure 3.3. Notice that the energy becomes positive; the satellite acquires escape velocity. If we lower the time step down to $\tau = 0.005$ yr we obtain better results, as shown in Figure 3.4. The results are still not perfect; the orbit has a noticeable spurious drift.

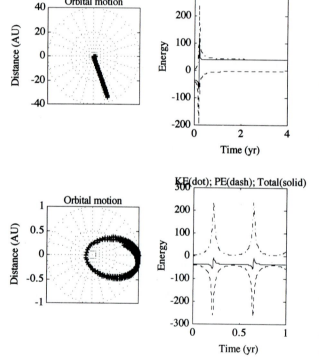

Figure 3.3   Graphical output from the orbec program (Euler–Cromer method). Initial radial distance is 1 AU and the initial tangential velocity is π AU/yr. The time step is $\tau = 0.02$ yr. Notice that the total energy becomes positive, so that the comet has escape velocity.

Figure 3.4   Graphical output from the orbec program (Euler–Cromer method). Initial radial distance is 1 AU and the initial tangential velocity is π AU/yr. The time step is $\tau = 0.005$ yr. Notice that the orbit has a spurious precession.

At this point you may be asking yourself, "Why are we studying this problem? The analytic solution is well known." It is true that there are more interesting celestial mechanics problems (e.g., the effect of perturbations on the orbit, the three-body problem). However, before doing the complicated cases we should always check our algorithms on known problems. Suppose that we introduced a small drag force on the comet. We might be fooled into believing that the precession in Figure 3.4 was a physical phenomenon rather than a numerical artifact.

Clearly, the Euler–Cromer method does an unacceptable job of tracking the more elliptical orbits. The results improve if we drop the time step, but then it takes forever to do a few orbits. Suppose that we wanted to track comets for possible Earth impacts. It has been suggested that a comet striking Earth could do more damage than a swarm of ICBMs. Many comets have extremely elliptical orbits and periods on the order of 50 to 100 years! This threat from outer space motivates our study of more advanced methods for solving ODEs.

EXERCISES

1. Suppose that a planet suddenly lost all its orbital velocity; of course it would plunge directly into the Sun. Show that Earth would reach the core of the Sun in about 65 days. (Hint: You don't need to use the computer; it's a two-line calculation.)

2. Prove that for the Kepler problem the Euler–Cromer method conserves angular momentum exactly.

3. Modify the orbec program so that instead of running for a fixed number of time steps, the program stops when the satellite completes one orbit.
   (a) Have the program compute the period, eccentricity, semimajor axis, and perihelion distance of the orbit; test it with circular and slightly elliptical orbits. Compare the measured eccentricity with

   $$\varepsilon = \sqrt{1 + \frac{2\,EL^2}{G^2 M^2 m^3}}$$

   (b) Show that your program confirms Kepler's third law.
   (c) Confirm that $\langle K \rangle = -\langle V \rangle / 2$, where $\langle K \rangle$ and $\langle V \rangle$ are the time-average kinetic and potential energy (virial theorem).

4. For an ellipse, the radial position varies with angle as

   $$r(\theta) = \frac{a(1 - \varepsilon^2)}{1 - \varepsilon \cos\theta}$$

   Modify your program for Exercise 3.3 to compute and plot the absolute fractional error in $r(\theta)$ over a single orbit. Obtain results for an initial radial distance of 1 AU, an initial tangential velocity of π AU/yr, and time steps of $\tau = 0.01$, 0.005, and 0.001.

5. Modify the orbec program to use the Verlet method. Compare the performance of the two methods.

6. Use your program from Exercise 3.3 to determine how small a time step is needed for the more elliptic orbits. Using an initial radial distance of 35 AU (Halley's comet) and various values for the aphelion velocity, find the largest value of τ for which the total energy is conserved to about 1% per orbit. Assemble a graph of τ versus initial velocity and estimate the time step needed to track Halley's comet.

7. The Lorentz force on a charged particle in a given electric and magnetic field is $\mathbf{F} = q(\mathbf{E} + \mathbf{v} \times \mathbf{B})$. Write a program to simulate the motion of an electron in uniform, perpendicular electric, and magnetic fields. Show that the motion is helical in form, with a pitch that depends on the initial particle velocity and with a drift velocity $\mathbf{u}_{\text{drift}} = \mathbf{E} \times \mathbf{B}/B^2$.

3.2 RUNGE–KUTTA METHODS

Second-Order Runge–Kutta

We now look at one of the most popular methods for numerically solving ODEs: Runge–Kutta. We will first work out the general Runge–Kutta formulas and then apply them specifically to our comet problem. In this way it will be easy to use the Runge–Kutta method for other physical systems.

Our general ODE takes the form

$$\frac{d\mathbf{x}}{dt} = \mathbf{f}(\mathbf{x}(t), t) \tag{3-11}$$

where the vector $\mathbf{x}(t) = [x_1(t) \ldots x_N(t)]$ is the desired solution. In the Kepler problem we have

$$\mathbf{x}(t) = [r_x(t) \quad r_y(t) \quad v_x(t) \quad v_y(t)] \tag{3-12a}$$

$$\mathbf{f}(\mathbf{x}(t), t) = [v_x(t) \quad v_y(t) \quad F_x(t)/m \quad F_y(t)/m] \tag{3-12b}$$

where r_x, v_x, and F_x are the x-components of position, velocity, and force, respectively (and similarly for the y-components). Notice that in the Kepler problem, the function $\mathbf{f}$ does not depend explicitly on time; rather, it only depends on $\mathbf{x}(t)$.

Our starting point is the simple Euler formula; in vector form it may be written as

$$\mathbf{x}(t + \tau) = \mathbf{x}(t) + \tau \mathbf{f}(\mathbf{x}(t), t) \tag{3-13}$$

The first Runge–Kutta formula we consider is

$$\mathbf{x}(t + \tau) = \mathbf{x}(t) + \tau \mathbf{f}(\mathbf{x}^*(t + \tfrac{1}{2}\tau), t + \tfrac{1}{2}\tau) \tag{3-14}$$

where

$$\mathbf{x}^*(t + \tfrac{1}{2}\tau) \equiv \mathbf{x}(t) + \tfrac{1}{2}\tau \mathbf{f}(\mathbf{x}(t), t) \tag{3-15}$$

To see where this formula comes from, consider for a moment the one-variable case. We know that the Taylor expansion

$$x(t + \tau) = x(t) + \tau \frac{dx(\zeta)}{dt}$$

$$= x(t) + \tau f(x(\zeta), \zeta) \tag{3-16}$$

is exact for some value of ζ between t and $t + \tau$ (see Section 1.5). The Euler formula takes $\zeta = t$; Euler–Cromer uses $\zeta = t$ in the velocity equation and $\zeta = t + \tau$ in the position equation. Runge–Kutta tries to use $\zeta = t + \tau/2$ since this is probably a better guess. However, $x(t + \tau/2)$ is not known so we approximate it in the simplest way possible: Use an Euler step to compute $x^*(t + \tau/2)$ and use this as our estimate of $x(t + \tau/2)$. Figure 3.5 illustrates this idea.

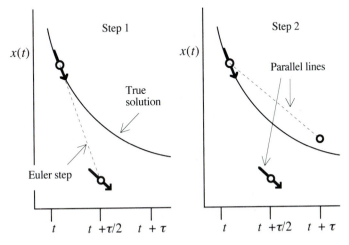

Figure 3.5 Graphical illustration of a simple, two-step Runge–Kutta formula. The arrows on the points have slope $dx/dt = f(x(t), t)$.

Let's walk through a simple illustration of using the Runge–Kutta formula. Take the equation

$$\frac{dx}{dt} = -x; \qquad x(t = 0) = 1 \tag{3-17}$$

The solution of (3-17) is $x(t) = e^{-t}$. Using Euler's method with a time step of $\tau = 0.1$, we get

$$x(0.1) = 1 + 0.1(-1) = 0.9$$

$$x(0.2) = 0.9 + 0.1(-0.9) = 0.81$$

$$x(0.3) = 0.81 + 0.1(-0.81) = 0.729$$

$$x(0.4) = 0.729 + 0.1(-0.729) = 0.6561$$

Now let's try Runge–Kutta. To make a fair comparison, we use a larger time step of $\tau = 0.2$ for Runge–Kutta since it makes twice as many evaluations of $f(x)$.

For the Runge–Kutta formula presented above,

$$x^*(0.1) = 1 + 0.1(-1) = 0.9$$

$$x(0.2) = 1 + 0.2(-0.9) = 0.82$$

$$x^*(0.3) = 0.82 + 0.1(-0.82) = 0.738$$

$$x(0.4) = 0.82 + 0.2(-0.738) = 0.6724$$

Compare this with the exact solution $x(0.4) = \exp(-0.4) \cong 0.6703$. Clearly, Runge–Kutta does much better than Euler; the absolute percent errors are 0.3% and 2.1%, respectively.

General Runge–Kutta Formulas

The formula discussed above is not the only second-order Runge–Kutta formula. Here is an alternative one:

$$\mathbf{x}(t + \tau) = \mathbf{x}(t) + \tfrac{1}{2}\tau[\mathbf{f}(\mathbf{x}(t), t) + \mathbf{f}(\mathbf{x}^*(t + \tau), t + \tau)] \qquad (3\text{-}18)$$

where

$$\mathbf{x}^*(t + \tau) \equiv \mathbf{x}(t) + \tau\mathbf{f}(\mathbf{x}(t), t) \qquad (3\text{-}19)$$

To understand this scheme, again consider the one variable case. In our original formula, we estimated $f(x(\zeta), \zeta)$ as $f(x^*(t + \tau/2), t + \tau/2)$. Our new formula is similar, but now we approximate $f(x(\zeta), \zeta)$ as $\tfrac{1}{2}[f(x, t) + f(x^*(t + \tau), t + \tau)]$.

These formulas were not pulled out of the air; you can work them out using the two-variable Taylor expansion,

$$f(x + h, t + \tau) = \sum_{i=0}^{\infty} \frac{1}{i!} \left(h\frac{\partial}{\partial x} + \tau\frac{\partial}{\partial t} \right)^i f(x, t) \qquad (3\text{-}20)$$

where all derivatives are evaluated at (x, t). For a general second-order Runge–Kutta formula, we want to obtain a formula of the form

$$x(t + \tau) = x(t) + w_1\tau f(x(t), t) + w_2\tau f(x^*, t + \alpha\tau) \qquad (3\text{-}21)$$

where

$$x^* \equiv x(t) + \beta\tau f(x(t), t) \qquad (3\text{-}22)$$

There are four unspecified coefficients: α, β, w_1, and w_2. Notice that we recover our earlier formulas with the values

$$w_1 = 0, \; w_2 = 1, \; \alpha = \tfrac{1}{2}, \; \beta = \tfrac{1}{2} \quad \text{(equations 14 and 15)}$$

$$w_1 = \tfrac{1}{2}, \; w_2 = \tfrac{1}{2}, \; \alpha = 1, \; \beta = 1 \quad \text{(equations 18 and 19)}$$

We want to pick these four coefficients such that we get second-order accuracy, that is, we want to match the Taylor series through the second derivative terms.

The details of the calculation are left as an exercise (see Exercise 3.9). I want to emphasize three points: (1) Any set of coefficients satisfying the relations $w_1 + w_2 = 1$, $\alpha w_2 = \tfrac{1}{2}$, and $\beta w_2 = \tfrac{1}{2}$ will give a second-order Runge–Kutta scheme. (2) It is not clear that one scheme is superior to another (actual error will vary from problem to problem). Unfortunately, the error term for these Runge–Kutta formulas is rather complicated. (3) The local truncation error is $O(\tau^3)$.

Fourth-Order Runge–Kutta

I presented the second-order Runge–Kutta formulas because it is easy to understand their construction. In practice, however, the most commonly used method is the following fourth-order Runge–Kutta formula:

$$\mathbf{x}(t + \tau) = \mathbf{x}(t) + \tfrac{1}{6}\tau[\mathbf{F}_1 + 2\mathbf{F}_2 + 2\mathbf{F}_3 + \mathbf{F}_4] \tag{3-23}$$

where

$$
\begin{aligned}
\mathbf{F}_1 &= \mathbf{f}(\mathbf{x}, t) \\
\mathbf{F}_2 &= \mathbf{f}(\mathbf{x} + \tfrac{1}{2}\tau\mathbf{F}_1, \, t + \tfrac{1}{2}\tau) \\
\mathbf{F}_3 &= \mathbf{f}(\mathbf{x} + \tfrac{1}{2}\tau\mathbf{F}_2, \, t + \tfrac{1}{2}\tau) \\
\mathbf{F}_4 &= \mathbf{f}(\mathbf{x} + \tau\mathbf{F}_3, \, t + \tau)
\end{aligned}
\tag{3-24}
$$

The following excerpt from *Numerical Recipes*[2] best summarizes the status that the above formula holds in the world of numerical analysis:

> For many scientific users, fourth-order Runge–Kutta is not just the first word on ODE integrators, but the last word as well. In fact, you can get pretty far on this old workhorse especially if you combine it with an adaptive stepsize algorithm. . . . Bulirsch–Stoer or predictor-corrector methods can be very much more efficient for problems where very high accuracy is a

---

[2] W. Press, B. Flannery, S. Teukolsky, and W. Vetterling, *Numerical Recipes in FORTRAN*, 2d ed. (Cambridge: Cambridge University Press, 1992), p. 706.

requirement. Those methods are the high-strung racehorses. Runge–Kutta is for ploughing the fields.

You may wonder, "Why fourth-order and not eighth- or twenty-third-order Runge–Kutta?" Well, the higher-order methods have better truncation error but also require more computation, that is, more evaluations of $f(x, t)$. There is a trade-off between doing more steps with a smaller τ using a low-order method as opposed to doing fewer steps with a larger τ using a high-order method. It turns out that the optimum, for Runge–Kutta methods, is the fourth-order scheme given above. By the way, the local truncation error for fourth-order Runge–Kutta is $O(\tau^5)$.

To implement the fourth-order Runge–Kutta method for our orbit problem, we use the function rk4 (see Listing 3.2). This function will accept as input the current state of the system, $\mathbf{x}(t)$, the time step to be used, τ, the current time, t, and the function $\mathbf{f}(\mathbf{x}(t), t)$. It returns the new state of the system, $\mathbf{x}(t + \tau)$, as computed by the Runge–Kutta method.

LISTING 3.2   Function rk4. Fourth-order Runge–Kutta routine.

```
1    function xout = rk4(x, t, tau, derivsRK, param)
2    %  Runge-Kutta integrator (4th order)
3    % Input arguments -
4    %    x = current value of dependent variable
5    %    t = independent variable (usually time)
6    %    tau = step size (usually timestep)
7    %    derivsRK = right hand side of the ODE; derivsRK is the
8    %               name of the function which returns dx/dt
9    %               Calling format derivsRK(x, t, param)
10   %    param = extra parameters passed to derivsRK
11   % Output arguments -
12   %    xout = new value of x after a step of size tau
13   half_tau = 0.5*tau;
14   F1 = feval(derivsRK, x, t, param);
15   t_half = t + half_tau;
16   xtemp = x + half_tau*F1;
17   F2 = feval(derivsRK, xtemp, t_half, param);
18   xtemp = x + half_tau*F2;
19   F3 = feval(derivsRK, xtemp, t_half, param);
20   t_full = t + tau;
21   xtemp = x + tau*F3;
22   F4 = feval(derivsRK, xtemp, t_full, param);
23   xout = x + tau/6.*(F1 + F4 + 2.*(F2+F3));
24   return;
```

Much of rk4 is comments; remember that if you type help rk4 from the command line, MATLAB will print lines 2–12. The one point I want to elaborate is the use of the MATLAB function feval. Our rk4 routine is written so that it may be used for different problems (as we'll do in later sections). As such, the equations of motion are given by another function; the *name* of this function is passed to rk4 in the derivsRK variable. For example, for the orbits problem we may use rk4 in our old orbe program by replacing the lines 21–23 with

```
state = rk4(state, time, tau, 'gravrk', GM);
r = [state(1) state(2)];
v = [state(3) state(4)];
```

and inserting the following line:

```
state = [r(1) r(2) v(1) v(2)];
```

just outside the main loop (above line 14). The function that defines the equations of motion is called gravrk (see Listing 3.3).

LISTING 3.3 Function gravrk. Used by rk4 to define the equations of motion for the Kepler problem.

```
1  function deriv = gravrk(s, time, GM)
2  %   The time is not used in this version
3  %   The vector s = [r(1) r(2) v(1) v(2)]
4  %   The vector deriv = [dr(1)/dt dr(2)/dt dv(1)/dt dv(2)/dt]
5  r = [s(1) s(2)]; % Unravel vector s into position & velocity
6  v = [s(3) s(4)];
7  accel = -GM*r/norm(r)^3;   % Gravity
8  deriv = [v(1) v(2) accel(1) accel(2)];
9  return;
```

To review how these routines work: Our main program knows the current values of **r** and **v**. It puts these together into a single vector called state, which is passed to the Runge–Kutta function. This routine computes the new value of state, that is, the new values of **r** and **v**. To do so, however, it needs to know the equations of motion, that is, the function $\mathbf{f}(\mathbf{x}, t)$. The main program tells rk4 that the equations of motion are given by the function gravrk. When rk4 executes line 14

```
F1 = feval(derivsRK, x, t, param);
```

this is equivalent to calling gravrk as

```
F1 = gravrk(x, t, param);
```

since the variable derivsRK contains the text string 'gravrk'. Notice that param is used to pass the value of GM.

The passing of a function name to a function is implemented in various ways by other languages. In FORTRAN, the EXTERNAL declaration indicates that a variable contains the name of a function. In C, a pointer to the function (i.e., a variable containing the memory address of the function) is passed to the routine that calls that function.

Numerical Results

Implementing the modifications described above, we create a new program, called orbrk, that uses the fourth-order Runge–Kutta scheme. For an initial radial distance of 1 AU, an initial tangential velocity of π AU/yr, and a time step of $\tau = 0.005$ yr, the program gives the results shown in Figure 3.6. Compare these results to those obtained in Section 3.1 using Euler–Cromer (Figure 3.4).

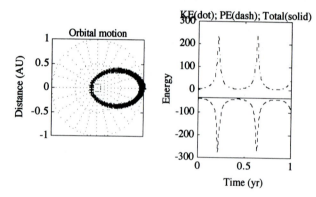

Figure 3.6 Graphical output from the orbrk program (Runge–Kutta method). The initial radial distance is 1 AU and the initial tangential velocity is π AU/yr. The time step is $\tau = 0.005$ yr. Theory predicts an elliptical orbit and constant total energy.

EXERCISES

8. Prove that any nth-order ODE of the form

$$\frac{d^n z}{dt^n} = f\left(z, \frac{dz}{dt}, \ldots, \frac{d^{n-1}z}{dt^{n-1}}\right)$$

may be written as a system of first-order ODEs.

9. (a) Use equation (3-20) to show that a second-order Runge–Kutta formula requires that $w_1 + w_2 = 1$, $\alpha w_2 = \frac{1}{2}$, and $\beta w_2 = \frac{1}{2}$.

(b) Show that the truncation error for second-order Runge–Kutta is

$$\tau^3 \left[\left(\frac{1}{6} - \frac{\alpha}{4}\right)\left(\frac{\partial}{\partial t} + f\frac{\partial}{\partial x}\right)^2 f + \frac{1}{6}\frac{\partial f}{\partial x}\left(\frac{\partial f}{\partial t} + f\frac{\partial f}{\partial x}\right) \right]$$

10. Modify the program used in Exercise 3.6 to use the fourth-order Runge–Kutta method and repeat that exercise. Compare the Euler–Cromer and Runge–Kutta methods.

11. Suppose that our comet is subjected to a constant force in one direction (e.g., gravitational attraction of a large but distant object).[3] Write a program that uses fourth-order Runge–Kutta to solve this problem. Set the strength of the perturbing force to be a few percent of the gravitational force. Show that an initially circular orbit is transformed into an elliptical orbit with the semimajor axis *perpendicular* to the perturbing force. Produce a graph of the angular momentum as a function of time.

12. The Wilberforce pendulum,[4] a popular demonstration device, is illustrated in Figure 3.7. The pendulum has two modes of oscillation: vertical and torsional motion. The Lagrangian for this system is

$$L = \tfrac{1}{2}m\left(\frac{dz}{dt}\right)^2 = \tfrac{1}{2}I\left(\frac{d\theta}{dt}\right)^2 - \tfrac{1}{2}kz^2 - \tfrac{1}{2}\delta\theta^2 - \tfrac{1}{2}\varepsilon z\theta$$

where m and I are the mass and rotational inertia of the bob, k and δ are the longitudinal and torsional spring constants, and ε is the coupling constant between the modes. Some typical values are $m = 0.5$ kg, $I = 10^{-4}$ kg $\cdot$ m^2, $k = 5$ N/m, $\delta = 10^{-3}$ N $\cdot$ m, and $\varepsilon = 10^{-2}$ N. Find the equations of motion and write a program to compute $z(t)$ and $\theta(t)$ using rk4. Try the initial conditions $z(0) \neq 0$, $\theta(0) = 0$ and $z(0) = 0$, and $\theta(0) \neq 0$. Show that when the longitudinal frequency, $f_z = 1/2\pi\sqrt{k/m}$, equals the torsional frequency, $f_\theta = 1/2\pi\sqrt{\delta/I}$, the motion periodically alternates between being purely longitudinal and purely torsional.

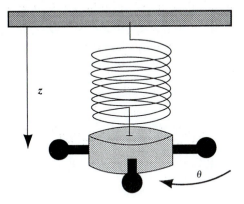

Figure 3.7 Wilberforce pendulum.

[3] A. Luehrmann, "Orbits in the Solar Wind—a Mini-Research Problem," *Am. J. Phys.*, **42**, 361–71 (1974).

[4] R. E. Berg and T. S. Marshall, "Wilberforce pendulum oscillations and normal modes," *Am. J. Phys.*, **59**, 32–38 (1991).

3.3 ADAPTIVE METHODS

Selection of a Time Step

Because the fourth-order Runge–Kutta method is more accurate (smaller truncation error), it does a better job with highly elliptical orbits. And yet, for an initial
aphelion distance of 1 AU and an initial aphelion velocity of $\pi/2$ AU/yr using a
time step as small as $\tau = 0.0005$ yr ($\cong 4\frac{1}{2}$ h), the total energy varies by over 7% per
orbit. If we think about the physics, we come to realize that the small time steps
are only needed when the comet makes its closest approach. At that point it
has its maximum velocity, and any small error in the trajectory as it rounds the
Sun causes a large deviation in the potential energy.

  The idea now is to design a program that would use smaller time steps when
the comet is near the Sun and larger time steps when it is far away. As it is, we
normally have only a rough idea of what τ should be; now we have to select a τ_{min}
and τ_{max} and a way to switch between them. If we have to do this by manual trial
and error, it could be worse than just doing the brute force calculation with a small
time step.

  Ideally, we wish to be completely freed of having to specify a time step. We
want to have the trajectory computed from some initial position up to some final
time with the assurance that the solution is correct to a specified accuracy.

Adaptive Time Step Programs

Adaptive programs continuously monitor the solution and modify the time step to
ensure that the user specified accuracy is maintained. These programs may do
some extra calculation to optimize the choice of τ, but in many cases this extra
work is worth it. Here is one way to implement this idea: Given the current state
$\mathbf{x}(t)$, the program computes $\mathbf{x}(t + \tau)$ as usual and then it repeats the calculation by
doing two steps, each with time step $\tau/2$. Visually, this is

$$
\begin{array}{ccccc}
 & & \text{big step} & & \\
\mathbf{x}(t) & \rightarrow & \rightarrow & \rightarrow & \mathbf{x}_b\,(t + \tau) \\
\hline
 & \text{small step} & & \text{small step} & \\
\mathbf{x}(t) & \rightarrow & \mathbf{x}(t + \tfrac{1}{2}\tau) & \rightarrow & \mathbf{x}_s(t + \tau)
\end{array}
$$

The difference between the two answers, $\mathbf{x}_b(t + \tau)$ and $\mathbf{x}_s(t + \tau)$, estimates the
local truncation error. If the error is tolerable, then the computed value is ac-
cepted and a larger value of τ is used on the next iteration. On the other hand, if
the error is too large, the answer is rejected, the time step is reduced, and the
procedure is repeated until an acceptable answer is obtained.

  A simple way to implement an adaptive scheme would be to compute $\mathbf{x}_b(t +$
$\tau)$ and $\mathbf{x}_s(t + \tau)$ as described above and halve τ each time the truncation error is

above the desired accuracy. On the other hand, if the error is well below the desired accuracy, we might double the time step for the next calculation. We shouldn't be to eager to increase τ, however, since we waste computer effort every time we reject an answer and need to reduce the time step.

A more sophisticated approach would estimate by how much we should increase or decrease τ. Here is how such a method can be implemented for our fourth-order Runge–Kutta scheme: Call Δ the local truncation error; we know that $\Delta \propto \tau^5$ for fourth-order Runge–Kutta. Suppose that the current time step τ gave an error of $\Delta_c = \mathbf{x}_b(t + \tau) - \mathbf{x}_s(t + \tau)$; this is our estimate for the truncation error. Given that we want the error to be less than or equal to the user specified ideal error, call it Δ_i, then the new time step should be

$$\tau_{\text{new}} \approx \tau \left| \frac{\Delta_i}{\Delta_c} \right|^{1/5} \tag{3-25}$$

Since this is only an estimate, we include a safety factor, call it $S_1 < 1$, that makes us overestimate the change when we lower τ and underestimate when we raise it. Our new time step is computed as

$$\tau_{\text{new}} = S_1 \tau \left| \frac{\Delta_i}{\Delta_c} \right|^{1/5} \tag{3-26}$$

We should also put in a second safety factor to be sure that the program is not too enthusiastic about precipitously raising or lowering the time step. For this reason, we add the following to our calculation of the new time step:

$$\tau_{\text{new}} = \max\left(\tau_{\text{new}}, \tau/S_2\right)$$
$$\tau_{\text{new}} = \min\left(\tau_{\text{new}}, S_2\tau\right) \tag{3-27}$$

where $S_2 > 1$ is a constant. These two lines of code ensure that our new estimate for τ never increases or decreases by more than a factor of S_2. Of course, this new τ may be found to be insufficiently small and we may have to continue reducing the time step, but at least we know it will not happen in an uncontrolled way.

Adaptive Runge–Kutta Program

An adaptive Runge–Kutta routine called rka that puts the above ideas into practice is given in Listing 3.4. Notice that the inputs in the calling sequence are the same as for rk4 except for the addition of the parameter err. The output from the routine is the new state of the system, xout, at the new time, t, and the new time step, tau.

LISTING 3.4   Function `rka`. Adaptive Runge–Kutta routine. Uses `rk4` (Listing 3.2).

```
1    function [x, t, tau] = rka(x,t,tau,err,derivsRK,param)
2    % Adaptive Runge-Kutta routine
3    % Input arguments -
4    %    x = current value of dependent variable
5    %    t = independent variable (usually time)
6    %    tau = step size (usually timestep) to be attempted
7    %    err = desired fractional local truncation error
8    %    derivsRK = right hand side of the ODE; derivsRK is the
9    %               name of the function with returns dx/dt
10   %               Calling format derivsRK(x,t,param).
11   %    param = extra parameters passed to derivsRK
12   % Output arguments -
13   %    x = new value of the dependent variable
14   %    t = new value of the independent variable
15   %    tau = suggested step size to be used on next call to rka
16   tsave = t; xsave = x;
17   safe1 = .9; safe2 = 4.;   % Safety factors
18   maxtry = 100; % Maximum number of attempts
19   for itry=1:maxtry
20     %% Do the two small steps %%
21     half_tau = 0.5 * tau;
22     xtemp = rk4(xsave,tsave,half_tau,derivsRK,param);
23     t = tsave + half_tau;
24     x = rk4(xtemp,t,half_tau,derivsRK,param);
25     %% Do the single large step %%
26     t = tsave + tau;
27     xtemp = rk4(xsave,tsave,tau,derivsRK,param);
28     %% Compute the error %%
29     scale = .5*(abs(x) + abs(xtemp))*err;
30     xdiff = x - xtemp;
31     errmax = max(abs(xdiff)./(scale + eps));
32     %% Estimate new tau value if errmax unacceptable %%
33     tau_old = tau;
34     tau = safe1*tau_old*errmax^(-0.20);
35     %% Never decrease tau by more than a factor of safe2 %%
36     tau = max(tau,tau_old/safe2);
37     if errmax < 1,   % If error is acceptable, break out of loop
38       break;
39     end
40   end
41   %% Never increase tau by more than a factor of safe2 %%
42   tau = min(tau,safe2*tau_old);
```

The parameter err gives the largest acceptable fractional error; on line 29 we use it to calculate a scale as the average value of $x(t + \tau)$ times the acceptable error. On line 31 we compute the ratio of the actual error in $\mathbf{x}(t + \tau)$, xdiff, to our allowed error, scale. If the largest ratio, errmax, is less than 1, we accept the time step; otherwise we repeat with a new time step. No matter what, we estimate the new time step as described above [see (3-26) and (3-27)].

To use our adaptive Runge–Kutta routine, we can replace lines 21–24 in orbe with

```
[state, time, tau] = rka(state,time,tau,err,'gravrk',GM);
r = [state(1) state(2)];
v = [state(3) state(4)];
```

Just outside the main loop of orbe (above line 11) we add the lines

```
state = [ r(1) r(2) v(1) v(2) ];
err = 1.e-3;
```

Finally, to see the orbit better we reduce nstep to 40. This new program is called orbrka. For an initial radial distance of 1 AU and an initial tangential velocity of $\pi/2$, the program gives the results shown in Figure 3.8. Notice that the program takes many more steps at perihelion (closest approach) than at aphelion.

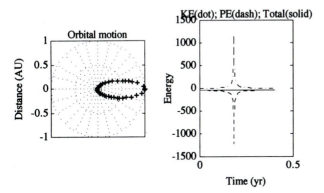

Figure 3.8   Results from the orbrka program. The initial radial distance is 1 AU and the initial tangential velocity is $\pi/2$ AU/yr.

It is interesting to plot the time step as a function of time (Figure 3.9); notice that τ varies by nearly three orders of magnitude. Interestingly, the graph of the time step versus radial distance (Figure 3.10) shows an approximate power law relation of the form $\tau \propto r^{3/2}$. Of course this dependence reminds us of Kepler's third law [equation (3-10)]. We expect some scatter in the points since our adaptive routine only estimates the optimum time step.

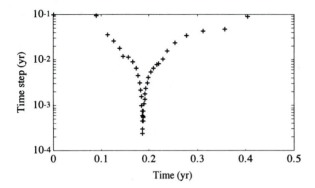

Figure 3.9   Time step, τ, as a function of time from the orbrka program (adaptive Runge–Kutta). Parameters are the same as for Figure 3.8. Notice how the time step varies by more than two orders of magnitude.

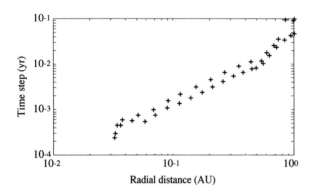

Figure 3.10   Time step, τ, as a function of radial distance from the orbrka program (adaptive Runge–Kutta). Parameters are the same as for Figure 3.8.

EXERCISES

13. Consider the central force

$$\mathbf{F(r)} = -\frac{GMm}{r^3}\left(1 - \frac{\alpha}{r}\right)\mathbf{r}$$

where α is a constant. Write a program, using adaptive Runge–Kutta, to compute the motion of an object under this force law. Show empirically that the orbit precesses $360(1 - a)/a$ degrees per revolution where $a = \sqrt{1 + GMm\alpha/L^2}$ and L is the angular momentum.

14. Modify orbrka to add a drag force on the comet such as if it were moving through a stationary medium (see Section 2.1). Show that the average kinetic energy (averaged over an orbit) *increases* with time. You may have to play with the parameters on the drag force to get a measurable effect, but keep it small enough that the perturbation per orbit is small.

15. In Rutherford scattering, an alpha particle is deflected as it passes near the nucleus of a heavy atom. Write a program using adaptive Runge–Kutta to simulate Rutherford scattering. Find the scattering angle for a 5-MeV alpha particle striking a gold nucleus at an impact parameter of 10 femtometers.

16. The adaptive Runge–Kutta routine, rka, uses a generic method for estimating the maximum error, errmax (see lines 28–31). Write a modified version of rka that accepts a user-specified function that computes errmax. For the comet problem, write a function that computes errmax from the error in the total energy. Test your routines and compare with the original version of rka.

17. Write an adaptive Runge–Kutta program to simulate a pendulum system consisting of a bob of mass m and a massless rod with rest length L. The rod acts like a stiff spring with spring constant k. Assume that the motion is in the xy plane. Obtain plots for the motion for the values $m = 0.1$ kg, $L = 1.0$ m, and spring constants in the range $k = 10^2$ N/m (rubber) to $k = 10^6$ N/m (metal wire). How does the average time step adopted by the algorithm vary with the spring constant?

18. Consider a double pendulum, as shown in Figure 3.11.
    (a) Show that its Lagrangian may be written as

$$\frac{m_1 + m_2}{2} L_1^2 \dot{\theta}_1^2 + \frac{m_2}{2} L_2^2 \dot{\theta}_2^2 + m_2 L_1 L_2 \dot{\theta}_1 \dot{\theta}_2 \cos(\theta_1 - \theta_2)$$

$$+ (m_1 + m_2)gL_1 \cos \theta_1 + m_2 g L_2 \cos \theta_2$$

where $\dot{\theta} \equiv d\theta/dt$.
    (b) Use this Lagrangian to find the equations of motion.

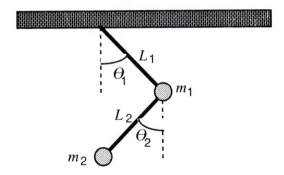

Figure 3.11 Double pendulum.

19. Write a program that uses adaptive Runge–Kutta to simulate the motion of a double pendulum (see Exercise 3.18). Take $g = 9.81$ m/s, $m_1 = m_2$, and $L_1 = L_2 = 0.1$ m; compute examples of the motion for various initial conditions. Show that in some cases, the lower mass spins completely around with an aperiodic motion.

3.4 LORENZ MODEL*

Unwinding the Mechanical Universe

Newton's success in solving the Kepler problem had effects far beyond physics. It inspired the mechanistic picture of the universe, a philosophy developed by Laplace and others. The orbits of the planets had the regularity of a well-made clock.

Even long-term events, such as solar eclipses and cometary returns, were predictable to high accuracy.

For centuries it was believed that other physical phenomena, such as weather, were unpredictable only because of the large number of variables in the problem. With the arrival of modern computers it was believed that long-range weather prediction would soon be within our grasp.

In the early 1960s, however, an MIT meteorologist named Ed Lorenz saw that it would not be so. He found that the weather was intrinsically unpredictable, not because of its complexity but because of the nonlinear nature of the governing equations. Lorenz formulated a simple model of the global weather, reducing the problem to a 12-variable system of nonlinear ODEs. What he observed was aperiodic behavior that was extremely sensitive to the initial conditions.[5]

To study this effect more easily, he introduced an even simpler model with only three variables. The Lorenz model is[6]

$$\frac{dx}{dt} = \sigma(y - x)$$

$$\frac{dy}{dt} = rx - y - xz \qquad\qquad (3\text{-}28)$$

$$\frac{dz}{dt} = xy - bz$$

where σ, r, and b are positive constants. These simple equations were originally developed as a model for buoyant convection in a fluid. Their derivation is beyond the scope of this text but, briefly, x measures the rate of convective overturning and y and z measure the horizontal and vertical temperature gradients. The parameters σ and b depend on the fluid properties and the geometry of the container. Commonly the values $\sigma = 10$ and $b = \frac{8}{3}$ are used. The parameter r is proportional to the applied temperature gradient.

Lorenz Model Program

A program, called `lorenz`, that solves this model using our adaptive Runge–Kutta method is given in Listing 3.5. Some notes about this program: Since it takes a few minutes to run, a message is periodically printed as to the progress (lines 20–22). On lines 28–31 three points called steady states are defined; these

---

[5] For a historical account see J. Gleick, *Chaos, Making a New Science* (New York: Viking Press, 1987).

[6] C. Sparrow, *The Lorenz Equations: Bifurcations, Chaos, and Strange Attractors* (New York: Springer-Verlag, 1982).

LISTING 3.5   Program `lorenz`. Computes the time evolution of the Lorenz model. Uses `rka` (Listing 3.4) and `lorzrk` (Listing 3.6).

```
1    % lorenz - Program to compute the trajectories of the Lorenz
2    % equations using the adaptive Runge-Kutta method.
3    clear; help lorenz;   % Clear memory; print header
4    state = input('Enter the initial position [x y z] - ');
5    r = input('Enter the parameter r - ');
6    s = 10.;   % Parameter sigma
7    b = 8./3.; % Parameter b
8    param = [r s b];   % Vector of parameter values passed to rka
9    tau = 1;   % Initial guess for the timestep
10   err = 1.e-3;    % Error tolerance
11   %%%% Main Loop %%%%
12   time = 0;
13   nstep = 300;   % Total number of steps
14   for istep=1:nstep
15     [state, time, tau] = rka(state,time,tau,err,'lorzrk',param);
16     x = state(1);  y = state(2);  z = state(3);
17     %% Record the orbit for plotting %%
18     tplot(istep) = time;   tauplot(istep) = tau;
19     xplot(istep) = x;  yplot(istep) = y;  zplot(istep) = z;
20     if ( rem(istep,10) < 1 )
21       fprintf('Finished %g steps out of %g\n',istep,nstep);
22     end
23   end
24   % Graph the time series
25   plot(tplot,xplot,'-',tplot,zplot,'--')
26   xlabel('Time'); ylabel('x (solid) and z (dashed)')
27   pause;   % Pause between plots; strike any key to continue
28   % Mark the location of the steady states
29   x_ss(1) = 0;            y_ss(1) = 0;         z_ss(1) = 0;
30   x_ss(2) = sqrt(b*(r-1));   y_ss(2) = x_ss(2); z_ss(2) = r-1;
31   x_ss(3) = -sqrt(b*(r-1));  y_ss(3) = x_ss(3); z_ss(3) = r-1;
32   subplot(121)
33     plot(xplot,zplot,'-',x_ss,z_ss,'*')
34     xlabel('x');   ylabel('z')
35   subplot(122)
36     plot(xplot,yplot,'-',x_ss,y_ss,'*')
37     xlabel('x');   ylabel('y')
38   subplot(111)
```

points are marked by asterisks on the plots. The meaning of these steady states is discussed in Chapter 4.

The function `lorzrk`, which contains the Lorenz model equations, is given in Listing 3.6. It is not really necessary to decompose the vectors a and `param`;

we could eliminate lines 7 and 8, and lines 9–11 in `lorzrk` could be rewritten as

```
deriv(1) = param(2)*(a(2)-a(1));
deriv(2) = param(1)*a(1) - a(2) - a(1)*a(3);
deriv(3) = a(1)*a(2) - param(3)*a(3);
```

However, I prefer to sacrifice a small bit of speed to improve the readability of the code.

Although an adaptive scheme has many advantages, it may not always be the best method to use for a particular problem. Note that for the runs of the `lorenz` program described above, the value of the time step τ does not vary by much more than one order of magnitude. This is one argument for returning to our nonadaptive methods. Another is that nonadaptive methods automatically produce data points that are evenly spaced in time, the form required by most data analysis techniques. In an exercise you are asked to make a comparison between simple and adaptive Runge–Kutta for the Lorenz problem and judge for yourself.

LISTING 3.6   Function `lorzrk`. Used by program `lorenz`; defines equations of motion for the Lorenz model.

```
1    function deriv = lorzrk(a,time,param)
2    % Function to define the Lorenz model equations
3    % The vector a = [x y z]
4    % The time is not used in this function
5    % The vector param = [r s b]
6    % The vector deriv = [dx/dt dy/dt dz/dt]
7    x = a(1); y = a(2); z = a(3);
8    r = param(1); s = param(2); b = param(3);
9    deriv(1) = s*(y-x);        % dx/dt
10   deriv(2) = r*x - y - x*z;  % dy/dt
11   deriv(3) = x*y - b*z;      % dz/dt
12   return;
```

Numerical Results

The time series $x(t)$ and $z(t)$ as obtained by the `lorenz` program are shown in Figure 3.12. The values oscillate in a fashion that does not seem much more complicated than simple harmonic motion. However, in Chapter 5 we analyze the power spectra for these time series and find they have a complex structure.

Figure 3.13 shows similar time series $x(t)$ and $z(t)$ but for slightly different initial conditions. Comparing the two plots shows that the evolution is initially very similar, but later the two time series are completely different. This extreme sensitivity to initial conditions led Lorenz to speculate that, if weather obeyed a similar dynamics, long-term prediction was impossible. He termed this the *butterfly effect*: Even a single butterfly flapping its wings could, in the long run, influence

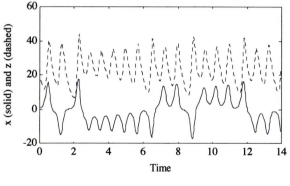

Figure 3.12 Time series for $x(t)$ (solid line) and $z(t)$ (dashed line) for the Lorenz model. The initial condition is $[x \quad y \quad z] = [1 \quad 1 \quad 20]$; the parameters are $\sigma = 10$, $b = 8/3$, and $r = 28$. Compare with Figure 3.13.

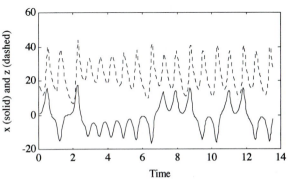

Figure 3.13 Time series for $x(t)$ (solid line) and $z(t)$ (dashed line) for the Lorenz model. The initial condition is $[x \quad y \quad z] = [1 \quad 1 \quad 20.01]$; the parameters are $\sigma = 10$, $b = 8/3$, and $r = 28$. Compare with Figure 3.12.

the world's weather. Because the trajectories of the Lorenz model are extremely sensitive to initial conditions the motion is considered *chaotic*.

Figure 3.14 shows the trajectory projected onto the xz and xy planes. In these phase space projections the motion looks far more interesting. The trajectory is said to lie on an *attractor*; you may think of this motion as a sort of aperiodic orbit. This picture helps us understand the butterfly effect and the origin of the chaotic motion. The center portion of the attractor mixes trajectories, sending some to the left lobe, some to the right. Trajectories with nearly identical

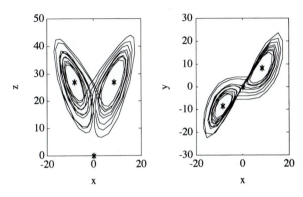

Figure 3.14 Phase space projections for the Lorenz model. The initial condition and parameters are the same as in Figure 3.12. Steady states are indicated by asterisks.

initial conditions are eventually separated in much the same way as adjacent particles of flour are separated in the kneading of bread.[7]

Although many celestial mechanics problems are accurately modeled using two-body interactions, objects moving in our solar system experience a gravitational attraction to all the planets. Specifically, the orbits of highly elliptical comets can be significantly influenced by the gas giants, especially Jupiter. It has recently been suggested that, given these perturbations, the motion of Halley's comet may actually be chaotic![8]

EXERCISES

20. Try running the `lorenz` program with the following values for the parameter r: (a) 0, (b) 1, (c) 14, (d) 20, (e) 100. Use the initial condition $[x\ y\ z] = [1\ 1\ 20]$. Describe the different types of behavior found and compare with Figure 3.14.

21. For $r = 28$, try the following initial conditions: $[x\quad y\quad z] =$ (a) $[0\ \ 0\ \ 0]$; (b) $[0\ \ 0\ \ 20]$; (c) $[0.01\quad 0.01\quad 0.01]$; (d) $[100\quad 100\quad 100]$; (e) $[8.5\quad 8.5\quad 27]$. Describe the different types of behavior found.

22. The following set of nonlinear ODEs is known as the Lotka–Volterra model:

$$\frac{dx}{dt} = (a - bx - cy)x; \qquad \frac{dy}{dt} = (-d + ex)y$$

where a, b, c, d, and e are positive constants.

(a) These equations model a simple ecological system of predators and prey.[9] For example, the variables x and y could represent the number of hares and foxes in a forest. Describe the physical meaning of each of the five parameters.

(b) Write a program using adaptive Runge–Kutta to compute the trajectory $(x(t), y(t))$ and plot $y(t)$ versus $x(t)$ for a variety of initial conditions using $a = 10$, $b = 10^{-5}$, $c = 0.1$, $d = 10$, and $e = 0.1$. Take $x(0) > 0$, $y(0) > 0$, since the number of animals should be positive.

23. Write a nonadaptive version of the `lorenz` program that uses `rk4`. Run the nonadaptive version using the minimum time step used by the adaptive version. Remember that `rka` is effectively using a time step of $\tau/2$ since this is the step size on the small steps. Modify the main loop so that the iteration stops at $t = 10$. Using the MATLAB flops command, count the number of floating point operations and determine the relative efficiency of the two methods.

---

[7] For a rigorous treatment of this geometric picture, see J. Guckenheimer and P. Holmes, *Nonlinear Oscillations, Dynamical Systems and Bifurcations of Vector Fields* (New York: Springer-Verlag, 1983).

[8] T. Y. Petrosky and R. Broucke, "Area-Preserving Mappings and Deterministic Chaos for Nearly Parabolic Motions," *Celestial Mech.*, **42**, 53–75 (1988); B. V. Chirikov and V. V. Vecheslavov, "Chaotic Dynamics of Comet Halley," *Astron. Astrophys.*, **221**, 146–54 (1989).

[9] E. C. Pielou, *An Introduction to Mathematical Ecology* (New York: Wiley, 1969).

24. One characteristic of chaotic dynamics is sensitivity to initial conditions. Write a nonadaptive version of the `lorenz` program that uses `rk4`, which simultaneously computes the trajectory for two different initial conditions. Plot the distance between the trajectories as a function of time. Use initial conditions that are very close together (e.g., [1 1 20] and [1 1 20.001]). Try plotting your data with normal and logarithmic scales. What can you say about how the distance varies with time?

25. Repeat Exercise 3.24 using the Lotka–Volterra equations (see Exercise 3.22).

BEYOND THIS CHAPTER

While adaptive fourth-order Runge–Kutta is a good general-purpose algorithm, for some problems it is useful to employ more advanced methods. Specifically, if the solution is smooth and you want to minimize the number of evaluations of $\mathbf{f}(\mathbf{x})$, you should consider trying Bulirsch–Stoer or predictor-corrector methods.[10] These are high-accuracy methods that, under the right conditions, allow you to use very large time steps. I especially recommend you try Bulirsch–Stoer if your computational budget is limited and the routine is available in a library package.

On some problems you may find that the adaptive Runge–Kutta method demands an extremely small time step. For example, suppose that you wanted to simulate a pendulum consisting of bob of mass m at the end of a massless rod of stiffness k and rest length L (see Exercise 3.17). The period of oscillation for a simple pendulum is $T_p = 2\pi\sqrt{L_p/g}$ where $L_p \cong L$ is the length of the pendulum. The period of vibration for a spring is $T_s = 2\pi\sqrt{m/k}$. If the rod is very stiff (large k), then $T_s \ll T_p$. The time step will have to be less than the period of vibration, so $\tau \lll T_p$. As you discover in Exercise 3.17, we may need to evaluate 10,000 time steps to simulate a single swing of the pendulum.

Systems of ordinary differential equations arising from physical problems with vastly different time scales, such as this spring-pendulum system, are said to be *stiff*. Another common example of such a system is a chemically reacting flow. The relaxation rates for the chemistry are often many orders of magnitude faster than the hydrodynamic time scales. Stiff ODEs are commonly solved using implicit schemes.[11]

APPENDIX 3A: FORTRAN LISTINGS

LISTING 3A.1 Program `orbe`. Computes the orbit of a comet about the Sun using the Euler method. By commenting lines in and out, this program may be modified into `orbec`, `orbrk` (Runge–Kutta), and `orbrka` (adaptive Runge–Kutta).

[10] J. Stoer and R. Bulirsch, *Introduction to Numerical Analysis* (New York: Springer-Verlag, 1980).

[11] C. W. Gear, *Numerical Initial Value Problems in Ordinary Differential Equations* (Englewood Cliffs, N.J.: Prentice-Hall, 1971).

```
         program orbe
! Program to compute the orbit of a comet using the Euler method
         parameter(nstep=200)
         real r(2),v(2),rplot(nstep),thplot(nstep),tplot(nstep)
         real kinetic(nstep),potential(nstep),norm_r
! orbrk and orbrka
!         real state(4),param(1),newstate(4)
!         external gravrk  ! Function which defines equations of
                           ! motion
!
         pi = 4*atan(1.) ! = 3.14159. . .
         write (*,*) 'Enter initial radial distance in au'
         read (*,*) r0
         r(1) = r0        ! Initial x-coordinate
         r(2) = 0.        ! Initial y-coordinate
         write (*,*) 'Enter initial tangential velocity in au/yr'
         read (*,*) v0
         v(1) = 0.        ! Initial x-velocity
         v(2) = v0        ! Initial y-velocity
         write (*,*) 'Enter time step in years, tau'
         read (*,*) tau
         GM = 4*pi**2     ! Grav. const. * Mass of Sun (au**3/yr**2)
         mass = 1.        ! Mass of the comet
!!!!! MAIN LOOP !!!!!
         time = 0.
! orbrk and orbrka !!!!!!!!!!!!
!         nstate = 4
!         state(1) = r(1)   ! Pack position and velocity information
!         state(2) = r(2)   ! into a single vector called ''state''
!         state(3) = v(1)
!         state(4) = v(2)
!         param(1) = GM     ! Parameter to pass into rk4 or rka
!!!!!!!!!!!!!!!!!!!!!!!!!!!!!!!!!!!
         do istep=1,nstep
           norm_r = sqrt(r(1)**2 + r(2)**2)
           rplot(istep) = norm_r               ! Record orbit for
           thplot(istep) = atan2(r(2),r(1))    ! polar plot
           tplot(istep) = time
           kinetic(istep) = 0.5*mass*(v(1)**2 + v(2)**2)
           potential(istep) = -GM*mass/norm_r  ! Record energies
! orbe !!!!!!!!!!!!!!!!!!!!!!!!
! Compute time evolution using Euler method
           do i=1,2
             accel = -GM*r(i)/norm_r**3 ! Gravity
             r(i) = r(i) + tau*v(i)      ! Euler step
             v(i) = v(i) + tau*accel
           end do
           time = time+tau
```

```
!!!!!!!!!!!!!!!!!!!!!!!!!!!!!!!
! orbrk
! Compute time evolution using 4-th order Runge-Kutta
!         call rk4(state,time,tau,gravRK,param,nstate,newstate)
!         time = time+tau
!!!!!!!!!!!!!!!!!!!!!!!!!!!!!!!
! orbrka
! Compute time evolution using adaptive 4-th order Runge-Kutta
!         call rka(state,time,tau,1e-3,gravRK,param,nstate,newstate)
!!!!!!!!!!!!!!!!!!!!!!!!!!!!!!!
! orbrk and orbrka
!         do i=1,4
!            state(i) = newstate(i)   ! Update state vector
!         end do
!         r(1) = state(1)      ! Decompose state vector into
!         r(2) = state(2)      ! x,y coordinates and velocities
!         v(1) = state(3)
!         v(2) = state(4)
!!!!!!!!!!!!!!!!!!!!!!!!!!!!!!!

      end do
! Print out the plotting variables -
!    rplot,thplot,tplot,kinetic,potential
!
      open(11,file='rplot.dat')
      open(12,file='thplot.dat')
      open(13,file='tplot.dat')
      open(14,file='kinetic.dat')
      open(15,file='potential.dat')
      do i=1,nstep
        write (11,*) rplot(i)
        write (12,*) thplot(i)
        write (13,*) tplot(i)
        write (14,*) kinetic(i)
        write (15,*) potential(i)
      end do
      stop
      end
```

LISTING 3A.2 Function rk4. Fourth-order Runge–Kutta routine.

```
      subroutine rk4(x,t,tau,derivsRK,param,nstate,xout)
! Runge-Kutta integrator (4th order)
      parameter(mstate = 20)    ! Maximum value for nstate
      real x(nstate),param(nstate),xout(nstate),xtemp(mstate)
      real F1(mstate),F2(mstate),F3(mstate),F4(mstate)
```

```
      half_tau = 0.5*tau
      call derivsRK(x,t,param,F1)              ! Compute F1
      t_half = t + half_tau
      do i=1,nstate
        xtemp(i) = x(i) + half_tau*F1(i)
      end do
      call derivsRK(xtemp,t_half,param,F2)     ! Compute F2
      do i=1,nstate
        xtemp(i) = x(i) + half_tau*F2(i)
      end do
      call derivsRK(xtemp,t_half,param,F3)     ! Compute F3
      t_full = t + tau
      do i=1,nstate
        xtemp(i) = x(i) + tau*F3(i)
      end do
      call derivsRK(xtemp,t_full,param,F4)     ! Compute F4
      do i=1,nstate
        xout(i) = x(i) + tau/6. * (F1(i) + F4(i) + 2*(F2(i)+F3(i)))
      end do
      return
      end
```

LISTING 3A.3   Function gravrk. Used by rk4 (Listing 3A.2) to define the equations of motion for the Kepler problem.

```
subroutine gravrk(x,time,param,deriv)
real x(4),param(1),deriv(4),r(2),v(2),accel(2),norm_r3

      r(1) = x(1)     ! Decompose the state vector x into
      r(2) = x(2)     ! x, y coordinates and velocities
      v(1) = x(3)
      v(2) = x(4)
      GM = param(1) ! Gravitational const. * Mass of Sun
      norm_r3 = sqrt(r(1)**2 + r(2)**2)**3
      accel(1) = -GM*r(1)/norm_r3   ! Acceleration in x direction
      accel(2) = -GM*r(2)/norm_r3   ! Acceleration in y direction
      deriv(1) = v(1)               ! dx/dt = v_x
      deriv(2) = v(2)               ! dy/dt = v_y
      deriv(3) = accel(1)           ! d(v_x)/dt = a_x
      deriv(4) = accel(2)           ! d(v_y)/dt = a_y
      return
      end
```

LISTING 3A.4   Function rka. Adaptive Runge–Kutta routine. Uses rk4 (Listing 3A.2).

```
      subroutine rka(x,t,tau,err,derivrk,param,nstate,xout)
      parameter(mstate = 20)   ! Maximum value of nstate
```

```fortran
      parameter (eps = 1e-16)    ! Epsilon
      real x(nstate),param(nstate),xout(nstate)
      real xsave(mstate),scale(mstate),xtemp(mstate),xdiff(mstate)
      external derivrk


      tsave = t
      do i=1,nstate
        xsave(i) = x(i)
      end do
      Safe1 = 0.9
      Safe2 = 4.0
      maxtry = 100
      do itry=1,maxtry
        !! Do the two small steps !!
        half_tau = 0.5 * tau
        call rk4(xsave,tsave,half_tau,derivrk,param,nstate,xtemp)
        t = tsave + half_tau
        call rk4(xtemp,t,half_tau,derivrk,param,nstate,x)
        !! Do the single large step !!
        t = tsave + tau
        call rk4(xsave,tsave,tau,derivrk,param,nstate,xtemp)
        !! Compute the error !!
        errmax = 0.
        do i=1,nstate
          scale(i) = 0.5*( abs(x(i) + xtemp(i)) )*err
          xdiff(i) = x(i) - xtemp(i)
          errmax = amax1(abs(xdiff(i))/(scale(i) + eps),errmax)
        end do
        !! Estimate new tau value if errmax unacceptable !!
        tau_old = tau
        tau = Safe1*tau_old*errmax**(-0.2)
        !! Never decrease tau by more than a factor of Safe2 !!
        tau = amax1(tau,tau_old/Safe2)
        if( errmax .lt. 1 ) then    ! If error is acceptable, break
                                    ! out
          goto 100
        end if
      end do
100   continue  ! Jump to here when error is acceptable
      !! Never increase tau by more than a factor of Safe2
      tau = amin1(tau,Safe2*tau_old)
      do i=1,nstate
        xout(i) = x(i)
      end do
      return
      end
```

LISTING 3A.5 Program `lorenz`. Computes the time evolution of the Lorenz model. Uses `rka` (Listing 3A.4) and `lorzrk` (Listing 3A.6).

```
      program lorenz
! Program to compute the trajectories of the Lorenz equations
! using adaptive Runge-Kutta method
      parameter( nstep = 300 )
      real state(3),param(3),newstate(3)
      real x_ss(3),y_ss(3),z_ss(3)
      real tplot(nstep),tauplot(nstep)
      real xplot(nstep),yplot(nstep),zplot(nstep)
      external lorzrk

      nstate = 3          ! Lorenz equations have 3 variables, x,y & z
      write (*,*) 'Enter the initial position x,y,z'
      read (*,*) state(1),state(2),state(3)
      write (*,*) 'Enter the parameter r'
      read (*,*) param(1)
      param(2) = 10.         ! Parameter sigma
      param(3) = 8./3.       ! Parameter b
      tau = 1.               ! Initial guess for the time step
      err = 1.e-3            ! Error tolerance
!!!!! MAIN LOOP !!!!!
      time = 0.
      do istep=1,nstep
        call rka(state,time,tau,err,lorzrk,param,nstate,newstate)
        do i=1,nstate
          state(i) = newstate(i)
        end do
        tplot(istep) = time        ! Record time for plotting
        tauplot(istep) = tau       ! Record time step for plotting
        xplot(istep) = state(1)    ! Record trajectory for plot
        yplot(istep) = state(2)
        zplot(istep) = state(3)
        if ( mod(istep,10) .lt. 1 ) then
          write (*,*) 'Finished ',istep,' steps out of ',nstep
        end if
      end do
      x_ss(1) = 0    ! Mark the locations of the steady states
      y_ss(1) = 0
      z_ss(1) = 0
      x_ss(2) = sqrt(param(3)*(param(1)-1.))
      y_ss(2) = x_ss(2)
      z_ss(2) = param(1)-1.
      x_ss(3) = -sqrt(param(3)*(param(1)-1.))
      y_ss(3) = x_ss(3)
      z_ss(3) = param(1)-1.
! Print out the plotting variables -
```

```fortran
!     tplot,tauplot,xplot,yplot,zplot,x_ss,y_ss,z_ss
!
      open(11,file='tplot.dat')
      open(12,file='tauplot.dat')
      open(13,file='xplot.dat')
      open(14,file='yplot.dat')
      open(15,file='zplot.dat')
      open(16,file='x_ss.dat')
      open(17,file='y_ss.dat')
      open(18,file='z_ss.dat')
      do i=1,nstep
        write (11,1001) tplot(i)
        write (12,1001) tauplot(i)
        write (13,1001) xplot(i)
        write (14,1001) yplot(i)
        write (15,1001) zplot(i)
      end do
      do i=1,3
        write (16,1001) x_ss(i)
        write (17,1001) y_ss(i)
        write (18,1001) z_ss(i)
      end do
      stop
1001  format(e14.7)
      end
```

LISTING 3A.6 Function `lorzrk`. Used by program `lorenz` (Listing 3A.5); defines equations of motion for the Lorenz model.

```fortran
      subroutine lorzrk(a,time,param,deriv)
!  Function to evaluate time derivatives for Lorenz model
      real a(3),param(3),deriv(3)
      x = a(1)                       ! State variables
      y = a(2)
      z = a(3)
      r = param(1)                   ! Parameters
      s = param(2)
      b = param(3)
      deriv(1) = s*(y-x)             ! Lorenz equations
      deriv(2) = r*x - y - x*z
      deriv(3) = x*y - b*z
      return
      end
```

Chapter 4
Solving Systems of Equations

In this chapter we learn how to solve systems of equations, both linear and nonlinear. You already know the basic algorithm for linear systems: Eliminate variables until you have a single equation in a single unknown. For nonlinear problems we develop an iterative scheme that at each step solves a linearized version of the equations. To maintain continuity, the discussion is motivated by the calculation of steady states, an important topic in ordinary differential equations.

4.1 LINEAR SYSTEMS OF EQUATIONS

Steady States of ODEs

In the last chapter we saw how to solve ordinary differential equations of the form

$$\frac{d\mathbf{x}}{dt} = \mathbf{f}(\mathbf{x}, t) \tag{4-1}$$

where $\mathbf{x} = [x_1 \quad x_2 \quad \ldots \quad x_N]$. Given the initial condition for the N variables, $x_i(t = 0)$, we can compute the time series $x_i(t)$ by a variety of methods (e.g., Runge–Kutta).

The examples we have studied so far have been autonomous systems where $\mathbf{f}(\mathbf{x}, t) = \mathbf{f}(\mathbf{x})$, that is, $\mathbf{f}$ does not depend explicitly on time. For autonomous systems, there often exists an important class of initial conditions for which $x_i(t) = x_i(0)$ for all i and t. These points in the N-dimensional space of our variables are called *steady states*. If we start at a steady state we stay there forever. Locating steady states for ODEs is important since they are used in bifurcation analysis.[1]

It is easy to see that $\mathbf{x}^* = [x_1^* \quad \ldots \quad x_N^*]$ is a steady state if and only if

$$\mathbf{f}(\mathbf{x}^*) = 0 \tag{4-2}$$

or

$$f_i(x_1^*, \ldots, x_N^*) = 0; \quad \text{for all } i \tag{4-3}$$

since this implies that $d\mathbf{x}^*/dt = 0$. Locating steady states now reduces to the problem of solving N equations in the N unknowns x_i^*. This problem is also called "finding the roots of $f(x)$." It should be clear that root finding has many more applications besides the computation of steady states.

As a simple example, consider a pendulum; the state is described by the angle θ and the angular velocity ω (see Section 2.2). The steady states are found by solving the nonlinear equations

$$-\frac{g}{L} \sin \theta^* = 0; \quad \omega^* = 0 \tag{4-4}$$

The roots are $\theta^* = 0, \pm\pi, \pm2\pi, \ldots$ and $\omega^* = 0$. Of course not all systems of equations are so easy to solve. In this chapter we consider a variety of ways to solve both linear and nonlinear systems. In this section and the next we consider the linear case, leaving the nonlinear problem to the latter part of the chapter.

Gaussian Elimination

The problem of solving $f_i(\{x_j\}) = 0$ is divided into two important classes. In this section we do the easier case of when $f_i(\{x_j\})$ is a linear function. The problem then reduces to solving a linear system of N equations with N unknowns

$$
\begin{aligned}
a_{11}x_1 + a_{12}x_2 + \ldots + a_{1N}x_N - b_1 &= 0 \\
a_{21}x_1 + a_{22}x_2 + \ldots + a_{2N}x_N - b_2 &= 0 \\
&\;\;\vdots \\
a_{N1}x_1 + a_{N2}x_2 + \ldots + a_{NN}x_N - b_N &= 0
\end{aligned}
\tag{4-5}
$$

[1] R. Seydel, *From Equilibrium to Chaos, Practical Bifurcation and Stability Analysis* (New York: Elsevier, 1988).

or in matrix form

$$\mathbf{Ax} - \mathbf{b} = 0 \qquad (4\text{-}6)$$

where

$$\mathbf{A} = \begin{bmatrix} a_{11} & a_{12} & \cdots \\ a_{21} & a_{22} & \cdots \\ \vdots & \vdots & \ddots \end{bmatrix}; \qquad \mathbf{x} = \begin{bmatrix} x_1 \\ x_2 \\ \vdots \end{bmatrix}; \qquad \mathbf{b} = \begin{bmatrix} b_1 \\ b_2 \\ \vdots \end{bmatrix} \qquad (4\text{-}7)$$

We learned how to solve linear sets of equations in grade school.[2] You keep combining equations to eliminate variables until you have an equation with only one unknown. Let's do a simple example to review how this procedure works. Take the equations

$$2x_1 + x_2 = 4$$
$$4x_1 - x_2 = 2 \qquad (4\text{-}8)$$

We want to eliminate x_1 from the second equation. To accomplish this we subtract twice the first equation from the second. This gives us

$$2x_1 + x_2 = 4$$
$$- 3x_2 = -6 \qquad (4\text{-}9)$$

This step is called *forward elimination*. For a larger set of equations the forward elimination procedure eliminates x_1 from the second equation, eliminates x_1 and x_2 from the third equation, and so on. The last equation will only contain the variable x_N.

Returning to the example, it is now trivial to solve the second equation for $x_2 = 2$. We can then substitute in for x_2 in the first equation to get

$$2x_1 + 2 = 4$$
$$x_2 = 2 \qquad (4\text{-}10)$$

so our answer is $x_1 = 1$, $x_2 = 2$. This second procedure is called *backsubstitution*. It should be clear how this works with larger systems of equations. Using the last

---

[2] These problems usually start with something like, "Johnny has twice as many apples as Suzy, and . . ."

equation to get x_N, this is plugged into the penultimate equation to obtain x_{N-1} and so forth.

This method of solving systems of linear equations is called *Gaussian elimination*.[3] It is a rote procedure that is simple for a computer to perform. Now that you understand how it works, let's write down the formal steps. For the system of equations (4-5), suppose that the matrix **A** and the vector **b** are the MATLAB variables a and b. Forward elimination is performed by the following segment of code:

```
1   % Forward elimination
2   for k=1:n-1            % Go column by column (k) operate
3     for i=k+1:n          % on the rows (i) below column k
4       coeff = a(i,k)/a(k,k);
5       for j=k+1:n
6         a(i,j) = a(i,j) - coeff*a(k,j);
7       end
8       a(i,k) = coeff;
9       b(i) = b(i) - coeff*b(k);
10    end
11  end
```

The best way to understand this code is to do a small example by hand (see the exercises). Notice that the elements in the lower triangular part of **A** (i.e., A_{ij} where $i > j$) are not explicitly set to zero. Instead, we simply skip these elements when performing backsubstitution:

```
1   % Backsubstitution
2   x(n) = b(n)/a(n,n);    % Start from the bottom
3   for i=n-1:-1:1         % and work upward (Note: This loop
4     sum = b(i);          % goes from n-1 to 1 in steps of -1)
5     for j=i+1:n          % Skip lower triangular part
6       sum = sum - a(i,j)*x(j);
7     end
8     x(i) = sum/a(i,i);
9   end
10  x = x'; % Convert x into column vector
```

These listings are provided for reference only. We will not be writing a MATLAB Gaussian elimination routine, for reasons discussed later.

---

[3] G. E. Forsythe and C. B. Moler, *Computer Solution of Linear Algebraic Systems* (Englewood Cliffs, N.J.: Prentice-Hall, 1967).

Pivoting

Gaussian elimination is a simple procedure, yet you should be aware of its pitfalls. To illustrate the first possible source of problems, consider the set of equations

$$\varepsilon x_1 + x_2 + x_3 = 5$$
$$x_1 + x_2 \qquad = 3 \qquad (4\text{-}11)$$
$$x_1 \qquad + x_3 = 4$$

In the limit $\varepsilon \to 0$, the solution is $x_1 = 1$, $x_2 = 2$, $x_3 = 3$. For these equations, the forward elimination step would start by multiplying the first equation by $(1/\varepsilon)$ and subtracting it from the second and third equations, giving

$$\varepsilon x_1 + \qquad x_2 \qquad + \qquad x_3 \qquad = \qquad 5$$
$$(1 - 1/\varepsilon)x_2 - \qquad (1/\varepsilon)x_3 \qquad = 3 - 5/\varepsilon \qquad (4\text{-}12)$$
$$-(1/\varepsilon)x_2 \quad + (1 - 1/\varepsilon)x_3 = 4 - 5/\varepsilon$$

Of course, if $\varepsilon = 0$ we have big problems, since the $1/\varepsilon$ factors blow up. Even if $\varepsilon \neq 0$ but is small, we are going to have serious round-off problems. Suppose that $1/\varepsilon$ is so large that $(C - 1/\varepsilon) \to -1/\varepsilon$ where C is of order unity. Our equations, after round-off, become

$$\varepsilon x_1 + \qquad x_2 \qquad + \qquad x_3 \qquad = \qquad 5$$
$$-(1/\varepsilon)x_2 - (1/\varepsilon)x_3 = -5/\varepsilon \qquad (4\text{-}13)$$
$$-(1/\varepsilon)x_2 - (1/\varepsilon)x_3 = -5/\varepsilon$$

At this point it is clear that we may not proceed since the second and third equations are now identical; we no longer have three independent equations. The next step of forward elimination would transform the third equation into the tautology $0 = 0$.

Fortunately, there is a simple fix; we can just interchange the order of the equations before doing the forward elimination. Exchanging the first and second equations,

$$x_1 + x_2 \qquad = 3$$
$$\varepsilon x_1 + x_2 + x_3 = 5 \qquad (4\text{-}14)$$
$$x_1 \qquad + x_3 = 4$$

The first step of forward elimination gives us the equations

$$
\begin{aligned}
x_1 + x_2 \qquad\qquad &= 3 \\
(1 - \varepsilon)x_2 + x_3 &= 5 - 3\varepsilon \\
-x_2 + x_3 &= 4 - 3
\end{aligned}
\tag{4-15}
$$

Round-off eliminates the ε terms, giving

$$
\begin{aligned}
x_1 + x_2 \qquad &= 3 \\
x_2 + x_3 &= 5 \\
-x_2 + x_3 &= 1
\end{aligned}
\tag{4-16}
$$

The second step of forward elimination removes x_2 from the third equation using the second equation,

$$
\begin{aligned}
x_1 + x_2 \qquad &= 3 \\
x_2 + x_3 &= 5 \\
2x_3 &= 6
\end{aligned}
\tag{4-17}
$$

You can easily check that backsubstitution gives the correct answer of $x_1 = 1$, $x_2 = 2$, and $x_3 = 3$.

Algorithms that rearrange the equations when they spot small diagonal elements are said to *pivot*. Even if all the elements of a matrix are initially of comparable magnitude, the forward elimination procedure may produce small elements on the main diagonal. The price of pivoting is just a little extra bookkeeping in the program, but it is essential to use pivoting for all but the smallest matrices. Even with pivoting, one cannot guarantee being safe from round-off problems when dealing with very large matrices.

Gaussian Elimination in MATLAB

Although Gaussian elimination with pivoting is straightforward, we will *not* write a MATLAB program to do it. Instead we will use the built-in matrix manipulation capabilities of MATLAB. Should you ever need to solve systems of linear equations outside of MATLAB, you should use the routines from a standard text such as *Numerical Recipes*[4], or from an established numerical library such as LIN-PACK[5]. In fact, MATLAB uses LINPACK routines to perform Gaussian elimina-

---

[4] W. Press, B. Flannery, S. Teukolsky, and W. Vetterling, *Numerical Recipes in FORTRAN*, 2d ed. (Cambridge: Cambridge University Press, 1992), chapter 2.

[5] J. J. Dongarra, J. R. Bunch, C. B. Moler, and G. W. Stewart, *LINPACK Users' Guide* (Philadelphia: Society for Industrial and Applied Mathematics (SIAM), 1979).

tion. A simple Gaussian elimination routine written in FORTRAN is listed in the appendix (Listing 4A.3).

The arguments for and against using canned routines come up frequently, so let me add my opinions. Virtually all scientists today are satisfied with using high-level languages (as opposed to writing in assembly code). Furthermore, I don't feel that I need to write my own version of the sqrt() function. It is important to understand the concept of square root and recognize possible pitfalls, especially computational ones [e.g., Does sqrt(-1) return an imaginary number or an error?]. As software technology advances, our concept of "primitive routines" changes. In MATLAB, Gaussian elimination is a primitive routine.

MATLAB implements Gaussian elimination using the slash, /, and backslash, \, operators. The linear system of equations, $\mathbf{xA} = \mathbf{b}$, where $\mathbf{x}$ and $\mathbf{b}$ are row vectors, is solved using the slash operator as $\mathbf{x} = \mathbf{b}/\mathbf{A}$. The linear system of equations, $\mathbf{Ax} = \mathbf{b}$, where $\mathbf{x}$ and $\mathbf{b}$ are column vectors, is solved using the back-slash operator as $\mathbf{x} = \mathbf{A} \backslash \mathbf{b}$. As an example, take equation (4-11) with $\varepsilon = 0$. We set it up in matrix form

$$
\begin{bmatrix} 0 & 1 & 1 \\ 1 & 1 & 0 \\ 1 & 0 & 1 \end{bmatrix} \begin{bmatrix} x_1 \\ x_2 \\ x_3 \end{bmatrix} = \begin{bmatrix} 5 \\ 3 \\ 4 \end{bmatrix} \tag{4-18}
$$

Using MATLAB interactively, the solution is illustrated below:

```
>>A=[0 1 1;1 1 0;1 0 1]

A =

        0       1       1
        1       1       0
        1       0       1

>>b=[5; 3; 4]

b =

        5
        3
        4

>>x=A\b

x =
        1
        2
        3
```

Clearly, MATLAB uses pivoting in this case. The slash and backslash operators may be used in the same way in programs. If MATLAB thinks that the computation is of questionable accuracy (for example, due to round-off) a warning message is issued.

In this chapter they are modest in size, but in later chapters we use large matrices. It is important for you to know that for an $N \times N$ matrix, the computation time for Gaussian elimination goes as N^3. Fortunately, if a matrix is sparse (most elements are zero), this calculation time can be greatly reduced. For example, if a matrix is tridiagonal, the computation time goes as N (see Chapter 8).

Determinants

It is easy to obtain the determinant of a matrix using Gaussian elimination. After completing forward elimination, one simply computes the product of the coefficients of the diagonal elements. Take our original example, equation (4-8); the matrix is

$$\mathbf{A} = \begin{bmatrix} 2 & 1 \\ 4 & -1 \end{bmatrix} \tag{4-19}$$

At the completion of forward elimination we obtained equation (4-9). The product of the coefficients of the diagonal elements is $(2)(-3) = -6$, which, you can check, is the determinant of $\mathbf{A}$. This is the algorithm used by the MATLAB function det (A). It should now be obvious that Cramer's rule is a computationally inefficient way to solve sets of linear equations. One final note: This method is slightly more complicated when pivoting is used. If the number of pivots is odd, the determinant is the *negative* of the product of the coefficients of the diagonal elements.

EXERCISES

1. Find the steady states of the one-variable Ginzburg–Landau equation,

$$\frac{dx}{dt} = -x + \lambda x^3$$

where x is a real number and λ is a real parameter. Plot the right-hand side of this equation as a function of x for a variety of values of λ (both positive and negative). Notice that the number of steady states varies with λ.

2. Show that the three steady states of the Lorenz model [equation (3-28)] are: $x^* = y^* = z^* = 0$, and $x^* = y^* = \pm\sqrt{b(r-1)}$, $z^* = r - 1$, if $\sigma \neq 0$. What are the steady states when $\sigma = 0$?

3. (a) Find the steady states of the Lotka–Volterra equations (see Exercise 3.22).
   (b) The trajectories in the Lotka–Volterra model are periodic cycles called *limit cycles*. Modify your program from Exercise 3.22 to compute the time-averaged values of x and y,

$$\langle x \rangle = \frac{1}{T} \int_0^T x(t) \, dt; \qquad \langle y \rangle = \frac{1}{T} \int_0^T y(t) \, dt$$

   where T is the period of the limit cycle. Show that $\langle x \rangle = x^*$, $\langle y \rangle = y^*$.

4. The Brusselator is a simple model for oscillatory chemical systems such as the Belousov–Zhabotinski reaction.[6] The time evolution of the concentration of two chemical species, x and y, is described by the ODEs,

$$\frac{dx}{dt} = A + x^2 y - (B + 1)x$$

$$\frac{dy}{dt} = Bx - x^2 y$$

   where $A > 0$, $B > 0$ are constants. Find the single steady state of this model. Write a program to compute $x(t)$ and $y(t)$ for a given initial condition $x(0) \geq 0$ and $y(0) \geq 0$. Plot a few trajectories in the xy plane and mark the location of the steady state. Investigate cases where $B/(1 + A^2)$ is less than, greater than, and equal to one.

5. The best way to understand an algorithm is to work out an example by hand.
   (a) Using Gaussian elimination, solve

$$\begin{bmatrix} 1 & 1 & 1 \\ 3 & 1 & 0 \\ -1 & 0 & 1 \end{bmatrix} \begin{bmatrix} x_1 \\ x_2 \\ x_3 \end{bmatrix} = \begin{bmatrix} 6 \\ 11 \\ -2 \end{bmatrix}$$

   Show the intermediate steps. Compute the determinant of the matrix using your forward elimination results.
   (b) Interchange the second and third equations and repeat part (a).

6. Using Kirchhoff's laws in circuit problems involves solving a set of simultaneous equations. Consider the simple circuit illustrated in Figure 4.1. Write a program that computes the currents given the resistances and voltages as inputs. Be sure to check your program by working out a simple case by hand. Have your program produce a graph of the power delivered to resistor 5 as a function of E_2 for the range of values $E_2 = 0$ V to $E_2 = 20$ V. For the other values, use $R_1 = R_2 = 1 \, \Omega$, $R_3 = R_4 = 2 \, \Omega$, $R_5 = 5 \, \Omega$, $E_1 = 2$ V, $E_3 = 5$ V.

7. Consider the Wheatstone bridge illustrated in Figure 4.2. There is no current flow through the ammeter when $R_1 = R_2 R_3 / R_4$ allowing us to measure R_1 if the other resistances are known (R_2 is a variable resistor). At first glance this suggests that it makes no difference whether $R_3 = R_4 = 1 \Omega$, 100Ω, or $10^4 \Omega$. Graph the current through the ammeter as a function of R_2 for each of these cases. Take $E = 6$V, $R_1 = 1 \Omega$, and the internal resistance of the ammeter to be $10^{-3} \Omega$. Given that the sensitivity of the ammeter is 10^{-4} amps, approximately how accurate is each bridge?

[6] G. Nicolis and I. Prigogine, *Self-Organization in Nonequilibrium Systems* (New York: Wiley, 1977) chapter 7.

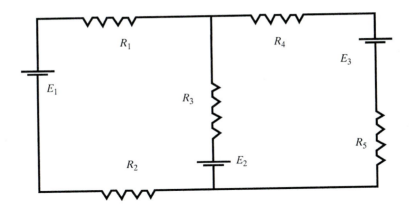

Figure 4.1 A simple resistor network.

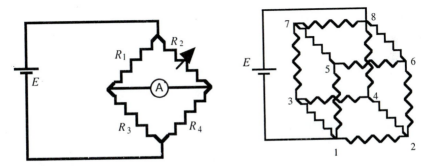

Figure 4.2 Wheatstone bridge.   **Figure 4.3** Resistor cube.

8. A problem you probably remember from freshman physics is the resistor cube (see Figure 4.3). If the resistances are all equal, then the equivalent resistance of the cube may be obtained by making use of the symmetry.

   (a) Write a program that uses Kirchhoff's laws to solve the resistor cube problem for the case where all the resistors are 1Ω except for the resistor between vertices 5 and 7. Plot the equivalent resistance as a function of this variable resistor.

   (b) Repeat but use 1–3 as the variable resistor.

   (c) Repeat but use 3–7 as the variable resistor.

4.2 MATRIX INVERSE

Matrix Inverse and Gaussian Elimination

In the previous section we reviewed how to solve a set of simultaneous equations by Gaussian elimination. However, if you are familiar with linear algebra, you would probably write the solution of

$$\mathbf{Ax} = \mathbf{b} \tag{4-20}$$

as

$$\mathbf{x} = \mathbf{A}^{-1}\mathbf{b} \tag{4-21}$$

where $\mathbf{A}^{-1}$ is the matrix inverse of $\mathbf{A}$. It should not be surprising that the calculation of a matrix inverse is related to the algorithm for solving a set of linear
equations. Usually the inverse of a matrix is computed by repeated applications of
Gaussian elimination (or a variant called LU decomposition).

The inverse of a matrix is defined by the equation

$$\mathbf{A}\mathbf{A}^{-1} = \mathbf{I} \tag{4-22}$$

where $\mathbf{I}$ is the *identity matrix*

$$\mathbf{I} = \begin{bmatrix} 1 & 0 & 0 & \cdots \\ 0 & 1 & 0 & \cdots \\ 0 & 0 & 1 & \cdots \\ \vdots & \vdots & \vdots & \ddots \end{bmatrix} \tag{4-23}$$

Defining the column vectors,

$$\mathbf{e}_1 = \begin{bmatrix} 1 \\ 0 \\ 0 \\ \vdots \end{bmatrix}; \quad \mathbf{e}_2 = \begin{bmatrix} 0 \\ 1 \\ 0 \\ \vdots \end{bmatrix}; \quad \cdots; \quad \mathbf{e}_N = \begin{bmatrix} \vdots \\ 0 \\ 0 \\ 1 \end{bmatrix} \tag{4-24}$$

we may write the identity matrix as a row vector of column vectors,

$$\mathbf{I} = [\mathbf{e}_1 \quad \mathbf{e}_2 \quad \cdots \quad \mathbf{e}_N] \tag{4-25}$$

If we solve the linear set of equations,

$$\mathbf{A}\mathbf{x}_1 = \mathbf{e}_1 \tag{4-26}$$

The solution vector $\mathbf{x}_1$ is the first column of the inverse $\mathbf{A}^{-1}$. If we proceed this
way with the other $\mathbf{e}$'s we will compute all the columns of $\mathbf{A}^{-1}$. In other words, our
matrix inverse equation $\mathbf{A}\,\mathbf{A}^{-1} = \mathbf{I}$ is solved by writing it as

$$\mathbf{A}[\mathbf{x}_1 \quad \mathbf{x}_2 \quad \cdots \quad \mathbf{x}_N] = [\mathbf{e}_1 \quad \mathbf{e}_2 \quad \cdots \quad \mathbf{e}_N] \tag{4-27}$$

After computing the **x**'s, we build $\mathbf{A}^{-1}$ as

$$\mathbf{A}^{-1} = [\mathbf{x}_1 \quad \mathbf{x}_2 \quad \cdots \quad \mathbf{x}_N] \tag{4-28}$$

We will not write a program to compute a matrix inverse since it is a simple variant of Gaussian elimination. In MATLAB the inverse of a matrix A may be computed using the built-in function `inv(A)`. It is possible to solve a system of linear equations using the matrix inverse, but usually this is overkill. An exception is the case where one wants to solve a number of similar problems in which the matrix **A** is fixed but the vector **b** takes many different values.

Here's a handy formula to keep around: The inverse of a 2×2 matrix is

$$\mathbf{A}^{-1} = \frac{1}{a_{11}a_{22} - a_{12}a_{21}} \begin{bmatrix} a_{22} & -a_{12} \\ -a_{21} & a_{11} \end{bmatrix} \tag{4-29}$$

For larger matrices the formulas very quickly become very messy.

Singular and Ill-Conditioned Matrices

Before getting back to physics, let's discuss another possible pitfall in solving linear systems of equations. Consider the equations

$$x_1 + x_2 = 1$$
$$2x_1 + 2x_2 = 2 \tag{4-30}$$

Notice that we really don't have two independent equations since the second is just twice the first. These equations do not have a unique solution. If we try to do forward elimination, we get

$$x_1 + x_2 = 1$$
$$0 = 0 \tag{4-31}$$

and are stuck.

Another way to look at this problem is to see that the matrix

$$\mathbf{A} = \begin{bmatrix} 1 & 1 \\ 2 & 2 \end{bmatrix} \tag{4-32}$$

has no inverse. A matrix without an inverse is said to be *singular*. A singular matrix also has a determinant of zero.

Here is what happens in MATLAB when we try to solve this problem:

```
>>A = [1 1; 2 2]

A =

        1       1
        2       2

>>b = [1;  2]

b =

        1
        2

>>x = A\b

Warning: Matrix is singular to working precision.

x =

        ∞
        ∞
```

Singular matrices are not always trivially spotted; would you have guessed that the matrix [1 2 3; 4 5 6; 7 8 9] was singular? Try inverting this matrix in MATLAB.

Sometimes a matrix is not singular but is so close to being singular that round-off may "push it over the edge." A trivial example would be

$$
\begin{bmatrix} 1 + \varepsilon & 1 \\ 2 & 2 \end{bmatrix}
\tag{4-33}
$$

where ε is a very small number. The *condition* of a matrix indicates how close it is to being singular; a matrix is said to be ill-conditioned if it is almost singular. Formally, the condition number is defined as the normalized "distance" between a matrix and the nearest singular matrix.[7]

There are a variety of ways to define this distance depending on the kind of norm used. The MATLAB function cond (A) returns the 2-norm condition number of a matrix A. The MATLAB function rcond (A) returns an estimate of the reciprocal of the 1-norm condition number of a matrix A. The latter function is more efficient but less reliable. A matrix with a large condition number is ill-conditioned. As a rule of thumb, $\log_{10}(\text{cond}(\mathbf{A}))$ is the number of significant digits (decimal) that you can expect to lose in Gaussian elimination.

---

[7] S. D. Conte and C. de Boor, *Elementary Numerical Analysis* (New York: McGraw-Hill, 1980).

Coupled Harmonic Oscillators

At the beginning of this chapter I discussed the problem of finding the steady states of ODEs. A canonical example of a system with linear interactions is the case of coupled harmonic oscillators. Consider the system shown in Figure 4.4; the spring constants are $k_1, \ldots, k_4$. The positions of the blocks, relative to the left wall, are x_1, x_2, and x_3. The distance between the walls is L_w and the unstretched lengths of the springs are $L_1, \ldots, L_4$. The blocks are of negligible width.

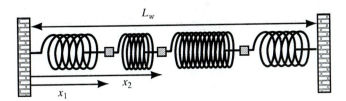

Figure 4.4 System of blocks coupled by springs.

The equation of motion for block i is

$$\frac{dx_i}{dt} = v_i; \qquad \frac{dv_i}{dt} = \frac{1}{m_i} F_i \tag{4-34}$$

where F_i is the net force on block i. At the steady state, the velocities, v_i, are zero and the net forces, F_i, are zero. This is just static equilibrium. Our job now is to find the rest positions of the masses. Explicitly, the net forces are

$$
\begin{aligned}
F_1 &= -k_1(x_1 - L_1) + k_2(x_2 - x_1 - L_2) \\
F_2 &= -k_2(x_2 - x_1 - L_2) + k_3(x_3 - x_2 - L_3) \\
F_3 &= -k_3(x_3 - x_2 - L_3) + k_4(L_w - x_3 - L_4)
\end{aligned}
\tag{4-35}
$$

or in matrix form,

$$
\begin{bmatrix} F_1 \\ F_2 \\ F_3 \end{bmatrix} =
\begin{bmatrix}
-k_1 - k_2 & k_2 & 0 \\
k_2 & -k_2 - k_3 & k_3 \\
0 & k_3 & -k_3 - k_4
\end{bmatrix}
\begin{bmatrix} x_1 \\ x_2 \\ x_3 \end{bmatrix}
$$
$$
- \begin{bmatrix}
-k_1 L_1 + k_2 L_2 \\
-k_2 L_2 + k_3 L_3 \\
-k_3 L_3 + k_4(L_4 - L_w)
\end{bmatrix}
\tag{4-36}
$$

For convenience, we abbreviate the above equation as $\mathbf{F} = \mathbf{K}\mathbf{x} - \mathbf{b}$. In matrix form the symmetries are clear, and one sees that it would not be difficult to extend the problem to larger systems of coupled oscillators. At static equilibrium the net forces are zero, so we obtain the rest positions of the masses by solving $\mathbf{K}\mathbf{x} = \mathbf{b}$, which you are asked to do in the exercises.

EXERCISES

9. To help you understand how the matrix inversion algorithm works, find $\mathbf{K}^{-1}$ by hand [see equation (4-36)]. Take $k_1 = k_2 = k_3 = k_4 = 1$. Check your result by showing that $\mathbf{K}\mathbf{K}^{-1} = \mathbf{I}$.

10. Write a program to solve for the rest positions of the masses in the harmonic oscillator system described above. Try your program with the following values:
    (a) $\mathbf{k} = [1\ \ 2\ \ 3\ \ 4]$; $\mathbf{L} = [1\ \ 1\ \ 1\ \ 1]$; $L_w = 4$
    (b) $\mathbf{k} = [1\ \ 2\ \ 3\ \ 4]$; $\mathbf{L} = [1\ \ 1\ \ 1\ \ 1]$; $L_w = 10$
    (c) $\mathbf{k} = [1\ \ 1\ \ 1\ \ 1]$; $\mathbf{L} = [2\ \ 2\ \ 1\ \ 1]$; $L_w = 4$
    (d) $\mathbf{k} = [1\ \ 1\ \ 1\ \ 0]$; $\mathbf{L} = [2\ \ 2\ \ 1\ \ 1]$; $L_w = 4$
    (e) $\mathbf{k} = [0\ \ 1\ \ 1\ \ 0]$; $\mathbf{L} = [2\ \ 2\ \ 1\ \ 1]$; $L_w = 4$
    Using general physical arguments, explain the results in each case.

11. The force on the right wall is $F_{rw} = -k_4(L_w - x_3 - L_4)$. Write a program to solve $\mathbf{K}\mathbf{x} = \mathbf{b}$, evaluate F_{rw}, and plot it as a function of L_w. For the other parameters, select any nontrivial values. Verify numerically that the force on the right wall is

$$F_{rw} = -k_o(L_w - L_o)$$

where $L_o = L_1 + L_2 + L_3 + L_4$ and

$$\frac{1}{k_o} = \frac{1}{k_1} + \frac{1}{k_2} + \frac{1}{k_3} + \frac{1}{k_4}$$

This is the law of equivalent springs. Since the matrix $\mathbf{K}$ is fixed, it is more efficient to use matrix inverse rather than Gaussian elimination. See Figure 4.4.

12. Gaussian elimination is not the only way to compute the inverse of a matrix. Consider the iterative scheme,

$$\mathbf{X}_{n+1} = 2\mathbf{X}_n - \mathbf{X}_n\mathbf{A}\mathbf{X}_n$$

where $\mathbf{X}_1$ is the initial guess for $\mathbf{A}^{-1}$. Write a computer program that uses this scheme to compute the inverse of a matrix. Test your program with a few matrices, including some that are singular, and comment on your results. How good does your initial guess have to be? When might this scheme be preferable to Gaussian elimination? For more details, see Exercise 4.21.

13. Consider the spring-mass system in Figure 4.5 (a simple model of a short polymer molecule). The springs connecting adjacent blocks have spring constant k_1, while the two outer springs have stiffness k_2. All the springs have a rest length of one.
    (a) Write the matrix equation for the equilibrium positions of the blocks.
    (b) Write a program that plots the total length of the system as a function of k_2/k_1.

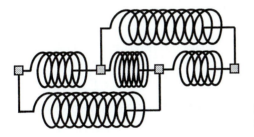

Figure 4.5   Spring-mass system with next-nearest neighbor couplings.

14. Consider a system of coupled masses (such as in Figure 4.4) with N blocks. The spring constants are
    (a) $k_1 = k_2 = k_3 = \ldots = k_{N+1} = 1$
    (b) $k_1 = 2k_2 = 3k_3 = \ldots = (N + 1)k_{N+1} = 1$
    (c) $k_1 = 4k_2 = 9k_3 = \ldots = (N + 1)^2 k_{N+1} = 1$
    (d) $k_1 = 2k_2 = 4k_3 = \ldots = 2^N k_{N+1} = 1$
    Write a program to compute the condition number of $\mathbf{K}$ and plot it as a function of N. Estimate the value of N for which cond($\mathbf{K}$) exceeds 10^{12}.

4.3 NONLINEAR SYSTEMS OF EQUATIONS*

One-Variable Newton's Method

Now that we know how to solve systems of linear equations, we proceed to the more general (and more challenging) case of solving systems of nonlinear equations. This problem is difficult enough that it is worthwhile to first consider the single variable case. We want to solve for x^* such that

$$f(x^*) = 0 \tag{4-37}$$

where $f(x)$ is now some general function.

There are a number of different methods available for one-variable root finding. If you've had a numerical analysis course you are probably familiar with bisection, the secant method, and perhaps other algorithms. The MATLAB toolbox has a function f zero that finds roots; open it up and see how it works. There are also specialized algorithms for when $f(x)$ is a polynomial. Instead of going through all these schemes, we will concentrate on Newton's method because it is the most useful for the N-variable case.

Newton's method[8] is based on the Taylor expansion of $f(x)$ around the root x^*. Suppose that we make an initial guess as to the location of the root, call this

---

[8] F. S. Acton, *Numerical Methods that Work* (New York: Harper and Row, 1970) chapters 2 and 14.

guess x_1. Our error may be written as $\delta x = x_1 - x^*$ or $x^* = x_1 - \delta x$. Writing out the Taylor expansion of $f(x^*)$,

$$f(x^*) = f(x_1 - \delta x) = f(x_1) - \delta x \frac{df(x_1)}{dx} + O(\delta x^2) \tag{4-38}$$

Notice that since x^* is a root, $f(x^*) = 0$, so we may solve for δx as

$$\delta x = \frac{f(x_1)}{f'(x_1)} + O(\delta x^2) \tag{4-39}$$

We drop the $O(\delta x^2)$ term (this will be our truncation error) and use the resulting expression for δx to "correct" our initial guess. The new guess is

$$\begin{aligned} x_2 &= x_1 - \delta x \\ &= x_1 - \frac{f(x_1)}{f'(x_1)} \end{aligned} \tag{4-40}$$

This procedure may be iterated to improve our guess further until the desired accuracy is obtained.

A few notes about Newton's method: First, when it converges it finds a root very quickly, but unfortunately it sometimes diverges and fails. For well-behaved functions, Newton's method is guaranteed to converge if we get "close enough" to the root. For this reason it is sometimes combined with a slower but surer root-finding algorithm such as bisection. Second, if there are multiple roots, the one it converges to depends on the initial guess (and it may not be the root you want). There are procedures (e.g., "deflation") for finding multiple roots using Newton's method. Finally, the method is slower when finding tangent roots, such as for $f(x) = x^2$.

The iterative procedure described above may best be understood graphically (see Figure 4.6). Notice that at each step we use the derivative of $f(x)$ to draw the tangent line of the function at that point. Effectively, we are linearizing $f(x)$ and solving the linear problem. If the function is well behaved, then it will be approximately linear in some neighborhood near the root x^*.

Multivariable Newton's Method

It is not difficult to generalize Newton's method to N-variable problems. Now our unknown $\mathbf{x} = [x_1 \quad \cdots \quad x_N]$ is a row vector, and we want to find the zeros (roots) of the row vector function

$$\mathbf{f}(\mathbf{x}) = [f_1(\mathbf{x}) \quad \cdots \quad f_N(\mathbf{x})] \tag{4-41}$$

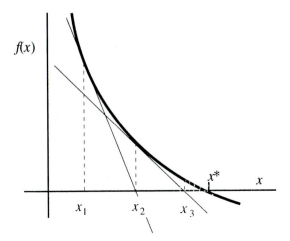

Figure 4.6 Graphical representation of Newton's method.

Again, we make an initial guess as to the location of the root, call this guess x_1. Our error may be written as $\delta x = x_1 - x^*$ or $x^* = x_1 - \delta x$. Using the Taylor expansion of $f(x_1)$,

$$
\begin{aligned}
f(x^*) &= f(x_1 - \delta x) \\
&= f(x_1) - \delta x D(x_1) + O(\delta x^2)
\end{aligned}
\tag{4-42}
$$

where the Jacobian matrix, **D**, is defined as

$$
D_{ij}(x) = \frac{\partial f_j(x)}{\partial x_i}
\tag{4-43}
$$

Since x^* is a root, $f(x^*) = 0$, so we may solve for δx as before. Dropping the error term, we may write (4-42) as

$$
f(x_1) = \delta x D(x_1)
\tag{4-44}
$$

or

$$
\delta x = f(x_1) D^{-1}(x_1)
\tag{4-45}
$$

Our new guess is

$$
x_2 = x_1 - \delta x = x_1 - f(x_1) D^{-1}(x_1)
\tag{4-46}
$$

This procedure may be iterated to improve our guess further until the desired accuracy is obtained.

Newton's Method Program

A program to find the roots of a system of equations using Newton's method, called newtn, is given in Listing 4.1. The function fnewt, on line 10, returns both the value of $f(x)$ and the matrix D. Since D changes at each iteration it would be wasteful to compute its inverse. Instead, we use Gaussian elimination on line 11 to solve for δx [see equation (4-44)]. After nstep iterations of Newton's method, the routine displays the current estimate for the root along with a plot to show the approach to the root.

LISTING 4.1   Program newtn. Finds steady states of the Lorenz model using Newton's method. Uses fnewt (Listing 4.2).

```
1   % newtn - Program to solve a system of nonlinear equations
2   % using Newton's method.  Equations defined by fnewt.
3   clear;  help newtn;   % Clear memory; print header
4   x = input('Enter the initial guess (row vector) - ');
5   a = input('Enter the parameter a - ');
6   nstep = 10;   % Number of iterations before stopping
7   %%%% MAIN LOOP %%%%
8   for istep=1:nstep
9     xplot(:,istep) = x(:);  % Save plot variables
10    [f D] = fnewt(x,a);    % fnewt returns value of f and D
11    dx = f/D;              % Find dx by Gaussian elimination
12    x = x - dx;            % Newton iteration for new x
13  end
14  xplot(:,nstep+1) = x(:); % Save plot variables
15  fprintf('After %g iterations the root is\n',nstep);
16  disp(x);
17  subplot(121)   % Plot iteration steps, x(1) vs. x(2)
18    plot(xplot(1,:),xplot(2,:),'o',...
19         xplot(1,:),xplot(2,:),'-',x(1),x(2),'*');
20    xlabel('x');  ylabel('y');
21  subplot(122)   % Plot iteration steps, x(1) vs. x(3)
22    plot(xplot(1,:),xplot(3,:),'o',...
23         xplot(1,:),xplot(3,:),'-',x(1),x(3),'*');
24    xlabel('x'); ylabel('z');
25  subplot(111)
```

To test the newtn program we use it to find the steady states of the Lorenz model (see Section 3.4). From (3-28), our equations are

$$\sigma(y - x) = 0$$

$$rx - y - xz = 0 \tag{4-47}$$

$$xy - bz = 0$$

For $r = 28$, $\sigma = 10$, and $b = 8/3$, the three roots are $[x \quad y \quad z] = [0 \quad 0 \quad 0]$, $[6\sqrt{2} \quad 6\sqrt{2} \quad 27]$, and $[-6\sqrt{2} \quad -6\sqrt{2} \quad 27]$. For the Lorenz equations, the function fnewt is given in Listing 4.2.

LISTING 4.2 Function fnewt. Used by program newtn; defines the equations of the Lorenz model along with the Jacobian matrix.

```
1    function [f,D] = fnewt(x,a)
2    %  Function used by the N-variable Newton's method
3    %  The variable a is contains parameters
4    %  Lorenz model x=[x y z], a=[r sigma b]
5    f(1) = a(2)*(x(2)-x(1));
6    f(2) = a(1)*x(1)-x(2)-x(1)*x(3);
7    f(3) = x(1)*x(2)-a(3)*x(3);
8    D(1,1) = -a(2);          % df(1)/dx(1)
9    D(1,2) = a(1)-x(3);      % df(2)/dx(1)
10   D(1,3) = x(2);           % df(3)/dx(1)
11   D(2,1) = a(2);           % df(1)/dx(2)
12   D(2,2) = -1;             % df(2)/dx(2)
13   D(2,3) = x(1);           % df(3)/dx(2)
14   D(3,1) = 0;              % df(1)/dx(3)
15   D(3,2) = -x(1);          % df(2)/dx(3)
16   D(3,3) = -a(3);          % df(3)/dx(3)
17   return;
```

An example of the output from newtn is given below; notice that the program obtains the root $[6\sqrt{2} \quad 6\sqrt{2} \quad 27]$.

```
>>newtn
    newtn - Program to solve a system of nonlinear equations
    using Newton's method.  Equations defined by function
    fnewt.

    Enter the initial guess (row vector) - [50 50 50]
    Enter the parameter a - [28 10 8/3]
  After 10 iterations the root is
      8.4853     8.4853    27.0000
```

The plot showing the convergence to the root for this run is illustrated in Figure 4.7. We converge to different roots when we use different initial conditions. For example, starting from $[2 \quad 2 \quad 2]$ we converge to $[0 \quad 0 \quad 0]$; starting from $[5 \quad 5 \quad 5]$ we converge to $[-6\sqrt{2} \quad -6\sqrt{2} \quad 27]$. Interestingly, starting from $[4 \quad 4 \quad 15]$ the method converges to a root only after making an incredibly distant excursion away from the origin.

A more sophisticated N-variable root finder is supplied by the built-in MATLAB function fsolve. It also uses Newton's method but is modified to better

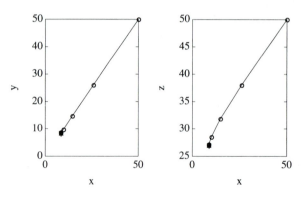

Figure 4.7 Graph of successive estimates of the root as obtained by newtn using the initial guess [50 50 50].

handle poor initial guesses. The fsolve function will also compute the Jacobian matrix using finite differences if an analytic form is not supplied.

Continuation

The primary difficulty with Newton's method is the necessity of providing a good initial guess for the root. Often our problem is of the form

$$\mathbf{f}(\mathbf{x}^*, \lambda) = 0 \tag{4-48}$$

where λ is some parameter in the problem. Suppose that we know a root $\mathbf{x}_o^*$ for the value λ_o but need to find a root $\mathbf{x}_a^*$ for a different value λ_a. Intuitively, if $\lambda_a \approx \lambda_o$, then $\mathbf{x}_o^*$ would be a good initial guess for finding $\mathbf{x}_a^*$ using Newton's method. But what if $\lambda_a \neq \lambda_o$; is there any way to make use of our known root?

   The answer is yes and the technique is known as *continuation*. The idea is to sneak up on λ_a by defining the following sequence of λ's:

$$\lambda_i = \lambda_o + (\lambda_a - \lambda_o)\frac{i}{N} \tag{4-49}$$

for $i = 1, \ldots, N$. Using Newton's method, we solve $\mathbf{f}(\mathbf{x}_1^*, \lambda_1) = 0$ with $\mathbf{x}_o^*$ as the initial guess. If N is sufficiently large, then $\lambda_1 \approx \lambda_o$ and the method should converge quickly. We then use $\mathbf{x}_1^*$ as an initial guess to solve $\mathbf{f}(\mathbf{x}_2^*, \lambda_2) = 0$; the sequence continues until we reach our desired value of λ_a. Continuation has the added benefit that we are often interested in obtaining not just a single root but actually want to know how $\mathbf{x}^*$ varies with λ.

EXERCISES

15. Write a program that uses Newton's method to find the roots of functions. Test your program on the following cases (in each case I indicate the initial guess):
   (a) $f(x) = \sin(x)$; $x_1 = 1$ (b) $f(x) = \sin(x)$; $x_1 = 2$

(c) $f(x) = x^{10}$; $x_1 = 1$ (d) $f(x) = \tanh(x)$; $x_1 = 1$

(e) $f(x) = \tanh(x)$; $x_1 = 3$ (f) $f(x) = \ln(x)$; $x_1 = 3$

Have the program make a plot of x_i versus i. Think carefully about how you want the program to decide when to stop the iteration process.

16. One of the fundamental problems in celestial mechanics is solving the Kepler equation,[9] $E - e\sin(E) = M$ where E is the eccentric anomaly, M is the mean anomaly, and e is the eccentricity. Write a program to find E using Newton's method. Run your program for a variety of values $0 \le e \le 1$ and plot E as a function of M for $0 \le M \le 2\pi$.

17. Consider a particle in a quantum square well potential of depth V and half-width a. The energy eigenvalues, E, are given by the transcendental equation

$$\sqrt{E} = \sqrt{V - E} \tan\left[\frac{a}{\hbar}\sqrt{2m(V - E)}\right]$$

for the even states and

$$\sqrt{E} = -\sqrt{V - E}\ \mathrm{ctn}\left[\frac{a}{\hbar}\sqrt{2m(V - E)}\right]$$

for the odd states.[10] Write a program to obtain the first 10 energy eigenvalues of an electron for $V = 13.6$ eV and $a = 20a_o$ where a_o is the Bohr radius.

18. For blackbody radiation, the radiant energy per unit volume in the wavelength range λ to $\lambda + d\lambda$ is[11]

$$u(\lambda)\ d\lambda = \frac{8\pi}{\lambda^5}\frac{hc}{\exp(hc/\lambda kT) - 1}\ d\lambda$$

where T is the temperature of the body, c is the speed of light, h is Planck's constant, and k is Boltzmann's constant. Show that the wavelength at which $u(\lambda)$ is maximum may be written as $\lambda_{\max} = \alpha hc/kT$ where α is a constant. Determine the value of α numerically from the resulting transcendental equation.

19. (a) Modify fnewt to solve for the steady states of the Lotka–Volterra model (see Exercise 3.22). Try a variety of initial conditions.

(b) Modify fnewt to solve for the steady state of the Brusselator model (see Exercise 4.4). Try a variety of initial conditions.

20. The program newtn is not very sophisticated since it always performs the same number of iterations. Modify the program so that it stops when the answer is within some user-specified tolerance or if the procedure appears to be diverging.

21. Newton's method gives us an iterative technique for finding the inverse of a matrix. Consider the function $\mathbf{F}(\mathbf{X}) = \mathbf{A} - \mathbf{X}^{-1}$; the root is $\mathbf{X}^* = \mathbf{A}^{-1}$. Show that Newton's method gives us the iterative scheme

$$\mathbf{X}_{n+1} = 2\mathbf{X}_n - \mathbf{X}_n\mathbf{A}\mathbf{X}_n$$

where $\mathbf{X}_1$ is the initial guess for $\mathbf{A}^{-1}$. When might this scheme be preferable to Gaussian elimination?

[9] J. M. A. Danby, *Fundamentals of Celestial Mechanics*, 2d ed. (Richmond, Va.: William-Bell Inc., 1988), section 6.6.

[10] D. Saxon, *Elementary Quantum Mechanics* (San Francisco: Holden-Day, 1968).

[11] W. Rosser, *An Introduction to Statistical Physics* (Chichester: Ellis Horwood, 1986).

22. Write a version of newtn that does not require the user to supply the matrix **D**. Instead, the program estimates it from $f(x)$ by centered finite difference. The program should determine the appropriate value for the grid spacing h (possibly with the user supplying an initial estimate).

23. For the simple, two-dimensional spring-mass system illustrated in Figure 4.8, the potential energy may be written as

$$V(x, y) = \tfrac{1}{2}k_1(\sqrt{x^2 + y^2} - L_1)^2$$
$$+ \tfrac{1}{2}k_2[\sqrt{(x - D)^2 + y^2} - L_2]^2 - mgy$$

where k_1, k_2 are the spring constants, L_1, L_2 are the rest lengths of the springs, m is the mass of the object, and $g = 9.81$ m/s^2. At static equilibrium, $\mathbf{F} = -\nabla V = 0$. Write a program to solve for the positions x and y at static equilibrium. What is the equilibrium position for $k_1 = 10$ N/m, $k_2 = 20$ N/m, $L_1 = L_2 = 0.1$ m, $D = 0.1$ m, $m = 0.1$ kg, and $g = 9.81$ m/s^2?

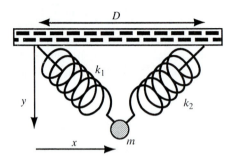

Figure 4.8   Mass-spring system for Exercise 4.23.

24. Modify your program from the previous exercise to use continuation. Using the spring constant k_2 as the variable parameter ($k_2 = 0$ to 20 N/m), obtain a graph of the equilibrium position for $k_1 = 10$ N/m, $L_1 = L_2 = 0.1$ m, $D = 0.1$ m, $m = 0.1$ kg, and $g = 9.81$ m/s^2.

BEYOND THIS CHAPTER

As you will see in the latter half of this book, many numerical schemes make use of linear algebra. For this reason it is one of the richest branches of numerical analysis. Many facets of matrix computations are covered by Golub and Van Loan.[12] Gaussian elimination and pivoting strategies are discussed in detail by Forsythe and Moler.[13]

Three principal concerns in numerical linear algebra are storage, speed, and singularity. In computational physics we often employ very large matrices, and

[12] G. H. Golub and C. F. Van Loan, *Matrix Computations* (Baltimore: Johns Hopkins University Press, 1983).

[13] G. E. Forsythe and C. B. Moler, *Computer Solution of Linear Algebraic Systems* (Englewood Cliffs, N.J.: Prentice Hall, 1967).

memory allocation can be a concern. Often these matrices are sparse, that is, most of the elements are zero, and full storage can be avoided. The computational expense of Gaussian elimination increases quickly with the size of the matrix. Again, if a matrix is sparse, the computation cost can often be reduced significantly. Sparse matrix techniques are discussed in Section 8.3.

For solving very large systems of equations, iterative techniques, such as *Gauss–Seidel*, provide alternatives to Gaussian elimination. These iterative techniques are described in the context of solving elliptic partial differential equations in Section 7.1. For further discussion, see Varga.[14]

Ill-conditioned matrices are common in numerical analysis (e.g., Vandermonde matrices in least squares curve fitting). *Singular value decomposition* (SVD) is a general technique for solving ill-conditioned problems.[15] It may also be used to solve overdetermined (more equations than unknowns) or underdetermined (more unknowns than equations) problems.

There are many schemes besides Newton's method for finding roots; you'll find them described in practically any numerical analysis textbook.[16] There are also specialized techniques for finding the roots of polynomials.[17] However, if you are finding the roots of the characteristic polynomial of a matrix to get the eigenvalues, you're barking up the wrong tree. There are much better ways to get eigenvalues and eigenvectors.

One complication in using Newton's method is the necessity of supplying the derivative, $f'(x)$. A natural solution is to evaluate this derivative numerically; the secant method is a variant based on this idea. For multidimensional problems, secant methods are not always suitable since the resulting estimated Jacobian matrix can be singular. An alternative is the class of schemes known as *quasi-Newton* methods; see Broyden[18] for a survey.

APPENDIX 4A: FORTRAN LISTINGS

LISTING 4A.1   Program newtn. Finds steady states of the Lorenz model using Newton's method. Uses fnewt (Listing 4A.2) and ge (Listing 4A.3).

```
      program newtn
! Program to solve a system of nonlinear equations using Newton's
! method
```

[14] R. S. Varga, *Matrix Iterative Analysis* (Englewood Cliffs, N.J.: Prentice Hall, 1962).

[15] G. E. Forsythe, M. A. Malcolm, and C. B. Moler, *Computer Methods for Mathematical Computations* (Englewood Cliffs N.J.: Prentice Hall, 1977), chapter 9.

[16] For example, R. L. Burden and J. D. Faires, *Numerical Analysis*, 4th ed. (Boston: PWS-Kent, 1989), chapter 2.

[17] W. Press, B. Flannery, S. Teukolsky, and W. Vetterling, *Numerical Recipes in FORTRAN*, 2d ed. (Cambridge: Cambridge University Press, 1992), section 9.5.

[18] C. G. Broyden, in *Numerical Methods for Unconstrained Optimization*, edited by W. Murray (New York: Academic Press, 1972).

```
! Equations defined by function fnewt
      parameter( nvar = 3, nparam = 3 )    ! Number of variables,
                                           ! parameters
      parameter( nstep = 10 ) ! Number of iterations before
                              ! stopping
      real x(nvar),xplot(nvar,nstep+1),a(nparam)
      real f(nvar),D(nvar,nvar),dx(nvar)

      write (*,*) 'Enter the initial guess for (x,y,z)'
      read (*,*) x(1),x(2),x(3)
      write (*,*) 'Enter the parameters (r,sigma,b)'
      read (*,*) a(1),a(2),a(3)
!!!!! MAIN LOOP !!!!!
      do istep=1,nstep
        do i=1,nvar
          xplot(i,istep)=x(i)                ! Save plot variables
        end do
        call fnewt(x,nvar,a,nparam,f,D)    ! fnewt returns f and D
        do i=1,nvar
          do j=i+1,nvar
            temp = D(i,j)
            D(i,j) = D(j,i)                ! Take the transpose of the
                                           ! matrix D
            D(j,i) = temp
          end do
        end do
        call ge(D,f,nvar,nvar,dx) ! Gaussian elimination to get
                                  ! dx
        do i=1,nvar
          x(i) = x(i) - dx(i)            ! Newton iteration for new x
        end do
      end do
      do i=1,nvar
        xplot(i,nstep+1)=x(i)         ! Save the last iteration
      end do
      write (*,*) 'After ',nstep,' iterations the value of x is'
      write (*,*) (x(i),i=1,nvar)
! Print out the plotting variables -
!   xplot
!
      open(11,file='xplot.dat')
      do j=1,nstep+1
        do i=1,nvar
          write (11,1001) xplot(i,j)
        end do
        write (11,1002)
      end do
1001  format(e12.6,' ',$)    ! The $ suppresses the carriage
```

```
                                    ! return
1002   format(/)                    ! New line
       stop
       end
```

LISTING 4A.2 Function `fnewt`. Used by program `newtn`; defines the equations of the Lorenz model along with the Jacobian matrix.

```
       subroutine fnewt(x,nvar,a,nparam,f,D)
! Subroutine used by the N-variable Newton's method
! The variable a is a parameter
! Lorenz model   x = (x,y,z),  a = (r,sigma,b)
       real x(nvar),f(nvar),D(nvar,nvar),a(nparam)

       f(1)  = a(2)*(x(2)-x(1))
       f(2)  = a(1)*x(1)-x(2)-x(1)*x(3)
       f(3)  = x(1)*x(2)  - a(3)*x(3)
       D(1,1) = -a(2)          ! df(1)/dx(1)
       D(1,2) = a(1)-x(3)      ! df(2)/dx(1)
       D(1,3) = x(2)           ! df(3)/dx(1)
       D(2,1) = a(2)           ! df(1)/dx(2)
       D(2,2) = -1.            ! df(2)/dx(2)
       D(2,3) = x(1)           ! df(3)/dx(2)
       D(3,1) = 0.             ! df(1)/dx(3)
       D(3,2) = -x(1)          ! df(2)/dx(3)
       D(3,3) = -a(3)          ! df(3)/dx(3)
       return
       end
```

LISTING 4A.3 Program `ge`. Performs Gaussian elimination with pivoting.

```
       subroutine ge(aa,bb,n,np,x)
! Perform Gaussian elimination to solve aa*x = bb
! Matrix aa is physically np by np but only n by n is used (n <=
! np)
       parameter( nmax = 100 )
       real aa(np,np),bb(np),x(np)
       real a(nmax,nmax),  b(nmax)
       integer index(nmax)
       real scale(nmax)

       if( np .gt. nmax ) then
         print *, 'ERROR - Matrix is too large for ge routine'
         stop
       end if
```

```
do i=1,n
 b(i) = bb(i)            ! Copy vector
 do j=1,n
   a(i,j) = aa(i,j)      ! Copy matrix
 end do
end do
```

```
!!!!! Forward elimination !!!!!
```

```
do i=1,n
  index(i) = i
  scalemax = 0.
  do j=1,N
    scalemax = amax1(scalemax,abs(a(i,j)))
  end do
  scale(i) = scalemax
end do
```

```
do k=1,N-1
  ratiomax = 0.
  do i=k,n
    ratio = abs(a(index(i),k))/scale(index(i))
    if( ratio .gt. ratiomax ) then
      j=i
      ratiomax = ratio
    end if
  end do
  indexk = index(j)
  index(j) = index(k)
  index(k) = indexk
  do i=k+1,n
    coeff = a(index(i),k)/a(indexk,k)
    do j=k+1,n
      a(index(i),j) = a(index(i),j) - coeff*a(indexk,j)
    end do
    a(index(i),k) = coeff
    b(index(i)) = b(index(i)) - a(index(i),k)*b(indexk)
  end do
end do
```

```
!!!!! Back substitution !!!!!
```

```
x(n) = b(index(n))/a(index(n),n)
do i=n-1,1,-1
  sum = b(index(i))
  do j=i+1,n
    sum = sum - a(index(i),j)*x(j)
  end do
```

```
      x(i) = sum/a(index(i),i)
    end do
    return
    end
```

LISTING 4A.4 Subroutine inv. Computes the inverse of a matrix with pivoting.

```
      subroutine inv(aa,n,np,ainv)
! Compute inverse of n by n complex matrix aa
! Physical size of aa is np by np (np >= n); inverse is ainv
      parameter( nmax = 100 )
      real aa(np,np),ainv(np,np)
      real a(nmax,nmax), b(nmax,nmax)
      integer index(nmax)
      real scale(nmax)

      if( np .gt. nmax ) then
        print *, 'ERROR - Matrix is too large for inv routine'
        stop
      end if

      do i=1,n
      do j=1,n
          a(i,j) = aa(i,j)      ! Copy matrix
          b(i,j) = 0.           ! Create identity matrix
        end do
        b(i,i) = 1.
      end do

!!!!! Forward elimination !!!!!

      do i=1,n
        index(i) = i
        scalemax = 0.
        do j=1,N
          scalemax = amax1(scalemax, abs(a(i,j)))
        end do
        scale(i) = scalemax
      end do

      do k=1,N-1
        ratiomax = 0.
        do i=k,n
          ratio = abs(a(index(i),k))/scale(index(i))
          if( ratio .gt. ratiomax ) then
            j=i
            ratiomax = ratio
```

```
            end if
          end do
          indexk = index(j)
          index(j) = index(k)
          index(k) = indexk
          do i=k+1,n
            coeff = a(index(i),k)/a(indexk,k)
            do j=k+1,n
              a(index(i),j) = a(index(i),j) - coeff*a(indexk,j)
            end do
            a(index(i),k) = coeff
            do jj=1,n
              b(index(i),jj)
    &=b(index(i),jj)-a(index(i),k)*b(indexk,jj)
            end do
          end do
        end do

!!!!! Back substitution !!!!!

        do jj=1,n
          ainv(n,jj) = b(index(n),jj)/a(index(n),n)
          do i=n-1,1,-1
            sum = b(index(i),jj)
            do j=i+1,n
              sum = sum - a(index(i),j)*ainv(j,jj)
            end do
            ainv(i,jj) = sum/a(index(i),i)
          end do
        end do
        return
        end
```

Chapter 5
Analysis of Data

It is often remarked that simulating physical systems on a computer is similar to experimental work. The reason this analogy is made is that computer simulations produce data in much the same way as laboratory experiments. We know that in experimental work one often needs to analyze the output, and it is the same with some numerical simulations. This chapter covers several topics in data analysis including curve fitting and Fourier transforms.

5.1 CURVE FITTING

Global Warming

At present, it appears that accurate long-range weather prediction will never be achieved. The reason is that the governing equations are highly nonlinear, and their solutions are extremely sensitive to initial conditions (see the Lorenz model discussed in Section 3.4). On the other hand, general predictions of Earth's climate are still possible. Forecasters can predict whether or not there will be drought conditions in Africa next year, although not the amount of precipitation on any given day.

Global warming is an important and hotly debated topic in climate research. The warming is blamed on greenhouse gases, such as carbon dioxide, in the atmosphere. These gases warm Earth by being transparent to the short-wave radiation arriving from the Sun but opaque to the infrared radiation from the ground.

Scientists and policy makers are still debating the threat of global warming.[1] However, no one questions that concentrations of greenhouse gases are increasing. Specifically, carbon dioxide levels have been rising steadily since the Industrial Revolution. Figure 5.1 shows the carbon dioxide concentration over the past decade, as measured in Mauna Loa, Hawaii.

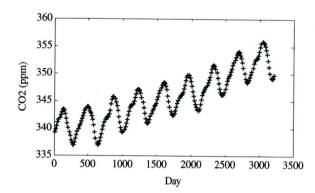

Figure 5.1   Carbon dioxide (in parts per million) measured at Mauna Loa, Hawaii.

The study of global warming has produced vast amounts of data, from both measurements in the field and computer simulations of the world's climate. In this chapter we study some basic techniques for analyzing and reducing such data sets. For example, for the data shown in Figure 5.1, what is the estimated rate of increase in CO_2 concentration per year? This first question motivates our study of curve fitting.

General Theory

The simplest type of data analysis is curve fitting. Suppose that we have a data set of N points (x_i, y_i). We wish to fit this data to a function $Y(x; \{a_j\})$ where $\{a_j\}$ is a set of M adjustable parameters. Our objective is to find the values of these parameters for which the function best "fits" the data. Intuitively, we expect that if our curve fit is good, then a graph of the data set (x_i, y_i) and the function $Y(x; \{a_j\})$ will show the curve passing "near" the points.

We can quantify this statement by measuring the distance between a data point and the curve

$$\Delta_i = Y(x_i; \{a_j\}) - y_i \tag{5-1}$$

---

[1] A. Gore, *Earth in the Balance, Ecology and the Human Spirit* (New York: Plume, 1992).

Our curve fitting criterion will be that the sum of the square of the errors be a minimum, that is, we need to find $\{a_j\}$ that minimizes the function

$$D(\{a_j\}) = \sum_{i=1}^{N} \Delta_i^2 = \sum_{i=1}^{N} [Y(x_i; \{a_j\}) - y_i]^2 \qquad (5\text{-}2)$$

This technique will give us the *least squares fit*; it is not the only way to obtain a curve fit, but it is the most common.[2]

Often, our data points have an estimated error bar (or confidence interval), which we write as $y_i \pm \sigma_i$. In this case we should modify our fit criterion so as to give less weight to the points with the most error. In this spirit, we define

$$\chi^2(\{a_j\}) = \sum_{i=1}^{N} \left(\frac{\Delta_i}{\sigma_i}\right)^2 = \sum_{i=1}^{N} \frac{[Y(x_i; \{a_j\}) - y_i]^2}{\sigma_i^2} \qquad (5\text{-}3)$$

The chi-square function is the most commonly used fitting function because, if the errors are Gaussian distributed, we can make statistical statements concerning the goodness of the fit.

Before continuing, I should remark that we will only briefly discuss the validation of our curve fit. You can fit any curve to any data set, but this does not mean that the results are meaningful. To establish significance we have to ask the following statistical question: What is the probability that the data, given the experimental error associated with each data point, are described by the curve? Unfortunately, hypothesis testing occupies a significant portion of a statistics course and is outside the scope of this book.[3]

Linear Regression

We first consider fitting the data set with a straight line,

$$Y(x; \{a_1, a_2\}) = a_1 + a_2 x \qquad (5\text{-}4)$$

This type of curve fit is also known as linear regression. We want to determine a_1 and a_2 such that

$$\chi^2(a_1, a_2) = \sum_{i=1}^{N} \frac{1}{\sigma_i^2}(a_1 + a_2 x_i - y_i)^2 \qquad (5\text{-}5)$$

---

[2] The least squares method was first used by Gauss to determine the orbits of comets from observational data.

[3] See any standard statistics textbook, such as W. Mendenhall, R. L. Scheaffer, and D. D. Wackerly, *Mathematical Statistics with Applications* (Boston: Duxbury Press, 1981).

is minimized. The minimum is found by differentiating (5-5) and setting the derivatives to zero:

$$\frac{\partial \chi^2}{\partial a_1} = 2 \sum_{i=1}^{N} \frac{1}{\sigma_i^2} (a_1 + a_2 x_i - y_j) = 0 \tag{5-6a}$$

$$\frac{\partial \chi^2}{\partial a_2} = 2 \sum_{i=1}^{N} \frac{1}{\sigma_i^2} (a_1 + a_2 x_i - y_j) x_i = 0 \tag{5-6b}$$

or

$$a_1 S + a_2 \Sigma x - \Sigma y = 0 \tag{5-7a}$$

$$a_1 \Sigma x + a_2 \Sigma x^2 - \Sigma xy = 0 \tag{5-7b}$$

where

$$S \equiv \sum_{i=1}^{N} \frac{1}{\sigma_i^2}; \quad \Sigma x \equiv \sum_{i=1}^{N} \frac{x_i}{\sigma_i^2}; \quad \Sigma y \equiv \sum_{i=1}^{N} \frac{y_i}{\sigma_i^2};$$

$$\Sigma x^2 \equiv \sum_{i=1}^{N} \frac{x_i^2}{\sigma_i^2}; \quad \Sigma xy \equiv \sum_{i=1}^{N} \frac{x_i y_i}{\sigma_i^2} \tag{5-8}$$

Since the sums may be computed directly from the data, they are known constants. We thus have a linear set of two simultaneous equations in the unknowns a_1 and a_2. These equations are easy to solve:

$$a_1 = \frac{\Sigma y \Sigma x^2 - \Sigma x \Sigma xy}{S \Sigma x^2 - (\Sigma x)^2}; \quad a_2 = \frac{S \Sigma xy - \Sigma y \Sigma x}{S \Sigma x^2 - (\Sigma x)^2} \tag{5-9}$$

Notice that if σ_i is a constant, that is, if the error is the same for all data points, then the σ's cancel out in equations (5-9). In this case, the parameters, a_1 and a_2, are independent of the error bar. It is not uncommon that a data set will not include the associated error bars. We may still use equations (5-9) to fit the data if we take σ_i to be constant ($\sigma_i = 1$ being the simplest choice).

Next, we want to obtain an associated error bar, $\sigma_{a_j}^2$, for the curve fit parameter a_j. Using the law of the propagation of errors,

$$\sigma_{a_j}^2 = \sum_{i=1}^{N} \left(\frac{\partial a_j}{\partial y_i} \right)^2 \sigma_i^2 \tag{5-10}$$

and inserting equations (5-9), after a bit of algebra we obtain

$$\sigma_{a_1} = \sqrt{\frac{\Sigma x^2}{S\Sigma x^2 - (\Sigma x)^2}}; \quad \sigma_{a_2} = \sqrt{\frac{S}{S\Sigma x^2 - (\Sigma x)^2}} \tag{5-11}$$

If the error bars on the data are constant ($\sigma_i = \sigma_o$), the error in the parameters is

$$\sigma_{a_1} = \sigma_o\sqrt{\frac{\Sigma'x^2}{N\Sigma'x^2 - (\Sigma'x)^2}}; \quad \sigma_{a_2} = \sigma_o\sqrt{\frac{N}{N\Sigma'x^2 - (\Sigma'x)^2}} \tag{5-12}$$

where

$$\Sigma'x \equiv \sum_{i=1}^{N} x_i; \quad \Sigma'x^2 \equiv \sum_{i=1}^{N} x_i^2 \tag{5-13}$$

Notice that σ_{a_j} is independent of y_i.

Finally, if our data set does not have an associated set of error bars, we may estimate σ_o from the sample variance of the data,

$$\sigma_o^2 \cong s^2 = \frac{1}{N-2} \sum_{i=1}^{N} [y_i - (a_1 + a_2x_i)]^2 \tag{5-14}$$

where s is the sample standard deviation. Notice that this sample variance is normalized by $N - 2$ since we have already extracted two parameters, a_1 and a_2, from the data.

Many nonlinear curve fitting problems may be transformed into linear problems by a simple change of variable. For example,

$$Z(x; \{\alpha, \beta\}) = \alpha e^{\beta x} \tag{5-15}$$

may be written as equation (5-4) using the change of variable

$$\ln Z = Y; \quad \ln \alpha = a_1; \quad \beta = a_2 \tag{5-16}$$

Similarly, to fit a power law of the form

$$Z(t; \{\alpha, \beta\}) = \alpha t^\beta \tag{5-17}$$

we use the change of variable

$$\ln Z = Y; \quad \ln t = x; \quad \ln \alpha = a_1; \quad \beta = a_2 \tag{5-18}$$

These transformations should be familiar since you use them whenever you plot data using semilog or log–log scales.

General Linear Least Squares Fit

The least squares fit procedure is easy to generalize to functions of the form

$$Y(x; \{a_j\}) = a_1 Y_1(x) + a_2 Y_2(x) + \ldots + a_M Y_M(x)$$

$$= \sum_{j=1}^{M} a_j Y_j(x)$$

(5-19)

To find the optimum parameters we proceed as before by finding the minimum of χ^2,

$$\frac{\partial \chi^2}{\partial a_j} = \frac{\partial}{\partial a_j} \sum_{i=1}^{N} \frac{1}{\sigma_i^2} \left(\sum_{k=1}^{M} a_k Y_k(x_i) - y_i \right)^2 = 0$$

(5-20)

or

$$\sum_{i=1}^{N} \frac{1}{\sigma_i^2} Y_j(x_i) \left(\sum_{k=1}^{M} a_k Y_k(x_i) - y_i \right) = 0$$

(5-21)

or

$$\sum_{i=1}^{N} \sum_{k=1}^{M} \frac{Y_j(x_i) Y_k(x_i)}{\sigma_i^2} a_k = \sum_{i=1}^{N} \frac{Y_j(x_i) y_i}{\sigma_i^2}$$

(5-22)

for $j = 1, \ldots, M$. This set of equations is known as the *normal equations* of the least squares problem.

The normal equations are easier to work with in matrix form. First, we define the design matrix, $\mathbf{A}$, as

$$\mathbf{A} = \begin{bmatrix} Y_1(x_1)/\sigma_1 & Y_2(x_1)/\sigma_1 & \cdots \\ Y_1(x_2)/\sigma_2 & Y_2(x_2)/\sigma_2 & \cdots \\ \vdots & \vdots & \ddots \end{bmatrix}$$

(5-23)

or

$$A_{ij} = \frac{Y_j(x_i)}{\sigma_i}$$

(5-24)

Notice that the design matrix does not depend on y_i, the data values, but does depend on x_i, that is, on the design of the experiment.

Using the design matrix we may write equation (5-22) as

$$\sum_{i=1}^{N}\sum_{k=1}^{M} A_{ij}A_{ik}a_k = \sum_{i=1}^{N} A_{ij}\frac{y_i}{\sigma_i} \tag{5-25}$$

or in matrix form,

$$(\mathbf{A}^{\mathrm{T}}\mathbf{A})\mathbf{a} = \mathbf{A}^{\mathrm{T}}\mathbf{b} \tag{5-26}$$

where the vector $\mathbf{b}$ is defined as $b_i = y_i/\sigma_i$ and $\mathbf{A}^{\mathrm{T}}$ is the transpose of the design matrix. This equation is easy to solve for the parameter vector $\mathbf{a}$,

$$\mathbf{a} = (\mathbf{A}^{\mathrm{T}}\mathbf{A})^{-1}\mathbf{A}^{\mathrm{T}}\mathbf{b} \tag{5-27}$$

Using this result and the law of the propagation of errors, equation (5-10), we find the estimated error in the parameter a_j to be

$$\sigma_{a_j} = \sqrt{C_{jj}} \tag{5-28}$$

where $\mathbf{C} \equiv (\mathbf{A}^{\mathrm{T}}\mathbf{A})^{-1}$.

Goodness of Fit

We can easily fit every data point if the number of parameters M equals the number of data points N. In this case we have either built a Rube Goldberg theory or have not taken enough data.[4] Instead, let's assume the more common scenario in which $N >> M$. Because each data point has an error, we don't expect the curve to exactly pass through the data. However, we ask, "With the given error bars, how likely is it that the curve actually describes the data?" Of course, if we are not given any error bars there is nothing we can say concerning the goodness of the fit.

Common sense suggests that if the fit is good, then on average the difference should be approximately equal to the error bar

$$|y_i - Y(x_i)| \approx \sigma_i \tag{5-29}$$

Putting this into our definition for χ^2, equation (5-3), we get the relation

$$\chi^2 \approx N \tag{5-30}$$

---

[4] As a cartoonist, Rube Goldberg created absurdly elaborate machines that performed trivial tasks.

Yet we know that the more parameters we use, the easier it is to get the curve to match the data; the fit can be perfect if $M = N$. This suggests that we modify our rule of thumb for a good fit to be

$$\chi^2 \approx N - M \tag{5-31}$$

Of course, this is only a crude indicator, but it is better than just "eye-balling" the curve. A complete analysis would use the χ^2-statistic to assign a probability that the data are fit by the curve.

   If we find that $\chi^2 \gg N - M$, then either we are not using an appropriate function, $Y(x)$, for our curve fit or the error bars, σ_i, are too small. On the other hand, if $\chi^2 \ll N - M$, then the fit is so spectacularly good that we may suspect that the error bars are actually too large.

Linear Regression Routine

Having worked out the general least squares problem, let's step back for a moment to the special case of fitting a straight line. Linear regression is such a common task that it is worth having a function especially designed to perform it. The function linreg (Listing 5.1) performs linear regression given the data vectors x, y and the error bars sigma. The routine returns the parameters a(1) and a(2) along with their associated estimated errors, sig_a(1) and sig_a(2). The curve fit is returned in the vector yy; the value of χ^2 is returned in the variable chisqr.

LISTING 5.1   Function linreg. Fits a straight line to a data set.

```
1    function [a_fit, sig_a, yy, chisqr] = linreg(x,y,sigma)
2    % Function to perform linear regression (fit a line)
3    % Inputs -
4    %    x - Independent variable
5    %    y - Dependent variable
6    %    sigma - Estimated error in y
7    % Outputs -
8    %    a_fit - Fit parameters; a(1) is intercept, a(2) is slope
9    %    sig_a - Estimated error in the parameters a()
10   %    yy - Curve fit to the data
11   %    chisqr - Chi squared statistic
12   N = length(x);              % Length of data set
13   temp = sigma .^ (-2);
14   s = sum(temp);
15   sx = sum(x .* temp);
16   sy = sum(y .* temp);        % Compute various sums
17   sxy = sum(x .* y .* temp);
18   sxx = sum((x .^ 2) .* temp);
19   denom = s*sxx - sx^2;
```

```
20   a_fit(1) = (sxx*sy - sx*sxy)/denom;   % Curve fit coeff.
21   a_fit(2) = (s*sxy - sx*sy)/denom;
22   sig_a(1) = sqrt(sxx/denom);           % Estimated error bar
23   sig_a(2) = sqrt(s/denom);             % on curve fit coeff.
24   yy = a_fit(1)+a_fit(2)*x;             % Curve fit to the data
25   chisqr = sum( ((y-yy)./sigma).^2 );   % Chi square
26   return;
```

The function linreg obtains the parameters by simply evaluating equations (5-9) on lines 20–21. The estimated error, σ_{a_j}, is computed using equations (5-11) on lines 22–23. Two programming notes: (1) linreg makes considerable use of the MATLAB function sum() that returns the sum of the elements in a vector. (2) The array operators (.*, ./, and .^) are used; recall from Chapter 1 that these are element-by-element operations. For example, for the vectors a and b,

```
z = a.*b;
```

is equivalent to

```
for i=1:N
   z(i) = a(i)*b(i);
end
```

The use of these element-by-element operators eliminates the need for any explicit loops in linreg.

The program lsftest (Listing 5.2) is useful for testing and demonstrating the linear regression function. It constructs a data set using the equations

$$x_i = i; \qquad \sigma_i = \alpha$$
$$y_i = c_1 + c_2 x_i + c_3 x_i^2 + \alpha \mathcal{R}_i \qquad (5\text{-}32)$$

where $\mathcal{R}$ is a Gaussian distributed random number with unit variance (see Chapter 10). Note that all the data points have the same estimated error, α.

LISTING 5.2 Program lsftest. Generates a data set and fits a curve to the data; may be used to test the linreg (Listing 5.1) and pollsf (Listing 5.3) functions.

```
1   % lsftest - Program for testing least squares fit routines
2   clear; help lsftest; % Clear memory and print header
3   N = 50;              % Number of data points
4   rand('seed',0);      % Initialize random number generator
5   rand('normal');      % Gaussian generator with unit variance
6   fprintf('Enter the coefficients of the quadratic \n');
7   fprintf('   Y(x) = c(1) + c(2)*x + c(3)*x^2 \n');
8   c = input('as [c(1) c(2) c(3)] - ');
9   alpha = input('Enter estimated error bar - ');
```

```
10   x = 1:N;                    % x(i)=i, i=1,. . .,N
11   r = alpha*rand(1,N);   % Gaussian distributed random vector
12   y = c(1) + c(2)*x + c(3)*x.^2 + r;
13   sigma = alpha*ones(1,N);      % Constant error bar
14   %%%%% Linear regression (Straight line) fit %%%%
15   M = 2;   % Number of fit parameters (M=2 is straight line)
16   [a_fit sig_a yy chisqr] = linreg(x,y,sigma);
17   fprintf('Fit parameters:\n');
18   for i=1:M
19      fprintf(' a(%g) = %g +/- %g \n',i,a_fit(i),sig_a(i));
20   end
21   fprintf('Chi square = %g, N-M = %g \n',chisqr,N-M);
22   plot(x,y,'o',x,yy,'-');      % Plot data and curve fit
23   errorbar(x,y,sigma)          % Add error bars to plot
24   xlabel('x'); ylabel('y');
25   title('Data (circle); curve fit (solid line)');
```

Figures 5.2 and 5.3 show some typical curve fits for linear and quadratic curves. The data is marked by the circles; the I-bars on the data points indicate the estimated error bars. The solid line is the curve fit to the data. In the former figure

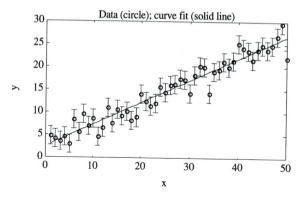

Figure 5.2 Curve fit result from lsftest using linreg. Input values are $c = [2.0 \quad 0.5 \quad 0.0]$, $\alpha = 2.0$. The fit gave the parameters $a = [2.63 \pm 0.57, 0.480 \pm 0.020]$ with $\chi^2 = 43.8$. The fit is good since $\chi^2 \approx N - 2 = 48$.

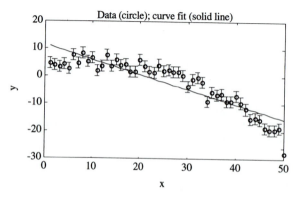

Figure 5.3 Curve fit result from lsftest using linreg. Input values are $c = [2.0 \quad 0.5 \quad -0.02]$, $\alpha = 2.0$. The fit gave the parameters $a = [11.47 \pm 0.57, -0.540 \pm 0.020]$ with $\chi^2 = 199.4$. The fit is poor since $\chi^2 \gg N - 2 = 48$.

the curve fit is good and $\chi^2 \approx N - 2$. In Figure 5.3 the function is attempting to fit a quadratic; the fit is poor and $\chi^2 >> N - 2$.

Notice that the estimated error in the parameters, σ_{a_j}, is the same in the two cases. Looking at equations (5-11) you will see that σ_{a_j} is independent of the values of the data vector y_i. The moral of the story is to use χ^2 to judge the fit and not the estimated errors, σ_{a_j}.

Polynomial Curve Fit Routine

The function `pollsf` uses the general formulation [equations (5-23) to (5-28)] to fit a polynomial to a data set (see Listing 5.3). Replacing lines 14–16 in `lsftest` with

```
14   %%%% Polynomial fit %%%%
15   M = 3;   % Number of fit parameters (M=3 is quadratic)
16   [a_fit sig_a yy chisqr] = pollsf(x,y,sigma,M);
```

allows us to fit a quadratic to the data. An example of its use is shown in Figure 5.4.

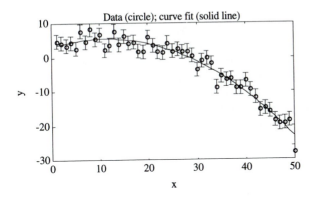

Figure 5.4 Curve fit result from `lsftest` using `pollsf`. Input values are $c = [2.0 \quad 0.5 \quad -0.02]$, $\alpha = 2.0$. The fit gave the parameters $a = [3.08 \pm 0.88, 0.427 \pm 0.080, -0.0190 \pm 0.0015]$ with $\chi^2 = 43.4$. The fit is good since $\chi^2 \approx N - 3 = 47$.

LISTING 5.3 Function `pollsf`. Fits a polynomial to a data set.

```
1    function [a_fit, sig_a, yy, chisqr] = pollsf(x, y, sigma, M)
2    % Function to fit a polynomial to data
3    % Inputs -
4    % x - Independent variable
5    % y - Dependent variable
6    % sigma - Estimate error in y
7    % M - Number of parameters used to fit data
8    % Outputs -
9    % a_fit - Fit parameters
```

```
10  % sig_a - Estimated error in the parameters a()
11  % yy - Curve fit to the data
12  % chisqr - Chi squared statistic
13  N = length(x); % Length of data set
14  b = y./sigma;  % Form the vector b
15  for i=1:N
16    for j=1:M
17      A(i,j) = x(i)^(j-1)/sigma(i);   % Form the design matrix
18    end
19  end
20  C = inv(A.' * A);         % Compute the correlation matrix
21  a_fit = C * A.' * b.';  % Compute the best fit parameters
22  for j=1:M
23    sig_a(j) = sqrt(C(j,j));  % Find the estimated error for
24  end                          % the fit parameters a(j)
25  % Compute chi-square statistic
26  yy = zeros(1,N);
27  for j=1:M
28    yy = yy + a_fit(j)*x.^(j-1);  % yy is the curve fit
29  end
30  chisqr = sum( ((y-yy)./sigma).^2 );  % Chi-square statistic
31  return;
```

Power corrupts; having a function like `pollsf` tempts one to fit data with all sorts of functions. However, one should not be too ambitious using too many parameters. First, the matrices encountered in curve fitting are notoriously ill-conditioned. The problem is rapidly aggravated if we try to fit a large number of parameters (e.g., a high-order polynomial). Second, although the function may pass near the data points, it may oscillate wildly between the data points (see Figure 5.5). Finally, can you really justify a theory that has 30 free parameters? In brief, you have better ways to apply your time.

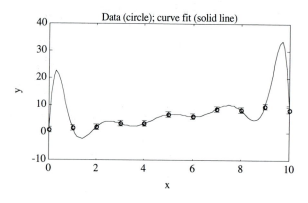

Figure 5.5 Curve fit of a tenth-order polynomial to 11 data points. Notice that the fit is good at the data points but oscillates wildly between the points. The data were generated as $y_i = x_i + \mathcal{R}$, where $\mathcal{R}$ is a Gaussian distributed random number with unit variance.

EXERCISES

1. Derive equations (5-11) using (5-9) and (5-10).

2. Two scientists want to measure the resistivity, $R(T)$, of a material as a function of temperature. They assume that $R(T) = a_1 + a_2 T$ where a_1 and a_2 are the constants to be determined. The first scientist measures R for temperatures $T = 0°, 10°, 20°, \ldots, 100°$. The second performs the measurement at $T = 0°, 100°, \ldots, 400°$. Assuming that they use the same instruments and that the measurement error is constant, which scientist will obtain more accurate estimates for the parameters?

3. Suppose that we want to fit a data set to the function $Y(x; \alpha) = \alpha x$, that is, a straight line that passes through the origin. Using the method of least squares, obtain an expression for α and its estimated error, σ_α.

4. Write a version of `linreg` that fits data sets which do not include error bars. What value do you get for χ^2? Have the function also return the sample standard deviation. Testing your function with `lsftest`, compare this standard deviation with the errors in the data.

5. (a) Repeat Exercise 1.22 using linear regression to confirm numerically that the truncation errors for the right and centered first derivatives go as h and h^2.
(b) Repeat Exercise 1.24 using linear regression to confirm numerically that the truncation error for the centered second derivative goes as h^2.

6. Repeat Exercise 2.10 and fit a quadratic to $T(\theta)$ for angles in the range $0 \le \theta \le 90°$. Compare your results with equation (2-29).

7. In Figure 3.10 we see that the time step selected by the adaptive Runge–Kutta routine varies with distance roughly as $\tau \approx r^{3/2}$. Use linear regression to fit this data to obtain a numerical estimate for the exponent including its estimated error.

8. In some problems we know, from physical grounds, that the curve must intercept the origin. Write a function that fits a data set to the equation

$$Y(x; \{a_1, a_2\}) = a_1 x + a_2 x^2$$

Use it to fit the trajectories obtained from `balle` for the initial condition $\mathbf{r}_i = [0\ 0]$ (see Chapter 2). Try a variety of values for the initial velocity, keeping the initial angle equal to 45°; include cases where air resistance is significant (see Figure 2.2). What is the largest velocity for which a parabola accurately fits the data?

9. Polynomial approximations are often used to evaluate special functions. Use the MATLAB function `bessel (n, x)` with $n = 0$ to obtain 300 values for $J_0(x)$ in the interval $0 \le x \le 3$. Fit this data to various polynomials up to twelfth order and compare your results with the polynomial approximation given in Abramowitz and Stegun.[5]

10. Carbon dioxide concentration (in parts per million) as measured in Mauna Loa, Hawaii (Figure 5.1), and Barrow, Alaska (Figure 5.6), is tabulated in Appendix 5B. The data

[5] M. Abramowitz and I. A. Stegun, *Handbook of Mathematical Functions* (New York: Dover, 1972).

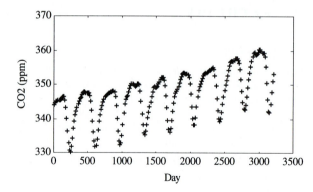

Figure 5.6 Carbon dioxide (in parts per million) measured at Barrow, Alaska.

were taken every 14 days, starting in 1981. Estimate error bars are $\sigma_o = 0.16$ ppm (Hawaii) and $\sigma_o = 0.27$ ppm (Alaska).

(a) Using linear regression, find the rate of increase of CO_2 in ppm per year at the two locations.

(b) Approximately when will the carbon dioxide concentration be 10% above its 1981 level?

5.2 SPECTRAL ANALYSIS

Discrete Fourier Transform

If our data set exhibits periodic oscillations, we probably want to fit it using trigonometric functions. This class of problems moves us from the regime of curve fitting to the topic of spectral analysis. Spectral analysis is a rich field that can easily fill a semester course (usually offered by engineers under the title "signal processing"). MATLAB has a full toolbox of spectral analysis routines. This section introduces some of the basic concepts, including the discrete Fourier transform and the power spectrum. For a more complete treatment, see Jenkins and Watts.[6]

Take a vector of N data points, $\mathbf{y} = [y_1 \quad y_2 \quad \ldots \quad y_N]$; we call this data set a time series since transform methods are often used in signal analysis. The data is evenly spaced in time, so $t_i = \tau (i - 1)$ where τ is the sampling interval, that is, the time increment between data points. We define the vector $\mathbf{Y}$, the discrete Fourier transform of $\mathbf{y}$, as

$$Y_{k+1} = \sum_{j=0}^{N-1} y_{j+1} e^{-2\pi ijk/N} \tag{5-33}$$

[6] G. Jenkins and D. Watts, *Spectral Analysis and Its Applications* (San Francisco: Holden-Day, 1968).

where $i = \sqrt{-1}$. Notice that j, $k = 0, \ldots, N-1$, while our vector indices start from one instead of zero. The inverse transform is

$$y_{j+1} = \frac{1}{N} \sum_{k=0}^{N-1} Y_{k+1} e^{2\pi i j k / N} \tag{5-34}$$

Note that texts (and numerical libraries) will vary slightly in how they define this transform, especially how it is normalized. If you use a library function from a numerical analysis package to evaluate a Fourier transform, check carefully how the transform is defined by that routine.

Each point Y_k of the transform has an associated frequency,

$$f_k = \frac{k-1}{\tau N} \tag{5-35}$$

The lowest (nonzero) frequency is $f_2 = 1/\tau N = 1/T$ where T is the length of the time series. To measure very low frequencies, we need to analyze long time series. The highest frequency is $f_N \approx 1/\tau$, so to measure very high frequencies we need to use a short sampling rate.[7]

The program $\mathtt{sftdemo}$ (short for slow Fourier transform demo) is given in Listing 5.4. The program creates a time series

$$y_{j+1} = \sin(2\pi f_s j\tau + \varphi_s) \tag{5-36}$$

for $j = 0, \ldots, N-1$. This signal is a sine wave of frequency f_s and phase φ_s (lines 8–10). The discrete Fourier transform of this time series is computed by the function $\mathtt{sft}$ (Listing 5.5) using equation (5-33). The original time series and its discrete Fourier transform are plotted side by side.

LISTING 5.4 Program $\mathtt{sftdemo}$. Demonstrates the discrete Fourier transform using the $\mathtt{sft}$ function (Listing 5.5).

```
1   % sftdemo - Discrete Fourier transform demonstration program
2   % This version uses the slow way of computing the tranform
3   clear; help sftdemo; % Clear memory and print header
4   N = input('Enter the number of points N - ');
5   freq = input('Enter frequency of the sine wave - ');
6   phase = input('Enter phase of the sine wave - ');
7   tau = 1; % Time increment
8   % Build the data set
9   t = (0:(N-1))*tau;    % t(i) = i*tau, i=0,...,N-1
10  y = sin(2*pi*t*freq + phase);   % Data set is sine wave
```

---

[7] For a real time series, the highest frequency is actually $1/(2\tau)$; see the discussion below on the Nyquist frequency.

```
11  flops(0);  % Reset the flops counter to zero
12  % Compute discrete Fourier transform
13  yt = sft(y);
14  fprintf('Total number of flops = %g\n',flops);
15  % Compute frequency vector
16  f = (0:(N-1))/(N*tau);    % f(i) = i/(N*tau),i=0,...,N-1
17  subplot(121)
18    plot(t,y);
19    title('Original data set');
20    ylabel('Amplitude');
21    xlabel('Time')
22  subplot(122)
23    plot(f,real(yt),'-',f,imag(yt),'--');
24    title('Real(solid); Imag(dash)');
25    ylabel('Fourier transform')
26    xlabel('Frequency')
27  subplot(111)
```

LISTING 5.5 Function sft. Computes the discrete Fourier transform directly from its definition; this is the slow algorithm.

```
1   function yt=sft(y)
2   % Slow Fourier transform function
3   % yt = sft(y)
4   % Input
5   %    y - Time series
6   % Output
7   %    yt - Discrete Fourier transform
8   N = length(y);   % Length of the time series
9   twopiN = -2*pi*sqrt(-1)/N;
10  for k=0:N-1
11    temp = exp(twopiN*(0:N-1)*k);
12    yt(k+1) = sum(y .* temp);
13  end
14  return;
```

The sftdemo program uses the MATLAB flops command to count the number of flops executed by sft.[8] On line 11 the counter is set to zero; line 14 prints the number of flops executed in the calculation of the transform. Computational efficiency is an important issue, so we want to establish a standard for its measurement (see the fast Fourier transform discussion below). Other languages provide various types of "clocking" routines.

---

[8] A flop is a floating point operation; each addition, subtraction, multiplication, or division counts as a single flop.

Let's walk through a few examples using `sftdemo`; in each case the sampling interval $\tau = 1$. For $N = 50$ data points, a signal frequency of $f_s = 0.2$, and phase of $\varphi_s = 0$, we obtain the sine wave and transform shown in Figure 5.7. Notice that the real part of the transform is zero and that the imaginary part has spikes at the frequencies $f = 0.2$ and 0.8. Using $N = 50$ data points, a signal frequency of $f_s = 0.2$, and phase of $\varphi_s = \pi/2$, we obtain the cosine wave (because of the phase) and transform shown in Figure 5.8. Notice that the imaginary part of the transform is zero and that the real part has spikes at the frequencies $f = 0.2$ and $f = 0.8$.

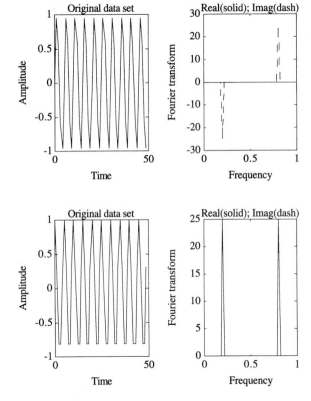

Figure 5.7 Time series and its Fourier transform from `sftdemo` for $N = 50$ data points, signal frequency $f_s = 0.2$, and phase $\varphi_s = 0$.

Figure 5.8 Time series and its Fourier transform from `sftdemo` for $N = 50$ data points, signal frequency $f_s = 0.2$, and phase $\varphi_s = \pi/2$.

Power Spectrum

Next, let's try a frequency that does not fall on a grid point, that is, $f_s \neq f_k$ for all k. Using the values $N = 50$, $f_s = 0.2123$, and $\varphi_s = 0$, we obtain the results shown in Figure 5.9. Notice that we still have a peak around the frequency of the sine wave, but the structure is more complicated.

In this example, because the frequency of the signal is not equal to a multiple of $1/\tau N = 1/50$, our Fourier transform is not a simple spike. For this reason, it is

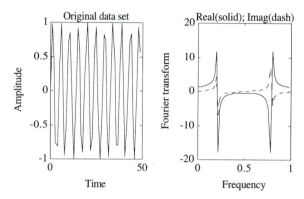

Figure 5.9 Time series and its Fourier transform from sftdemo for $N = 50$ data points, signal frequency $f_s = 0.2123$, and phase $\varphi_s = 0$.

often useful to compute the (unnormalized) power spectrum,

$$S_k = |Y_k|^2 = Y_k Y_k^* \tag{5-37}$$

where Y^* is the complex conjugate of Y. To obtain the power spectrum, we may add the following lines to the bottom of the sftdemo program:

```
for k=1:N
    spect(k)  =  yt(k) * conj(yt(k));
end
```

An equivalent but slicker way to compute the spectrum is to use the .* operator,

```
spect = yt .* conj(yt);
```

The power spectrum from our previous example ($N = 50, f_s = 0.2123$, and $\varphi_s = 0$) is shown in Figure 5.10. This spectrum shows two well-defined spikes; the first peak in the spectrum is between $f = 0.20$ and $f = 0.22$. If you are wondering about the second spike at $f = 1 - f_s$, read on.

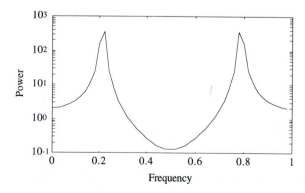

Figure 5.10 Plot of the power spectrum for $N = 50$ data points, signal frequency $f_s = 0.2123$, and phase $\varphi_s = 0$. Compare with the Fourier transform shown in Figure 5.9.

Aliasing and Nyquist Frequency

For our next example let's try a higher signal frequency, say $f_s = 0.8$. Using $N = 50$ data points and $\varphi_s = 0$ we obtain the results shown in Figure 5.11. Comparing with Figure 5.7 we see that the results for $f_s = 0.2$ and $f_s = 0.8$ are almost identical; the time series only differ by a phase shift of π. But how is this possible since these sine waves have completely different frequencies?

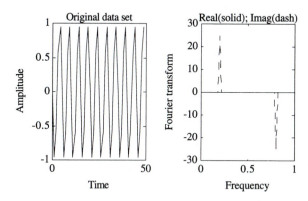

Figure 5.11 Plots from `sftdemo` for $N = 50$ data points, signal frequency $f_s = 0.8$, and phase $\varphi_s = 0$. Compare with Figure 5.7, where $f_s = 0.2$.

To help you understand why we obtain similar Fourier transforms, consider Figure 5.12. The two sine waves have frequencies $f_s = 0.2$ and $f_s = 0.8$; the former is shifted by $\varphi = \pi$. When the sampling interval is $\tau = 1$, the two data sets for these sine waves (the circles in Figure 5.12) are identical. This phenomenon is known as *aliasing*.

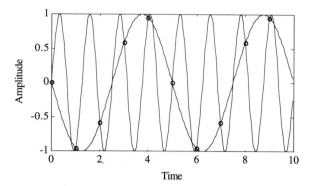

Figure 5.12 Illustration of aliasing. The two sine waves have frequencies $f_s = 0.2$ and $f_s = 0.8$; the former is shifted by $\varphi = \pi$. If the sampling interval is $\tau = 1$, then the two data sets (circles) are identical.

Not surprisingly, because of aliasing there is a limit as to how high a frequency we may resolve for a given sampling interval τ. This upper bound, called the *Nyquist frequency*, is

$$f_{Ny} = \frac{1}{2\tau} \tag{5-38}$$

For our examples above, $f_{Ny} = \frac{1}{2}$ since $\tau = 1$. Truncating our Fourier transform at this upper bound means that we discard the upper half of the vector Y.

Another way to understand this upper bound is to consider the following "information" argument. The (real) time series y contains N data points, but its Fourier transform contains N *complex* data points. However, the information content of the signal and its transform must be the same. Since the transform contains twice as many data points, it must contain a duplicate of the signal. This is why the Nyquist frequency cutoff truncates the transform vector in half.

Fast Fourier Transform

Consider again the definition of the discrete Fourier transform, as given by equation (5-33). The number of operations required to compute Y_{k+1} for a single value of k goes as N, the number of data points in our time series. To compute Y for all values of k (from 0 to $N - 1$) thus requires $O(N^2)$ operations. For many years spectral work was hobbled because it was computationally prohibitive to analyze large data sets.

In 1965, Cooley and Tukey introduced an algorithm that later became known as the *fast Fourier transform* (or FFT).[9] They showed that by cleverly rearranging the order in which the calculation was performed, the number of operations could be reduced to $O(N \log_2 N)$. Their original algorithm was limited to the case where $N = 2^n$, that is, the number of data points in the time series is a power of two. Modern implementations of the FFT (such as MATLAB's built-in version) can handle any value of N but are still most efficient when N is a power of two. If N is a prime number, the number of operations returns to $O(N^2)$.

To illustrate the use of the FFT, modify the sftdemo program by replacing line 13 with the line

```
yt = fft(y);
```

Repeating the examples presented above you should notice that (1) the output is *identical*, and (2) the run times (flop counts) are significantly shorter. Also notice that it actually takes less time to compute the transform if we use $N = 64$ (= 2^6) data points than if we use $N = 50$.

As long as you understand that the FFT does nothing more than evaluate the discrete Fourier transform in an efficient way, you don't really need to know how it works. If you're satisfied knowing what it does and don't care how, go on and skip to the next section. On the other hand, if you're one of those who can't stand using a black box without having some idea of what's inside, read on. The rest of

---

[9] J. W. Cooley and J. W. Tukey, "An Algorithm for the Machine Calculation of the Complex Fourier Series," *Math. Comput.* **19**, 297–301 (1965).

this section works through an example illustrating the fast Fourier transform algorithm. The discussion is adapted from Brigham;[10] see that excellent reference for a complete coverage of the FFT algorithm.

How the FFT Works

The discrete Fourier transform is defined as

$$Y_{k+1} = \sum_{j=0}^{N-1} y_{j+1} W^{kj} \tag{5-39}$$

where $W \equiv e^{-2\pi i/N}$. The easiest way to understand the FFT algorithm is to work through a simple example; we take $N = 8$ so $j,k = 0, \ldots, 7$. It is useful to decompose j and k into binary form,

$$\begin{aligned} j &= 4j_2 + 2j_1 + j_0 \\ k &= 4k_2 + 2k_1 + k_0 \end{aligned} \tag{5-40}$$

where $j_2, j_1, \ldots, k_1, k_0 = 0$ or 1.

To make the notation easier, define

$$y_{j+1} = y(j_2, j_1, j_0); \qquad Y_{k+1} = Y(k_2, k_1, k_0) \tag{5-41}$$

Using the binary notation defined above, the discrete Fourier transform may be written as

$$Y(k_2, k_1, k_0) = \sum_{j_0=0}^{1} \sum_{j_1=0}^{1} \sum_{j_2=0}^{1} y(j_2, j_1, j_0) W^{(4k_2+2k_1+k_0)(4j_2+2j_1+j_0)} \tag{5-42}$$

The first simplification comes from noticing that $W^8 = W^{16} = \ldots = 1$ since $e^{-i2\pi n} = 1$, so

$$\begin{aligned} W^{(4k_2+2k_1+k_0)4j_2} &= W^{4k_0 j_2} \\ W^{(4k_2+2k_1+k_0)2j_1} &= W^{(2k_1+k_0)2j_1} \end{aligned} \tag{5-43}$$

---

[10] E. O. Brigham, *The Fast Fourier Transform and Its Applications* (Englewood Cliffs, N.J.: Prentice Hall, 1988), chapter 8.

and

$$Y(k_2, k_1, k_0) = \sum_{j_0=0}^{1} W^{(4k_2+2k_1+k_0)j_0} \sum_{j_1=0}^{1} W^{(2k_1+k_0)2j_1}$$

$$\times \sum_{j_2=0}^{1} y(j_2, j_1, j_0) W^{4k_0j_2} \tag{5-44}$$

The sums may be further simplified by using $W^0 = 1$.

The nested sums are usually processed in layers. The inner sum over j_2 is evaluated in the first layer as

$$y_1(k_0, j_1, j_0) = y(0, j_1, j_0) + y(1, j_1, j_0) W^{4k_0} \tag{5-45}$$

for k_0, j_1 and $j_0 = 0, 1$. The subsequent layers are

$$y_2(k_0, k_1, j_0) = y_1(k_0, 0, j_0) + y_1(k_0, 1, j_0) W^{(4k_1+2k_0)} \tag{5-46}$$

for k_0, k_1 and $j_0 = 0, 1$ and

$$y_3(k_0, k_1, k_2) = y_2(k_0, k_1, 0) + y_2(k_0, k_1, 1) W^{(4k_2+2k_1+k_0)} \tag{5-47}$$

for k_0, k_1 and $k_2 = 0, 1$. Finally, the vector y_3 is "bit unscrambled" to give,

$$Y(k_2, k_1, k_0) = y_3(k_0, k_1, k_2) \tag{5-48}$$

where Y is the desired Fourier transform of y.

Note that processing each layer requires eight complex multiplications and eight complex additions. Since there are three ($= \log_2(8)$) layers, the total number of operations is 24 multiplies and adds. It is not difficult to extend the above example to any value N that is a power of 2. The number of layers will be $\log_2(N)$, so the number of operations will be $O[N \log_2(N)]$. Another attractive feature of the FFT is that the operations may be performed "in place," that is, the vector Y may be created in the same space in memory originally occupied by the time series y. This feature can be important if memory is constrained.

EXERCISES

11. Watching a Western on television you may have noticed that sometimes the wheels on a fast stagecoach appear to be rotating backwards. Explain how aliasing causes this effect.

12. The fast Fourier transform is most efficient when the number of data points is a power of two. Unfortunately, we cannot always control the number of points in our time

series. A common workaround is to "pad" the time series with zeros, that is, add to the end of the data extra points whose value is zero. Write a program to demonstrate the effect of padding a time series, comparing the spectra from padded and nonpadded data.

13. The sftdemo program prints the number of flops (floating point operations) executed in computing the Fourier transform. Modify the program to use the fast Fourier transform and graph flop count versus N, number of points in the time series. Show that the lower bound goes as $N \log_2(N)$ while the upper bound goes as N^2.

14. (a) The following two lines of MATLAB generate a vector y of independent, Gaussian distributed random numbers,

$$\text{rand}('\text{normal}');$$
$$\text{y} = \text{rand}(1, N);$$

This time series is an example of *white noise*. Write a program to compute its power spectrum using the FFT; use at least 512 data points.

(b) Assemble a time series consisting of white noise plus a sinusoid of amplitude α. Compute its power spectrum for a variety of values of α. In your judgment, what is the minimum value of α for which the sinusoid is distinctly seen in the spectrum; does it depend on the number of data points?

15. (a) Consider the following simple digital filter.[11] We may smooth our time series y_i by averaging adjacent data points to create a new time series z_i,

$$z_i = \tfrac{1}{2}(y_i + y_{i+1})$$

where $y_{N+1} \equiv y_1$. Write a program that applies this simple smoothing to a signal composed of a sum of sine waves of various frequencies. Show that the averaging serves as a digital low-pass filter by plotting the power spectra of both y and z.

(b) Repeat part (a) but using the difference filter

$$z_i = \tfrac{1}{2}(y_i - y_{i+1})$$

Show that this is a high-pass filter.

(c) Repeat parts (a) and (b) using Gaussian white noise as the time series y_i (see Exercise 5.14).

16. Carbon dioxide concentration (in parts per million) as measured in Barrow, Alaska, and Mauna Loa, Hawaii, is tabulated in Appendix 5B. Besides the linear trend, the data show an annual cycle in CO_2 concentration.

(a) Write a program that removes the annual cycle from the data. Remove the cycle by transforming into the frequency domain, zeroing the appropriate values, and transforming back. For comparison, plot the filtered and unfiltered data together on one graph.

(b) Repeat part (a) but remove the linear trend from the data before filtering and then replace it after filtering. Compare and comment with your results from part (a).

---

[11] R. W. Hamming, *Digital Filters* (Englewood Cliffs, N.J.: Prentice Hall, 1977).

5.3 NORMAL MODES*

Coupled Mass System

In this section we use Fourier transforms to study oscillations in a simple spring-mass system. Consider the system introduced in the previous chapter (see Figure 4.4). It consists of three blocks (of mass m) coupled together and to opposite walls by four springs. To simplify the analysis, we assume that the springs are identical and take the spring constants and rest lengths of the springs equal to k and L, respectively. Given the distance between the walls as L_w, the rest position of the masses are $x_j^* = \frac{1}{4} j L_w$. In Chapter 4 we studied the more general problem for which finding $\mathbf{x}^*$ was more complicated.

We now want to analyze the oscillatory motion when the masses are not at rest. The equation of motion is

$$m \frac{d^2}{dt^2} \mathbf{x}(t) = \mathbf{K}\mathbf{x}(t) - \mathbf{b} \tag{5-49}$$

where

$$\mathbf{K} = -k \begin{bmatrix} 2 & -1 & 0 \\ -1 & 2 & -1 \\ 0 & -1 & 2 \end{bmatrix}; \qquad \mathbf{b} = -kL_w \begin{bmatrix} 0 \\ 0 \\ 1 \end{bmatrix} \tag{5-50}$$

Physical intuition suggests the periodic trial solution,

$$\mathbf{x}(t) = \mathbf{a}e^{i\omega t} + \mathbf{x}^* \tag{5-51}$$

where $\mathbf{a}$ is the (complex) amplitude vector. Inserting this trial solution into (5-49) gives us

$$-m\omega^2 \mathbf{a} = \mathbf{K}\mathbf{a} \tag{5-52}$$

This is an *eigenvalue* problem; the above equation will have solutions only for certain values of ω.

We may write equation (5-52) as

$$\begin{bmatrix} 2 & -1 & 0 \\ -1 & 2 & -1 \\ 0 & -1 & 2 \end{bmatrix} \mathbf{a} - \lambda \mathbf{a} = 0 \tag{5-53}$$

where $\lambda = m\omega^2/k$. If we further rearrange it as

$$\begin{bmatrix} 2 - \lambda & -1 & 0 \\ -1 & 2 - \lambda & -1 \\ 0 & -1 & 2 - \lambda \end{bmatrix} \mathbf{a} = 0 \tag{5-54}$$

it is clear that we have a nontrivial solution (i.e., $\mathbf{a} \neq 0$) only if the matrix has no inverse.

The matrix is singular if its determinant is zero,

$$\begin{vmatrix} 2 - \lambda & -1 & 0 \\ -1 & 2 - \lambda & -1 \\ 0 & -1 & 2 - \lambda \end{vmatrix} = 0 \tag{5-55}$$

or

$$(2 - \lambda)\{(2 - \lambda)^2 - 1\} - (2 - \lambda) = 0 \tag{5-56}$$

This equation is easy to factor,

$$[2 - \lambda][(2 + \sqrt{2}) - \lambda][(2 - \sqrt{2}) - \lambda] = 0 \tag{5-57}$$

The three eigenvalues are thus $\lambda = 2, 2 + \sqrt{2}$, and $2 - \sqrt{2}$. I should point out that while this analysis is suitable for small matrices, it is *not* the recommended way of computing eigenvalues numerically.

Our spring-mass system has three *normal modes* of oscillation; their angular frequencies are

$$\omega_0 = \sqrt{2\frac{k}{m}}; \qquad \omega_\pm = \sqrt{(2 \pm \sqrt{2})\frac{k}{m}} \tag{5-58}$$

Some further calculation gives us the associated eigenvectors,

$$\mathbf{a}_0 = \begin{bmatrix} 1/\sqrt{2} \\ 0 \\ -1/\sqrt{2} \end{bmatrix}; \qquad \mathbf{a}_\pm = \begin{bmatrix} 1/2 \\ \mp 1/\sqrt{2} \\ 1/2 \end{bmatrix} \tag{5-59}$$

Notice that the three vectors are orthogonal and normalized to unit length.

Our final solution may be constructed as a linear combination of the three normal modes,

$$\mathbf{x}(t) = c_0 \mathbf{a}_0 e^{i\omega_0 t} + c_+ \mathbf{a}_+ e^{i\omega_+ t} + c_- \mathbf{a}_- e^{i\omega_- t} + \mathbf{x}^* \tag{5-60}$$

We may prepare the system in one of these three states, in which case the motion is very simple. In general, however, an arbitrary initial condition excites all three modes (see the numerical results below).

Numerical Simulation

Now that we have done the problem analytically, let's solve the equations of motion numerically and compare the results. The program sprfft (Listing 5.6) simulates the coupled mass-spring system using fourth-order Runge–Kutta. Since we want our time series to have a constant time step, I chose not to use the adaptive routine.[12] The program is straightforward; lines 4–8 initialize variables. The evolution of the state of the system, $[x_1 \ \ x_2 \ \ x_3 \ \ v_1 \ \ v_2 \ \ v_3]$, is computed by the Runge–Kutta routine rk4 (line 14). The equations of motion are defined in the function sprrk (Listing 5.7). Notice that the program computes the displacements from the steady states, that is, all the x's are zero when the system is at rest.

LISTING 5.6 Program sprfft. Computes the evolution of a coupled mass-spring system; uses Fourier transforms to obtain the eigenfrequencies. Uses rk4 (Listing 3.2) and sprrk (Listing 5.7).

```
1    % sprfft - Program to compute the power spectrum of a
2    % coupled mass-spring system.
3    clear; help sprfft;  % Clear memory and print header
4    x = input('Enter initial displacement [x1 x2 x3] - ');
5    v = [0 0 0];          % Masses are initially at rest
6    state = [x v];        % Positions and velocities; used by rk4
7    tau = input('Enter timestep, tau - ');  % (sec)
8    k_over_m = 1;         % Ratio of spring const. over mass (1/sec^2)
9    %%%%%%%%%%%%% Main Loop %%%%%%%%%%%%%%%%
10   time = 0;
11   nstep = 256;          % Number of steps in the main loop
12   nprint = nstep/16;    % Number of steps between printing
                              progress
13   for istep=1:nstep
14     state = rk4(state,time,tau,'sprrk',k_over_m);
               % Runge-Kutta
15     time = time + tau;
16     xplot(istep,1:3) = state(1:3);    % Record positions for
                                            plots
17     tplot(istep) = time;              % Record time for plots
18     if ( rem(istep,nprint) < 1 )
19       fprintf('Finished %g out of %g steps/n',istep,nstep);
20     end
```

---

[12] However, with just a bit of extra coding you can get the adaptive routine to deliver data that is evenly spaced in time.

```
21  end
22  f = (0:(nstep-1))/(tau*nstep);    % Frequency
23  x1 = xplot(:,1);           % Displacement of mass 1
24  x1fft = fft(x1);            % Fourier transform of displacement
25  spect = x1fft .* conj(x1fft); % Power spectrum of displacement
26  % Build the weight vector for the Hanning window
27  window = 0.5*(1-cos(2*pi*((1:nstep)-1)/(nstep-1)));
28  x1w = x1 .* window';          % Window the data
29  x1wfft = fft(x1w);            % Fourier transf. (windowed
                                              data)
30  spectw = x1wfft .* conj(x1wfft); % Power spectrum(windowed
                                              data)
31  % Graph the displacements of the masses
32  ipr = 1:nprint:nstep;
33  plot(tplot,xplot(:,1),'-',tplot(ipr),xplot(ipr,1),'o',. . .
34       tplot,xplot(:,2),'-.',tplot(ipr),xplot(ipr,2),'+',. . .
35       tplot,xplot(:,3),'--',tplot(ipr),xplot(ipr,3),'*')
36  title('Displacement of mass 1- o  2- +  3- *');
37  xlabel('Time'); ylabel('Displacement');
38  pause;  % Pause between plots; strike any key to continue
39  semilogy(f(1:(nstep/2)),spect(1:(nstep/2)),'-',. . .
40           f(1:(nstep/2)),spectw(1:(nstep/2)),'--');
41  title('Power spectrum (dashed is windowed data)');
42  xlabel('Frequency'); ylabel('Power');
```

LISTING 5.7 Function sprrk. Defines the equations of motion for sprfft (Listing 5.6).

```
1   function deriv = sprrk(a,time,param)
2   % Function to compute 3 mass-spring system
3   % using Runge-Kutta integrator
4   % The time is not used in this function
5   % The vector a is [x(1) x(2) x(3) v(1) v(2) v(3)]
6   % The vector param = k_over_m = (spring constant)/mass
7   % The vector deriv = [dx(1)/dt dx(2)/dt . . . dv(3)/dt]
8   deriv(1) = a(4);       % dx(1)/dt = v(1)
9   deriv(2) = a(5);       % dx(2)/dt = v(2)
10  deriv(3) = a(6);       % dx(3)/dt = v(3)
11  deriv(4) = -2*param*a(1) + param*a(2);          % dv(1)/dt
12  deriv(5) = -2*param*a(2) + param*(a(1)+a(3));   % dv(2)/dt
13  deriv(6) = -2*param*a(3) + param*a(2);          % dv(3)/dt
14  return;
```

If the initial displacement is one of the normal modes, then we have a simple uniform oscillation with a single frequency (see Figure 5.13). On the other hand, for an arbitrary initial displacement of the masses we have a linear combination of the three modes. To the untrained eye, the masses appear to oscillate in a chaotic fashion with no discernible pattern (see Figure 5.14).

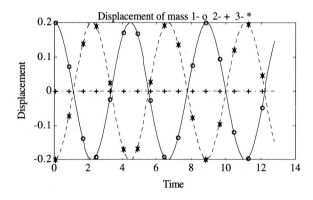

Figure 5.13 Displacement of masses in the coupled spring system as a function of time as computed by sprfft. The initial displacement was [0.2 0 −0.2], which is the normal mode with frequency $\omega_o = \sqrt{2k/m}$. The time step used was $\tau = 0.05$.

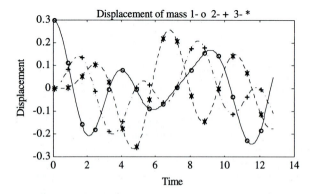

Figure 5.14 Displacement of masses in the coupled spring system as a function of time as computed by sprfft. The initial displacement was [0.3 0 0]; all three normal modes are excited. The time step used was $\tau = 0.05$.

Power Spectra

While the time series in Figure 5.14 may appear chaotic, the power spectrum (Figure 5.15) clearly reveals the three eigenfrequencies of the system. The peaks are not sharp because the length of our time series, $T = N\tau$, is not equal to an integer number of periods of any of the modes. This effect is called *leakage*. We see the same phenomenon in Figure 5.10; although the input is a pure tone, the output is a broad spike.

One way to compensate for our time series being finite is to apply a window. On line 28 of sprfft we apply the Hanning window,

$$x_i^w = [\tfrac{1}{2} - \tfrac{1}{2}\cos(2\pi(i - 1)/(N - 1))]x_i \qquad (5\text{-}61)$$

where x_i is the original time series and x_i^w is the windowed time series and $i = 1$, . . . , N. This window smoothly tapers the data at the ends, reducing spurious leakage effects. Notice in Figure 5.15 how the spikes in the windowed data (dashed line) are significantly more pronounced. The peaks become even narrower if we add more data points to the time series.

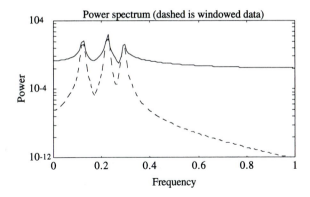

Figure 5.15 Power spectrum for the coupled spring system as computed by sprfft. The initial displacement was [0.3   0   0], which gives a combination of the three normal modes (notice the three spikes). The time step used was $\tau = 0.5$; frequencies are truncated at the Nyquist frequency. Dashed line is spectrum of windowed data.

The results from this coupled oscillator problem may remind you of the Lorenz model. Looking back on the time series from that model (Figure 3.12) we recall that they appear periodic. It is a simple matter to modify sprfft to compute the Lorenz model. Although in Chapter 3 we used the adaptive Runge–Kutta routine, we found that the time step, τ, did not vary significantly. Using rk4 with a fixed time step gives us a time series we can easily analyze.

The resulting power spectrum for $z(t)$ is shown in Figure 5.16 (the results are similar for the other two variables in the Lorenz model). While there is a peak around $f \cong 1.5$, on the whole the spectrum is quite complicated. A signal whose spectrum contains all frequencies in equal amplitude is called *white noise* in analogy with the spectrum of white light. Spectra such as that in Figure 5.16 are commonly referred to as *red noise* since all frequencies are present but with more power in the lower frequencies.

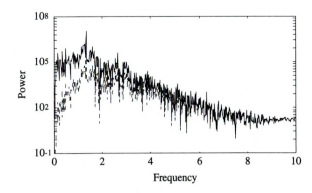

Figure 5.16 Power spectrum of $z(t)$ from the Lorenz model; dashed line is spectrum of windowed data. The parameters used are the same as in Figure 3.12 with time step, $\tau = 0.05$, and $N = 1024$ data points.

For a long time it was believed that a noisy spectrum indicated that a physical system contained many degrees of freedom. For example, a coupled oscillator system with many masses (e.g., atoms in a crystal) has a spectrum with many eigenfrequencies. The Lorenz model, however, is a simple, *nonlinear* system with only three degrees of freedom that produces a red noise spectrum. These results in nonlinear dynamics have revolutionized time series analysis and forecasting.

EXERCISES

17. Derive the eigenvectors (5-59) for the coupled spring-mass system.
18. (a) Find the normal mode frequencies for the Wilberforce pendulum (see Exercise 3.12).
    (b) Using your program from Exercise 3.12, compute the power spectrum and confirm your results from part (a).
19. (a) Find the normal mode frequencies for the spring-mass system in Exercise 4.13; assume $k_1 = k_2$.
    (b) Write a program to compute the time series for this system. Use the power spectrum to confirm your results from part (a).
20. (a) Using your program from Exercise 3.22, compute the power spectra for the Lotka–Volterra model.
    (b) Repeat part (a) for the Brusselator model (see Exercise 4.4).
21. Compute the power spectrum of the angular displacement $\theta(t)$ for the simple pendulum (see Section 2.2). Show that for small initial displacements (e.g., $\theta(0) = 10°$) the spectrum only has a single peak, but with large displacements (e.g., $\theta(0) = 170°$) multiple peaks appear. What is the relation between the frequencies of these multiple peaks?

BEYOND THIS CHAPTER

Curve fitting is such an important topic in scientific data analysis that entire books are dedicated to the subject.[13] Even linear least squares analysis can be complicated since the matrices in the normal equations are often ill-conditioned. Gaussian elimination can fail and the best approach is to use *singular value decomposition* (SVD).[14]

In this chapter we only consider fitting data to functions that are linear in the coefficients. In general, to fit nonlinear functions requires an iterative procedure for finding the minimum of χ^2. In some cases, the problem is notoriously difficult; the function

$$Y(x; \{a_1, \ldots, a_4\}) = a_1 e^{-a_2 x} + a_3 e^{-a_4 x} \qquad (5\text{-}62)$$

[13] P. Bevington, *Data Reduction and Error Analysis for the Physical Sciences*, 2d ed. (New York: McGraw-Hill, 1992); S. Brandt, *Statistical and Computational Methods in Data Analysis* (Amsterdam: North-Holland, 1970).

[14] G. E. Forsythe, M. A. Malcolm, and C. B. Moler, *Computer Methods for Mathematical Computations* (Englewood Cliffs, N.J.: Prentice-Hall, 1977), chapter 9.

is a famous example. There are, however, specialized algorithms for tackling these problems.[15]

Least squares analysis can give poor results when the data contains outliers or if the errors are not Gaussian distributed. So-called *robust* techniques have been developed to provide alternative ways to fit data.[16] It is often useful to apply both least squares and a robust algorithm to a data set; if the fit parameters differ significantly, you might want to investigate why. Robust techniques are also useful for identifying outliers.

Sometimes we want to draw a smooth curve through a data set but don't really care to specify the function. In this case it's simplest to assemble the curve using a piecewise function, the *cubic spline* being the most popular choice.[17] This is also a good approach when we want to interpolate between data points.

The Fourier transform has many applications besides estimating power spectra. Computing convolutions and correlations of data sets is most efficiently performed using FFTs. Chapter 7 covers spectral methods for solving partial differential equations. Specifically, the Fourier transform is used to numerically solve the Poisson equation. See Brigham[18] for an extensive summary of other applications.

An alternative technique for power spectrum estimation is the *maximum entropy* method (MEM).[19] This algorithm is useful when a data set has sharp peaks in the power spectrum, especially when trying to resolve peaks that are close together. Unfortunately, careless use of the method leads to spurious features (e.g., false peaks) in the spectrum. MEM can be significantly improved when combined with an adaptive filter.[20]

APPENDIX 5A: FORTRAN LISTINGS

LISTING 5A.1 Function linreg. Fits a straight line to a data set.

```
      subroutine linreg(x,y,sigma,N,a_fit,sig_a,yy,chisqr)
! Subroutine to perform linear regression (fit a line)
      real x(N),y(N),sigma(N),yy(N),a_fit(2),sig_a(2)
```

[15] D. M. Bates and D. G. Watts, *Nonlinear Regression Analysis and Its Applications* (New York: Wiley, 1988); G. A. F. Seber and C. J. Wild, *Nonlinear Regression* (New York: Wiley, 1989).

[16] P. J. Huber, *Robust Statistics* (New York: Wiley, 1981); P. J. Rousseeuw and A. M. Leroy, *Robust Regression and Outlier Detection* (New York: Wiley, 1987).

[17] C. de Boor, *A Practical Guide to Splines* (New York: Springer-Verlag, 1978); P. Lancaster and K. Salkavskas, *Curve and Surface Fitting* (London: Academic Press, 1986).

[18] E. O. Brigham, *The Fast Fourier Transform and Its Applications* (Englewood Cliffs, N.J.: Prentice Hall, 1988).

[19] *Modern Spectrum Analysis*, edited by D. G. Childers (New York: IEEE Press, 1978).

[20] C. Penland, M. Ghil, and K. M. Weickmann, "Adaptive Filtering and Maximum Entropy Spectra with Application to Changes in Atmospheric Angular Momentum," *J. Geo. Res.*, **96**, 659–71 (1991).

```
s = 0.
sx = 0.
sy = 0.
sxy = 0.
sxx = 0.
do i=1,N
  temp = 1./sigma(i)**2
  s = s + temp                    ! Evaluate the various sums
  sx = sx + x(i)*temp
  sy = sy + y(i)*temp
  sxy = sxy + x(i)*y(i)*temp
  sxx = sxx + x(i)**2*temp
end do
denom = s*sxx - sx**2
a_fit(1) = (sxx*sy - sx*sxy)/denom   ! Curve fit
                                     ! coefficients
a_fit(2) = (s*sxy - sx*sy)/denom
sig_a(1) = sqrt(sxx/denom)            ! Error in coefficients
sig_a(2) = sqrt(s/denom)
chisqr = 0.                           ! Chi square statistic
do i=1,N
  yy(i) = a_fit(1) + a_fit(2)*x(i)   ! Curve fit to the data
  chisqr = chisqr + ((y(i)-yy(i))/sigma(i))**2  ! Chi
                                                 ! square
end do
return
end
```

LISTING 5A.2   Program lsftest. Generates a data set and fits a curve to the data; may be used to test the linreg (Listing 5A.1) and pollsf (Listing 5A.3) functions. Also uses the random number generator randn (Listing 10A.8).

```
      program lsftest
! Program to test least squares fit routines
      parameter( Nmax = 1000, Mmax = 10 ) ! Max number of data,
                                          ! paramaters
      real x(Nmax),y(Nmax),sigma(Nmax),yy(Nmax)
      real c(3),a_fit(Mmax),sig_a(Mmax)
      integer*4 iseed

      N = 50                   ! Number of data points to plot
      iseed = 123456789   ! Seed for random number generator
      write (*,*) 'Enter the coefficients of the quadratic'
      write (*,*) '   Y(x) = c(1) + c(2)*x + c(3)*x**2'
      write (*,*) 'as c(1),c(2),c(3)'
```

```
      read (*,*) c(1),c(2),c(3)
      write (*,*) 'Enter estimated error bar'
      read (*,*) alpha
      do i=1,N              ! Construct data vectors
        x(i) = i
        r = alpha*randn(iseed)   ! Gaussian distrib. random number
        y(i) = c(1) + c(2)*x(i) + c(3)*x(i)**2 + r
       sigma(i) = alpha         ! Constant error bar
      end do
!!!!! Linear regression (Straight line) fit
!!!!!!!!!!!!!!!!!!!!!!!!!
      M = 2       ! Number of parameters is M=2 for linear fit
      call linreg(x,y,sigma,N,a_fit,sig_a,yy,chisqr)
!!!!!!!!!!!!!!!!!!!!!!!!!!!!!!!!!!!!!!!!!!!!!!!!!!!!!!!!!!!!!!!!!
!!!!! Polynomial least squares fit
!!!!!!!!!!!!!!!!!!!!!!!!!!!!!!!!!!!
!      M = 3     ! M=3 is a quadratic polynomial fit
!      call pollsf(x,y,sigma,N,M,a_fit,sig_a,yy,chisqr)
!!!!!!!!!!!!!!!!!!!!!!!!!!!!!!!!!!!!!!!!!!!!!!!!!!!!!!!!!!!!!!!!!!
      write (*,*) 'Fit parameters'
      do i=1,M
        write (*,*) 'a(',i,') = ',a_fit(i),' +/- ',sig_a(i)
      end do
      write (*,*) 'Chi square = ',chisqr,'  N-M = ',N-M
! Print out the plotting variables -
!   x,y,yy,sigma
!
      open(11,file='x.dat')
      open(12,file='y.dat')
      open(13,file='yy.dat')
      open(14,file='sigma.dat')
      do i=1,N
        write (11,*) x(i)
        write (12,*) y(i)
        write (13,*) yy(i)
        write (14,*) sigma(i)
      end do
      stop
      end
```

LISTING 5A.3 Function `pollsf`. Fits a polynomial to a data set. Uses `inv` (Listing 4A.4).

```
      subroutine pollsf(x,y,sigma,N,M,a_fit,sig_a,yy,chisqr)
! Subroutine to fit a polynomial to data
      parameter( Nmax = 1000, Mmax = 20 )   ! Maximum number of
```

```
                                            ! parameters
      real x(N),y(N),sigma(N),a_fit(M),sig_a(M),yy(N)
      real b(Nmax), A(Nmax,Mmax)
      real C_inv(Mmax,Mmax),C(Mmax,Mmax)

      do i=1,N
        b(i) = y(i)/sigma(i)       ! Form the vector b
        do j=1,M
          A(i,j) = x(i)**(j-1)/sigma(i)    ! Form the design matrix
        end do
      end do
      do j=1,M
       do jj=1,M
         temp = 0.
         do i=1,N
           temp = temp + A(i,j)*A(i,jj)
         end do
         C_inv(j,jj) = temp      ! Set up matrix to be inverted
       end do
      end do

      call inv(C_inv,M,Mmax,C)   ! Calculate C using matrix inverse

      do j=1,M
        temp = 0
        do jj=1,M
          do jjj=1,N
            temp = temp + C(j,jj)*A(jjj,jj)*b(jjj)
          end do
        end do
        a_fit(j) = temp          ! Compute best fit parameter
        sig_a(j) = sqrt(C(j,j))  ! Estimated error in fit
                                 ! parameter
      end do

      do i=1,N
        yy(i) = 0
        do j=1,M                 ! yy() is the fitted curve
          yy(i) = yy(i) + a_fit(j)*x(i)**(j-1)
        end do
      end do

      chisqr = 0    ! Compute chi square statistic
      do i=1,N
        chisqr = chisqr + ( (y(i)-yy(i))/sigma(i) )**2
      end do
      return
      end
```

LISTING 5A.4　Program `sftdemo`. Demonstrates the discrete Fourier transform using the `sft` function (Listing 5A.5) or the `fft` function (Listing 5A.8).

```fortran
      program sftdemo
! Discrete Fourier transform demonstration program
      parameter ( Nmax = 4096 )
      real t(Nmax), f(Nmax), spect(Nmax)
      complex y(Nmax), yt(Nmax)

      pi = 4.*atan(1.)   ! = 3.1415926. . .
!!!!! Use these lines for SFT !!!!!
!     write (*,*) 'Enter the number of points N'
!     read (*,*) N
!!!!!!!!!!!!!!!!!!!!!!!!!!!!!!!!!!!!!!!
!!!!! Use these lines for FFT !!!!!
      write (*,*) 'Enter power of 2 for number of data points'
      read (*,*) npow2
      N = 2**npow2
!!!!!!!!!!!!!!!!!!!!!!!!!!!!!!!!!!!!!!!
      write (*,*) 'Enter frequency and phase of sine wave'
      read (*,*) freq,phase
      tau = 1.   ! Sampling interval (time between samples)
      ! Build the data set
      do i=1,N
        t(i) = (i-1)*tau
        y(i) = sin(2.*pi*t(i)*freq + phase)
      end do
      do i=1,N
        yt(i) = y(i)           ! Make copy of y(i) to send to sft
      end do
!!!!!!!!!!!!!!!!!!!
!     call sft(yt,N)      ! Slow fourier transform
!!!!!!!!!!!!!!!!!!!
      call fft(yt,npow2) ! Fast fourier transform (power of 2)
!!!!!!!!!!!!!!!!!!!
      ! Compute the frequency vector
      do k=0, (N-1)
        f(k+1) = k/(N*tau)
      end do
      ! Compute the power spectrum
      do k=1,N
        spect(k) = yt(k)*conjg(yt(k))
      end do
! Print out the plotting variables -
!    t,y,f,yt,spect
!
      open(11,file='t.dat')
      open(12,file='y.dat')
```

```
        open(13,file='f.dat')
        open(14,file='yt.dat')
        open(15,file='spect.dat')
        do i=1,N
          write (11,1001) t(i)
          write (12,1002) y(i)
          write (13,1001) f(i)
          write (14,1002) yt(i)
          write (15,1001) spect(i)
        end do
        stop
1001    format(e12.6)
1002    format(e12.6,2x,e12.6)
        end
```

LISTING 5A.5 Function sft. Computes the discrete Fourier transform directly from its definition; this is the slow algorithm.

```
        subroutine sft(y,N)
! Discrete fourier transform routine (slow version)
        parameter( Nmax = 2000 )   ! Max size for data vector
        complex y(N),yt(Nmax),ii

        twopiN = -6.283185307/N  ! = -2*pi/N
        ii = (0. , 1.)           ! ii = sqrt(-1)
        do k=0,(N-1)
          yt(k+1) = 0.   ! Compute transform using definition (sum)
          do j=0,(N-1)
            yt(k+1) = yt(k+1) + y(j+1)*( cos(twopiN*j*k)
        &                               + ii*sin(twopiN*j*k) )
            !! Notice we use exp(ii*x) = cos(x)+ii*sin(x)
          end do
        end do

        do i=1,N
          y(i) = yt(i)   ! Copy transform into y
        end do
        return
        end
```

LISTING 5A.6 Program sprfft. Computes the evolution of a coupled mass-spring system; uses Fourier transforms to obtain the eigenfrequencies. Uses rk4 (Listing 3A.2), sprrk (Listing 5A.7) and fft (Listing 5A.8).

```
        program sprfft
! Program to compute the power spectrum of a coupled
! mass-spring system
```

```fortran
      ! Notice that nstep is required to be a power of 2 for FFTs
      parameter( npow2 = 8, nstep = 2**npow2, nstate = 6 )
      real state(nstate),new_st(nstate)
      real xplot(nstep,3),f(nstep),tplot(nstep),k_over_m
      complex x1fft(nstep),x1wfft(nstep)
      real spect(nstep),spectw(nstep)
      external sprrk ! Function which defines equations of motion

      pi = 4*atan(1.)   ! = 3.14159. . .
      write (*,*) 'Enter initial displacements x1,x2,x3'
      read (*,*) state(1), state(2), state(3)
      state(4) = 0    ! Initial velocity of mass 1 is zero
      state(5) = 0    ! Initial velocity of mass 2 is zero
      state(6) = 0    ! Initial velocity of mass 3 is zero
      write (*,*) 'Enter time step, tau'
      read (*,*) tau
      k_over_m = 1    ! Ratio of spring const. / mass
!!!!! MAIN LOOP !!!!!
      do istep=1,nstep
        !! Compute time evolution of oscillators using
        !! Runge-Kutta
        call rk4(state,time,tau,sprrk,k_over_m,nstate,new_st)
        do i=1,nstate
          state(i) = new_st(i)      ! Reset state with new values
        end do
        time = time + tau           ! Update time
        do j=1,3
          xplot(istep,j) = state(j)   ! Record positions for plots
        end do
        tplot(istep) = time          ! Record time for plots
      end do
      do i=1,nstep
        f(i) = (i-1)/(tau*nstep)   ! Frequency
        x1fft(i) = xplot(i,1)        ! Record positions to send to FFT
        window = .5*(1.-cos(2.*pi*(i-1)/(nstep-1))) ! Hanning
                                                    ! window
        x1wfft(i) = xplot(i,1)*window  ! Windowed data
      end do
      call fft(x1fft,npow2)     ! Take FFT of position of mass 1
      call fft(x1wfft,npow2)    ! Ditto but using windowed data
      do i=1,nstep
        spect(i) = x1fft(i)*conjg(x1fft(i))      ! Power spectrum
        spectw(i) = x1wfft(i)*conjg(x1wfft(i))   ! Spectrum with
                                                 ! window
      end do
! Print out the plotting variables -
!     tplot,xplot,f,spect,spectw
!
```

```
open(11,file='tplot.dat')
open(12,file='xplot.dat')
open(13,file='f.dat')
open(14,file='spect.dat')
open(15,file='spectw.dat')
do i=1,nstep
  write (11,*) tplot(i)
  write (12,*) (xplot(i,j),j=1,3)
  write (13,*) f(i)
  write (14,*) spect(i)
  write (15,*) spectw(i)
end do
stop
end
```

LISTING 5A.7 Function sprrk. Defines the equations of motion for sprfft.

```
      subroutine sprrk(a,time,param,deriv)
!  Function to evaluate time derivatives for 3 mass-spring system
      real a(6),param,deriv(6)

      deriv(1) = a(4)
      deriv(2) = a(5)
      deriv(3) = a(6)
      param2 = -2*param
      deriv(4) = param2*a(1) + param*a(2)
      deriv(5) = param2*a(2) + param*(a(1)+a(3))
      deriv(6) = param2*a(3) + param*a(2)

      return
      end
```

LISTING 5A.8 Function fft. Computes the discrete Fourier transform using the FFT algorithm.

```
      subroutine fft(A,M)
! Routine to compute discrete Fourier transform using FFT
! algorithm
! The complex vector A contains a data vector with 2**M points
! On output the vector A contains the transform of the input
      complex A(*), u, w, temp

      Pi = 4.*atan(1.)
      N = 2**M          ! Number of data points
      N_half = N/2
      Nm1 = N-1
```

```
      j=1
      do i=1,Nm1
        if( i .lt. j ) then
          temp = A(j)
          A(j) = A(i)
          A(i) = temp
        end if
        k = N_half
1       if( k .ge. j ) goto 2  ! while loop
          j=j-k
          k=k/2
          goto 1
2       continue
        j = j+k
      end do

      do k=1,M
        ke = 2**k
        ke1 = ke/2
        u = (1.0, 0.0)
        angle = -Pi/ke1
        w = cmplx(cos(angle),sin(angle))
        do j=1,ke1
          do i=j,N,ke
            ip = i + ke1
            temp = A(ip)*u
            A(ip) = A(i)-temp
            A(i) = A(i)+temp
          end do
          u = u*w
        end do
      end do
      return
      end
```

LISTING 5A.9 Function `ifft`. Computes the discrete inverse Fourier transform using the FFT algorithm. Uses `fft` (Listing 5A.8).

```
      subroutine ifft(A,npow)
! Routine to compute the inverse Fourier transform
! Uses the fft routine
      complex A(*)

      N = 2**npow     ! Number of data points
      do i=1,N
        A(i) = conjg(A(i))
      end do
```

```
call fft(A,npow)
do i=1,N
   A(i) = conjg(A(i))/N
end do

return
end
```

APPENDIX 5B: CARBON DIOXIDE DATA

TABLE 5.1   Carbon dioxide (in parts per million) measured at Mauna Loa, Hawaii

339.35	339.96	340.59	341.17	341.67
342.13	342.61	343.10	343.49	343.60
343.34	342.72	341.90	341.01	340.18
339.41	338.66	337.93	337.32	337.00
337.07	337.52	338.21	338.96	339.60
340.10	340.51	340.89	341.32	341.84
342.39	342.92	343.40	343.78	343.99
343.96	343.69	343.28	342.85	342.36
341.68	340.69	339.45	338.24	337.36
337.01	337.17	337.69	338.37	339.11
339.84	340.56	341.28	341.79	342.07
342.15	342.25	342.64	343.43	344.46
345.37	345.86	345.87	345.51	344.95
344.23	343.34	342.27	341.16	340.21
339.60	339.40	339.54	339.91	340.40
341.00	341.71	342.48	343.26	343.81
344.15	344.36	344.61	345.06	345.70
346.41	346.99	347.28	347.25	346.96
346.46	345.77	344.91	343.91	342.85
341.89	341.22	340.98	341.19	341.69
342.30	342.89	343.39	343.79	344.14
344.42	344.78	345.30	345.93	346.55
347.07	347.53	347.96	348.34	348.53
348.41	347.93	347.14	346.12	345.01
343.99	343.19	342.69	342.47	342.52
342.83	343.36	344.02	344.69	345.26
345.71	345.97	346.15	346.36	346.74
347.35	348.11	348.88	349.48	349.80
349.79	349.43	348.75	347.85	346.90
346.01	345.24	344.56	343.97	343.54
343.43	343.78	344.53	345.45	346.23
346.74	347.05	347.30	347.62	348.01
348.46	348.98	349.61	350.32	351.00
351.48	351.66	351.52	351.07	350.33
349.38	348.34	347.40	346.70	346.31
346.20	346.35	346.77	347.42	348.17
348.83	349.29	349.65	350.02	350.52
351.09	351.63	352.12	352.63	353.19
353.72	354.07	354.14	353.95	353.55
352.98	352.21	351.25	350.19	349.26
348.68	348.54	348.75	349.11	349.51
349.96	350.55	351.27	352.07	352.55
352.77	352.87	353.07	353.51	354.22
355.02	355.66	355.97	355.92	355.63
355.19	354.59	353.74	352.60	351.30
350.12	349.36	349.13	349.37	349.91

Notes: Measurements were taken every 14 days starting in 1981. Values in the table are ordered left to right and then down (i.e., the second value is 339.96). Estimated error is $\sigma_0 = 0.16$ ppm.

TABLE 5.2   Carbon dioxide (in parts per million) measured at Barrow, Alaska

344.20	344.83	345.20	345.37	345.41
345.42	345.52	345.79	346.21	346.56
346.47	345.51	343.42	340.24	336.50
333.06	330.85	330.45	331.84	334.40
337.20	339.52	341.17	342.39	343.49
344.52	345.39	345.95	346.44	347.01
347.57	347.88	347.82	347.62	347.56
347.67	347.60	346.87	345.18	342.59
339.41	336.15	333.46	332.02	332.24
334.01	336.66	339.31	341.39	342.83
343.91	344.91	345.89	346.55	346.89
347.08	347.29	347.55	347.76	347.86
347.93	348.12	348.30	347.98	346.60
343.92	340.29	336.55	333.67	332.37
332.93	335.17	338.43	341.78	344.35
345.82	346.48	346.99	347.87	348.96
349.92	350.41	350.38	350.05	349.69
349.57	349.77	350.19	350.45	349.98
348.29	345.33	341.66	338.24	336.01
335.45	· 336.40	338.28	340.36	342.16
343.62	345.01	346.53	348.04	349.23
349.53	349.38	349.29	349.57	350.19
350.92	351.57	352.05	352.30	352.15
351.36	349.66	346.92	343.38	339.78
337.13	336.24	337.24	339.58	342.34
344.78	346.52	347.60	348.28	348.85
349.59	350.48	351.54	352.56	353.29
353.57	353.47	353.26	353.19	353.23
353.07	352.24	350.37	347.40	343.72
340.30	338.30	338.47	340.61	343.70
346.52	348.44	349.53	350.29	351.11
352.02	352.82	353.25	353.44	353.61
353.87	354.12	354.27	354.36	354.57
354.89	355.03	354.46	352.76	349.83
346.14	342.57	340.10	339.33	340.20
342.16	344.50	346.69	348.52	350.02
351.40	352.83	354.41	355.73	356.60
356.95	357.01	357.10	357.36	357.65
357.77	357.65	357.27	356.38	354.58
351.72	348.20	344.98	343.04	342.84
344.22	346.56	349.18	351.57	353.58
355.32	356.92	358.35	359.40	359.55
359.18	358.86	359.07	359.75	360.34
360.37	359.90	359.43	359.33	359.26
358.23	355.34	350.71	345.74	342.40
342.01	344.35	347.96	351.22	353.33

Notes: Measurements were taken every 14 days starting in 1981. Values in the table are ordered left to right and then down (i.e., the second value is 344.83). Estimated error is $\sigma_o = 0.27$ ppm.

Chapter 6
Partial Differential Equations I: Explicit Methods

Up to now, most of the physical systems we have simulated were formulated using ordinary differential equations. However, much of physics involves working with partial differential equations (PDEs). We have the Schrödinger equation in quantum mechanics, Maxwell's equations in electricity and magnetism, and the wave equation in optics and acoustics. The next three chapters cover how to treat such equations numerically. This chapter starts by discussing the various types of PDEs and covers some of the explicit marching methods.

6.1 INTRODUCTION TO PDEs

Classification of PDEs

In solving ordinary differential equations we developed some general methods, such as Runge–Kutta, that could be applied to any problem. The situation is different with partial differential equations (PDEs). The classification of the equation determines the type of method that should be used.

Instead of introducing the classification of PDEs in the abstract, let's discuss some familiar, concrete examples. There are three PDEs that are model equations of our classification scheme and, fortunately, you have already seen them in your other physics courses. The first is the one-dimensional diffusion equation:

$$\frac{\partial}{\partial t} T(x, t) = \kappa \frac{\partial^2}{\partial x^2} T(x, t) \tag{6-1}$$

161

This equation is used to describe many different diffusion processes. Here I write it as the Fourier equation from the theory of heat transport. The variable $T(x, t)$ is the temperature at position x and time t. The constant κ is the thermal diffusion coefficient. This first equation is an example of a *parabolic equation*. The time-dependent Schrödinger equation is another example (see Section 8.2).

The second important PDE that we study is the one-dimensional wave equation

$$\frac{\partial^2 A}{\partial t^2} = c^2 \frac{\partial^2 A}{\partial x^2} \tag{6-2}$$

where $A(x, t)$ is the wave amplitude and c is the wave speed. This equation is classified as a *hyperbolic equation*.

Our third partial differential equation is Poisson's equation from electrostatics. In two dimensions, it takes the form

$$\frac{\partial^2 \Phi(x, y)}{\partial x^2} + \frac{\partial^2 \Phi(x, y)}{\partial y^2} = -\frac{1}{\varepsilon_o} \rho(x, y) \tag{6-3}$$

where $\Phi(x, y)$ is the electrostatic potential, $\rho(x, y)$ is the charge density, and ε_o is the permittivity of free space. If $\rho = 0$, then we have Laplace's equation. The Poisson and Laplace equations are classified as *elliptic equations*.

Notice that in each of these examples we have two independent variables, either (x, t) or (x, y). All the methods we discuss are extendible to higher-dimensional systems, but are much easier to understand when first used with only two variables. Formally, a second-order PDE in two independent variables of the form

$$a \frac{\partial^2 A}{\partial x^2} + b \frac{\partial^2 A}{\partial x \partial y} + c \frac{\partial^2 A}{\partial y^2} + d \frac{\partial A}{\partial x} + e \frac{\partial A}{\partial y} + fA(x, y) + g = 0 \tag{6-4}$$

is classified as hyperbolic if $b^2 - 4ac > 0$, parabolic if $b^2 - 4ac = 0$, and elliptic if $b^2 - 4ac < 0$.

Initial Value Problems

The diffusion equation and the wave equation are similar in that they are usually solved as *initial value problems*. For the diffusion equation, we might be given an initial temperature distribution $T(x, t = 0)$ and want to find $T(x, t)$ for all $t > 0$. Similarly, for the wave equation we could start with the initial amplitude $A(x, t = 0)$ and velocity $dA(x, t = 0)/dt$ of a wave pulse and be asked to find the shape of the wave pulse, $A(x, t)$, for all $t > 0$.

Besides initial conditions, we also need to specify boundary conditions. Say our solution is constrained to the region of space between $x = -L/2$ and $x = L/2$.

Boundary conditions are imposed at these end points. For the diffusion equation we could fix the temperature at the boundaries,

$$T(x = -L/2, t) = T_a; \qquad T(x = L/2, t) = T_b \qquad (6\text{-}5)$$

This is an example of *Dirichlet boundary conditions*. An alternative would be to fix the flux at the boundaries

$$-\kappa \left.\frac{dT}{dx}\right|_{x=-L/2} = F_a; \qquad -\kappa \left.\frac{dT}{dx}\right|_{x=L/2} = F_b \qquad (6\text{-}6)$$

In this case we have *Neumann boundary conditions*. A third type of boundary condition is common in numerical simulations. We can equate the function at the two ends,

$$T(x = -L/2, t) = T(x = L/2, t) \qquad (6\text{-}7\text{a})$$

$$\left.\frac{dT}{dx}\right|_{x=-L/2} = \left.\frac{dT}{dx}\right|_{x=L/2} \qquad (6\text{-}7\text{b})$$

This is called a *periodic boundary condition*. In one dimension, using periodic boundary conditions is equivalent to transforming our coordinates from a line to a circle.

   A useful way to picture initial value problems is to view them on the xt plane, as shown in Figure 6.1. Notice that the interior region is open ended; we may compute $T(x, t)$ or $A(x, t)$ as far into the future as we want.

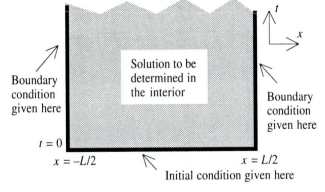

Boundary condition given here

Solution to be determined in the interior

Boundary condition given here

$t = 0$

$x = -L/2$

$x = L/2$

Initial condition given here

Figure 6.1 Schematic representation for an initial value problem.

   As with ODEs, we discretize the time as $t_n = (n - 1)\tau$ where τ is the time step and $n = 1, 2, \ldots$. Similarly, we discretize space as $x_i = (i - 1)h - L/2$ where h is the grid spacing and $i = 1, \ldots, N$. If the number of points in the x direction is N, the grid spacing is $h = L/(N - 1)$. A schematic of an initial value problem in discretized form is shown in Figure 6.2. Our job is to determine the unknown

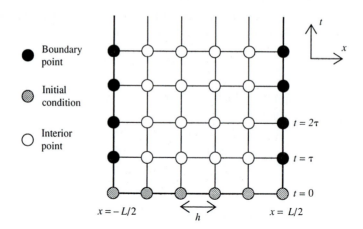

Figure 6.2 Schematic representation for a discretized initial value problem.

values in the interior (open circles) given the initial conditions (gray circles) and the boundary conditions (filled circles).

Initial value problems are often solved using *marching methods*. Starting from the initial condition, we compute the solution one time step into the future. Using this result, the solution at $t = 2\tau$ is computed. The algorithm proceeds in this manner and marches forward in time. Initial value problems are solved both analytically and numerically in the following sections of this chapter.

Boundary Value Problems

Elliptic equations, such as Laplace's equation in electrostatics, are not initial value problems but rather *boundary value problems*. For example, we may be told the potential on the four sides of a rectangle,

$$\Phi(x = 0, y) = \Phi_1; \qquad \Phi(x = L_x, y) = \Phi_2;$$
$$\Phi(x, y = 0) = \Phi_3; \qquad \Phi(x, y = L_y) = \Phi_4 \qquad (6\text{-}8)$$

and be asked to solve for $\Phi(x, y)$ at all points inside the rectangle (see Figure 6.3). We discretize space as $x_i = (i - 1)h_x$, $y_j = (j - 1)h_y$ where h_x and h_y are the x and y grid spacings. Our task is now to determine Φ at the interior points given the constraints specified by the boundary conditions (see Figure 6.4).

Algorithms for solving boundary value problems are sometimes called *jury methods*. The potential at an interior point is influenced by all the boundary points; the solution in the interior is a weighted result that reconciles all the

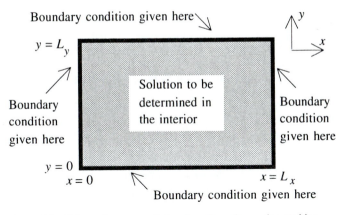

Figure 6.3 Schematic representation for a boundary value problem.

demands (constraints) imposed by the boundary. The methods for solving boundary value problems are covered in Chapter 7.

I don't want to dwell too much on the classification of equations. We could go on to discuss the uniqueness of solutions; instead, let's start looking at algorithms for solving our three model equations. One last note: Most equations encountered in real research do not fall neatly into one of our three categories. Instead, they are often hybrids. For example, consider acoustic wave propagation with attenuation. The equations are in part hyperbolic and in part parabolic. Still, the methods we develop here will be applicable to such hybrid problems.

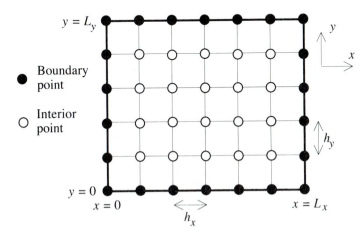

Figure 6.4 Schematic representation for a discretized boundary value problem.

6.2 DIFFUSION EQUATION

Method of Images

We'll start with the most cooperative of the three PDEs introduced in the previous section, the diffusion equation:

$$\frac{\partial}{\partial t} T(x, t) = \kappa \frac{\partial^2}{\partial x^2} T(x, t) \tag{6-9}$$

where $T(x, t)$ is the temperature at time t at location x and κ is the thermal diffusion coefficient. Before solving this equation numerically, let's obtain the analytical solutions to some simple initial value problems. There are several ways to solve the diffusion equation (e.g., separation of variables, integral transforms). In this section we use one of the lesser known techniques: the method of images.[1]

An important solution of this PDE is a Gaussian of the form

$$T_g(x, t) = \frac{1}{\sigma(t)\sqrt{2\pi}} \exp\left[\frac{-(x - x_o)^2}{2\sigma^2(t)}\right] \tag{6-10}$$

where x_o is the location of the maximum. The standard deviation, $\sigma(t)$, increases in time as

$$\sigma(t) = \sqrt{2\kappa t} \tag{6-11}$$

Notice that as $t \to 0$, $\sigma(t) \to 0$, so the width of the Gaussian tends to zero. Since the Gaussian is normalized,

$$\int_{-\infty}^{\infty} dx T_g(x, t) = 1 \tag{6-12}$$

then

$$\lim_{t \to 0} T_g(x, t) = \delta(x - x_o) \tag{6-13}$$

where $\delta(x)$ is the Dirac delta function. In fact, this Gaussian is one of the definitions of a delta function.[2] This Gaussian solution is important since it is also the Green's function for the infinite domain.[3]

---

[1] J. Mathews and R. Walker, *Mathematical Methods of Physics* (Menlo Park, Calif.: W. A. Benjamin. 1970).

[2] G. Arfken, *Mathematical Methods for Physicists* (New York: Academic Press, 1970).

[3] P. Morse and H. Feshbach, *Methods of Theoretical Physics*, vol. 1 (New York: McGraw-Hill, 1953).

The problem we want to solve is the following: Given the initial condition $T(x, t = 0) = \delta(x)$ and the Dirichlet boundary conditions,

$$T(x = -L/2, t) = T(x = L/2, t) = 0 \qquad (6\text{-}14)$$

find $T(x, t)$ for all x and t. Physically, this problem corresponds to the diffusion of heat in a bar of length L whose ends are held at temperature $T = 0$. If this zero temperature bothers you, remember that we may add an arbitrary constant to the solution. At time $t = 0$, an infinitesimal spot at the center of the bar is instantaneously heated to a very high (infinite) temperature. The total heat (or energy) in the bar remains finite since the integral of the temperature is finite.

By the method of images, we may construct an analytical solution as

$$T(x, t) = \sum_{n=-\infty}^{\infty} (-1)^n T_g(x + nL, t) \qquad (6\text{-}15)$$

To understand the construction, consider the schematic shown in Figure 6.5. The initial condition has images at $-L$ and L. But the image at L has an image of itself at $-2L$; similarly, the image at $-L$ has an image at $2L$. Of course, there are an infinity of images of images. The images are initially all delta functions of alternating signs. In time each spreads as a Gaussian. This set of images maintains the boundary conditions $x = \pm L/2$.

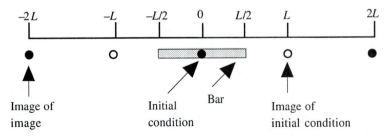

Figure 6.5 Method of images solution for the diffusion of a delta function initial condition. The images are positive or negative as indicated by the solid or open circles, respectively. There is an infinity of images at $\pm L$, $\pm 2L$, ...

For the values $\kappa = 1$, $L = 1$, the solution $T(x, t)$, as given by (6-15), is plotted in Figure 6.6. Notice that the solution looks somewhat like a Gaussian except that the values at $x = \pm L/2$ are fixed at zero by the boundary conditions.

The method of images is most useful when we are interested in the solution for small t. The solution is an infinite sum but for short times only the images near the origin contribute significantly. As t increases, more and more images contribute; in the case of large t it is better to use the solution obtained by separation of variables (see Chapter 7). That solution is also an infinite sum but the contributions from the higher-order terms decrease with time.

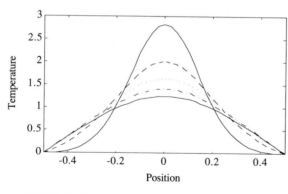

Figure 6.6 Plot of $T(x, t)$ as given by equation (6-15) at $t = 0.01, 0.02, 0.03, 0.04$, and 0.05. Only the seven leading terms of the infinite sum are evaluated (i.e., original pulse and three images on each side).

FTCS Scheme

Now let's try to numerically solve the diffusion equation. As mentioned earlier in the chapter, we discretize time and space, so it is useful to introduce the shorthand

$$T_i^n = T(x_i, t_n) \tag{6-16}$$

where $x_i = (i - 1)h - L/2$ and $t_n = (n - 1)\tau$ (see Figure 6.2). The index i denotes the spatial location of a grid point, while the index n indicates the temporal step. Note that the boundary points are T_1^n and T_N^n, so the grid spacing is $h = L/(N - 1)$.

The time derivative, discretized using the right derivative form, becomes

$$\frac{\partial T(x, t)}{\partial t} \Rightarrow \frac{T(x_i, t_n + \tau) - T(x_i, t_n)}{\tau} = \frac{T_i^{n+1} - T_i^n}{\tau} \tag{6-17}$$

The space derivative is discretized using the centered derivative form,

$$\frac{\partial^2 T(x, t)}{\partial x^2} \Rightarrow \frac{T(x_i + h, t_n) + T(x_i - h, t_n) - 2T(x_i, t_n)}{h^2}$$
$$= \frac{T_{i+1}^n + T_{i-1}^n - 2T_i^n}{h^2} \tag{6-18}$$

From the way we discretized the derivatives, this method takes the name forward time centered space (FTCS) scheme.

Using the above, our discretized diffusion equation is

$$\frac{T_i^{n+1} - T_i^n}{\tau} = \kappa \frac{T_{i+1}^n + T_{i-1}^n - 2T_i^n}{h^2} \tag{6-19}$$

or

$$T_i^{n+1} = T_i^n + \frac{\kappa\tau}{h^2} (T_{i+1}^n + T_{i-1}^n - 2T_i^n) \tag{6-20}$$

Notice that everything that depends on time step *n* is on the right-hand side, while only the future value of temperature is on the left. The FTCS scheme may remind you of the Euler scheme for ODEs (see Chapter 2).

FTCS Program

A MATLAB program, called dftcs, for solving the diffusion equation using the FTCS method is given in Listing 6.1. A few notes about the program: The FTCS scheme is implemented on lines 28–31. The temperature profile is periodically recorded (lines 32–37) and displayed as a function of *x* and *t* using a wire-mesh plot (lines 40–42) and a contour plot (lines 44–46).

LISTING 6.1 Program dftcs. Solves the diffusion equation using the FTCS scheme.

```
1    % dftcs - Program to solve the diffusion equation
2    % using the FTCS scheme
3    clear; help dftcs;   % Clear memory and print header
4    tau = input('Enter time step - ');
5    N = input('Enter the number of grid points - ');
6    L = 1.;   % The system extends from x=-L/2 to x=L/2
7    h = L/(N-1);   % Grid size
8    kappa = 1.;    % Diffusion coefficient
9    coeff = kappa*tau/h^2;
10   if( coeff < 1/2 )
11     disp('Solution is expected to be stable');
12   else
13     disp('WARNING: Solution is expected to be unstable');
14   end
15   %%%% Initial conditions and boundary conditions %%%%
16   tt = zeros(N,1); % Initialize temperature to zero
17   tt_new = zeros(N,1);  % Initialize temporary array
18   tt(N/2) = 1/h;    % Initial cond. is delta function in center
19   %% The boundary conditions are tt(1) = tt(N) = 0
20   %%%% Set up loop and plot variables %%%%
21   xplot = (0:N-1)*h - L/2;   % Record the x scale for plots
22   iplot = 1;                 % Counter used to count plots
23   nstep = 250;               % Maximum number of iterations
24   nplots = 25;               % Number of snapshots (plots)
25   plot_step = nstep/nplots;  % Number of time steps between
                                       plots
26   %%%% MAIN LOOP %%%%
27   for istep=1:nstep
28     % FTCS scheme
29     tt_new(2:(N-1)) = tt(2:(N-1)) + ...
```

```
30              coeff*(tt(3:N) + tt(1:(N-2)) - 2*tt(2:(N-1))));
31     tt = tt_new;           % Reset temperature to new values
32     if( rem(istep,plot_step) < 1 )    % Every plot_step steps
33        ttplot(:,iplot) = tt(:);        % record tt(i) for
                                                plotting
34        tplot(iplot) = istep*tau;       % Record time for plots
35        iplot = iplot+1;
36        fprintf('%g out of %g steps completed\n',istep,nstep);
37     end
38   end
39   subplot(121)    % Wire-mesh plot of the solution
40     mesh(ttplot,[-70 30]);
41     xlabel('Position');   ylabel('Time');
42     title('Diffusion of a delta spike');
43   subplot(122)    % Contour plot of the solution
44     cs = contour(rot90(ttplot),0:.5:5,xplot,tplot);
45     xlabel('Position');   ylabel('Time');  clabel(cs,0:5);
46     title('Contour plot');
47   subplot(111)
```

The initial condition we want to use is $T(x, 0) = \delta(x)$, but we cannot put a delta function in the program. To best represent the delta function numerically, we use

$$\Delta(x) = \begin{cases} (x + h)/h^2 & \text{for } -h < x \leq 0 \\ (h - x)/h^2 & \text{for } \quad 0 \leq x < h \\ 0 & \text{otherwise} \end{cases} \qquad (6\text{-}21)$$

Notice that

$$\lim_{h \to 0} \Delta(x) = \delta(x) \qquad (6\text{-}22)$$

The function $\Delta(x)$ is just a triangular spike with unit area. When we discretize $\Delta(x)$ we have $\Delta_i = 1/h$ at $i = N/2$ (center of the system) and zero elsewhere.

Numerical Stability

Before running the program, we should have some idea as to how to specify the time step. From equation (6-11), we know the width of our Gaussian will spread in time as

$$\sigma(t) = \sqrt{2\kappa t} \qquad (6\text{-}23)$$

Call t_σ the time it takes the width σ to increase from zero to h (i.e., by one grid spacing). From the above, we may write

$$t_\sigma = \frac{h^2}{2\kappa} \tag{6-24}$$

Notice that the FTCS scheme, equation (6-20), may be written in terms of t_σ as

$$T_i^{n+1} = T_i^n + \frac{\tau}{2t_\sigma} (T_{i+1}^n + T_{i-1}^n - 2T_i^n) \tag{6-25}$$

Our physical intuition tells us that we probably don't want to use a time step much larger than t_σ.

   In running the dftcs program for $N = 41$ and $\tau = 2.0 \times 10^{-4}$, we obtain the results shown in Figure 6.7. In this case $h = L/(N - 1) = 1/40$, so $t_\sigma = 6.25 \times 10^{-4}$. Qualitatively, the profile looks correct; in the exercises you are asked to compare it quantitatively with the analytical solution.

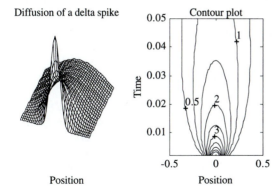

Figure 6.7 Graphical output from dftcs. Number of grid points is $N = 41$, time step $\tau = 2.0 \times 10^{-4}$.

   You will discover that the FTCS method is not stable for all values of τ and N. In fact, it is unstable when $\tau > t_\sigma$ (try Exercise 6.2). In Chapter 8 we will discover why this happens and will discuss some more advanced numerical methods that don't suffer from numerical instabilities. For now, we move on to our next PDE (after a few exercises, of course).

EXERCISES

1. Show that equation (6-10) is a solution to the diffusion equation (6-9).
2. Run the dftcs program with $N = 41$ and a variety of values for τ between 10^{-3} and 10^{-5}. For $\tau = 2.0 \times 10^{-4}$, try a variety of values for N. What do you observe?

3. Write a program to evaluate equation (6-15) numerically for $T(x, t)$ and reproduce Figure 6.6. Insert the code from this program into dftcs to produce a graph of $|T_a(x, t) - T_c(x, t)|$, the difference between the analytical temperature profile, T_a, and the profile obtained by the FTCS scheme, T_c.

4. Suppose that we replace our Dirichlet boundary conditions with the following Neumann boundary conditions:

$$\left.\frac{\partial T}{\partial x}\right|_{x = -L/2} = \left.\frac{\partial T}{\partial x}\right|_{x = L/2} = 0$$

Using the method of images, find the solution $T(x, t)$ for the same initial condition as before, $T(x, 0) = \delta(x)$.

5. Modify dftcs so that it uses periodic boundary conditions. With periodic boundary conditions the points T_1 and T_N are adjacent; T_1 is to the right of T_N and T_N is to the left of T_1 (see Figure 6.8 in the next section). Test your program with a variety of values for N and τ.

6. Discuss why the periodic boundary conditions in the previous exercise are equivalent to using the Neumann boundary conditions as described in Exercise 6.4. Quantitatively compare your results from Exercises 6.4 and 6.5. In general, are periodic and Neumann boundary conditions equivalent, or is there something special about this problem?

7. The DuFort–Frankel scheme for solving the diffusion equation uses the following discretization:

$$\frac{T_i^{n+1} - T_i^{n-1}}{2\tau} = \kappa \frac{T_{i+1}^n + T_{i-1}^n - (T_i^{n+1} + T_i^{n-1})}{h^2}$$

Note that this is a three time-level scheme, that is, the method uses T^{n+1}, T^n, and T^{n-1}, so the scheme is not self-starting. Write a program that uses this scheme to solve the diffusion problem discussed in this section. Use the FTCS scheme on the first step to get it started. Try a variety of values for τ and show that the method is unconditionally stable but that the accuracy of the solution decreases with increasing τ.

6.3 ADVECTION EQUATION

Wave and Advection Equations

We now look at hyperbolic equations, the paradigm of which is the familiar wave equation,

$$\frac{\partial^2 A}{\partial t^2} = c^2 \frac{\partial^2 A}{\partial x^2} \tag{6-26}$$

where $A(x, t)$ is the wave amplitude and c is the wave speed. In Chapters 2 and 3 the equations of motion for a particle were ODEs of the form

$$\frac{d^2\mathbf{r}}{dt^2} = f(\mathbf{r}, \mathbf{v}) \tag{6-27}$$

where **r** and **v** are the position and velocity **d** of the particle. To solve them numerically, we usually rewrote (6-27) as a pair of first-order equations,

$$\frac{d\mathbf{v}}{dt} = f(\mathbf{r}, \mathbf{v}); \qquad \frac{d\mathbf{r}}{dt} = \mathbf{v} \tag{6-28}$$

For the wave equation, we use a similar trick and introduce the variables

$$p = \frac{\partial A}{\partial t}; \qquad q = c \frac{\partial A}{\partial x} \tag{6-29}$$

The wave equation may now be written as the pair of equations

$$\frac{\partial p}{\partial t} = c \frac{\partial q}{\partial x}; \qquad \frac{\partial q}{\partial t} = c \frac{\partial p}{\partial x} \tag{6-30}$$

or

$$\frac{\partial \mathbf{a}}{\partial t} = c \mathbf{B} \frac{\partial}{\partial x} \mathbf{a} \tag{6-31}$$

where $\mathbf{a} = \begin{bmatrix} p \\ q \end{bmatrix}$ and $\mathbf{B} = \begin{bmatrix} 0 & 1 \\ 1 & 0 \end{bmatrix}$.

This suggests that even though the wave equation is the most familiar hyperbolic equation, it is not the simplest possible hyperbolic equation. When formulating and studying numerical methods it is best to first use them with the simplest, nontrivial problem. We will thus use as our model hyperbolic equation the *advection equation*

$$\frac{\partial a}{\partial t} = -c \frac{\partial a}{\partial x} \tag{6-32}$$

Physically, this equation describes the evolution of the passive scalar field, $a(x, t)$, carried along by a flow with constant velocity c. This equation is also known as the linear convection equation. In the wave equation we have left- and right-moving waves; with the advection equation waves only move in one direction. Of course, left- and right-moving waves may be described using (6-31), the vector advection equation.

The advection equation is the simplest example of a flux-conservation equation,

$$\frac{\partial \mathbf{p}}{\partial t} = -\nabla \cdot \mathbf{F}(\mathbf{p}) \tag{6-33}$$

which in one dimension is

$$\frac{\partial p}{\partial t} = -\frac{\partial}{\partial x} F(p) \qquad (6\text{-}34)$$

Equations of this form are ubiquitous in physics because if p is any conserved quantity (such as mass or energy), then $F(p)$ is the flux. The Navier–Stokes equations of fluid mechanics are all of this form.

Solution of the Advection Equation

The analytical solution of the advection equation is easy to obtain. For the initial condition

$$a(x, t = 0) = f_o(x) \qquad (6\text{-}35)$$

where f_o is an arbitrary function, the solution is

$$a(x, t) = f_o(x - ct) \qquad (6\text{-}36)$$

For example, suppose that our initial condition is a cosine-modulated Gaussian pulse,

$$a(x, t = 0) = \cos[k(x - x_o)] \exp\left[-\frac{(x - x_o)^2}{2\sigma^2}\right] \qquad (6\text{-}37)$$

where the constants x_o and σ give the location of the peak and the width of the pulse. The wave number $k = 2\pi/\lambda$ where λ is the wavelength of the modulation. The solution is

$$a(x, t) = \cos\{k[(x - ct) - x_o]\} \exp\left\{-\frac{[(x - ct) - x_o]^2}{2\sigma^2}\right\}$$
$$= \cos\{k[x - (x_o + ct)]\} \exp\left\{-\frac{[x - (x_o + ct)]^2}{2\sigma^2}\right\} \qquad (6\text{-}38)$$

Notice that the solution $a(x, t)$ is still a cosine-modulated Gaussian but with the location of the peak displaced to $x_o + ct$. Although the advection equation is simple to solve analytically, it makes an excellent test case for our numerical methods for hyperbolic equations. We will discover that even this simple equation is nontrivial to solve numerically.

FTCS Method for Advection Equation

Let's try to solve the advection equation numerically. We will start with the FTCS method from the previous section. The time derivative is replaced by its right (forward) discretized form

$$\frac{\partial a}{\partial t} \Rightarrow \frac{a(x_i, t_n + \tau) - a(x_i, t_n)}{\tau} = \frac{a_i^{n+1} - a_i^n}{\tau} \tag{6-39}$$

where $x_i = (i - 1)h - L/2$ and $t_n = (n - 1)\tau$ (see Figure 6.2). The index i denotes the spatial location of a grid point, while the index n indicates the temporal step.
  The space derivative is replaced by its centered discretized form,

$$\frac{\partial a}{\partial x} \Rightarrow \frac{a(x_i + h, t_n) - a(x_i - h, t_n)}{2h} = \frac{a_{i+1}^n - a_{i-1}^n}{2h} \tag{6-40}$$

We'll use periodic boundary conditions, so grid points x_1 and x_N are adjacent; the grid spacing is $h = L/N$ (see Figure 6.8). The discretized advection equation is

$$\frac{a_i^{n+1} - a_i^n}{\tau} = -c\,\frac{a_{i+1}^n - a_{i-1}^n}{2h} \tag{6-41}$$

The FTCS scheme is obtained by solving for a_i^{n+1},

$$a_i^{n+1} = a_i^n - \frac{c\tau}{2h}\,(a_{i+1}^n - a_{i-1}^n) \tag{6-42}$$

A program, called `aftcs`, that implements the FTCS scheme for the advection equation is given in Listing 6.2. A few notes about the program: The initial condition is a cosine-modulated Gaussian pulse [see equation (6-37)] that is set up on lines 10–14. The FTCS scheme is implemented on lines 24–28; notice that we handle the end points $i = 1$ and N separately.

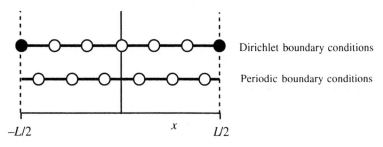

Dirichlet boundary conditions

Periodic boundary conditions

$-L/2$ x $L/2$

Figure 6.8 Grids used by Dirichlet and periodic boundary conditions. With periodic boundary conditions the boundary lies between the first and last grid points.

LISTING 6.2 Program `aftcs`. Solves the advection equation using the FTCS scheme.

```
1   % aftcs - Program to solve the advection equation
2   % using the FTCS scheme
3   clear;  help aftcs;  % Clear memory and print header
4   tau = input('Enter time step - ');
5   N = input('Enter number of grid points - ');
6   L = 1.;       % System size
7   h = L/N;      % Grid spacing
8   c = 1;        % Wave speed
9   coeff = -c*tau/(2.*h);  % Coefficient used by FTCS scheme
10  %%%% Initial condition is a Gaussian-cosine pulse%%%%
11  sigma = 0.1;         % Width of the Gaussian pulse
12  k_wave = pi/sigma;  % Wave number of the cosine
13  x = ((1:N)-1/2)*h - L/2;  % Record x for plotting
14  a = cos(k_wave*x) .* exp(-x.^2/(2*sigma^2));
15  %%%% Set up plot variables %%%%
16  iplot = 1;           % Plot counter
17  aplot(:,1) = a(:);  % Record the initial state
18  tplot(1) = 0;        % Record the time (t=0)
19  %%%% MAIN LOOP %%%%
20  nstep = floor(L/(c*tau)); % Wave circles system once
21  %%% nstep = 10*nstep;  % Increase steps for chap 8 plot
22  plot_step = ceil(nstep/50); % Number of steps between plots
23  for istep=1:nstep
24    % FTCS scheme
25    a_new(2:(N-1)) = a(2:(N-1)) + coeff*(a(3:N)-a(1:(N-2)));
26    a_new(1) = a(1) + coeff*(a(2)-a(N));    % Periodic boundary
27    a_new(N) = a(N) + coeff*(a(1)-a(N-1)); % conditions
28    a = a_new;  % Replace old amplitude with new values
29    if( rem(istep,plot_step) < 1 )  % Every plot_iter steps
30      iplot = iplot+1;
31      aplot(:,iplot) = a(:);          % Record a(i) for ploting
32      tplot(iplot) = tau*istep;
33      fprintf('%g out of %g steps completed\n',istep,nstep);
34    end
35  end
36  % Since graph is cluttered, show only initial and final
       states
37  plot(x,aplot(:,1),'-',x,a,'--');
38  xlabel('x');  ylabel('a(x, t)');
39  title('Initial and final amplitudes');
40  pause;  % Pause between plots; strike any key to continue
41  mesh(aplot,[-100 60]);   % Wire-mesh plot of solution
42  xlabel('Position');  ylabel('Time');
43  title('Amplitude versus x and t');
```

The number of iterations performed, nstep, is set to $L/(c\tau)$. With periodic boundary conditions, the pulse should move across and around the system once, returning to its starting point. Since the wave speed is c, the time it takes a wave to move a distance equal to the grid spacing, h, is $t_w = h/c$. This gives us a characteristic time scale for the problem.

The aftcs program was run with the following values: $\tau = 0.002$, $N = 50$; in this case, $h = 1/50$ and $t_w = 0.02$. Figure 6.9 shows the initial and final values for the wave amplitude, $a(x, t)$. The solid line is the initial condition; the dashed line shows the pulse after it evolves long enough to circle the system once. Clearly, the FTCS method failed; the pulse does not maintain its shape. The wire-mesh plot shown in Figure 6.10 clearly illustrates how the pulse distorts in time.

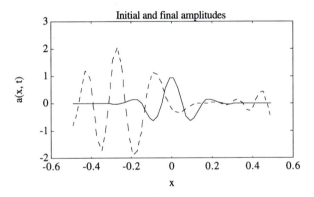

Figure 6.9 Initial and final shapes of the wave packet as obtained by the aftcs program. Notice that the wave does not correctly retain its shape. The time step is $\tau = 0.002$; the number of grid points is $N = 50$.

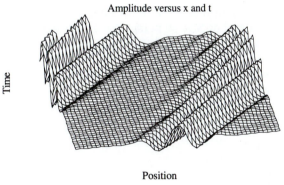

Figure 6.10 Output from the aftcs program. Notice how the wave packet moves to the right but incorrectly distorts with time. The parameters are as in Figure 6.9.

Lax Method for Advection Equation

Now for the bad news: For the advection equation the FTCS method is numerically unstable for all values of τ! For a smaller value of τ we can delay the problem but not escape it. Fortunately, the stability problem is simple to fix. We introduce the *Lax method*, defined by the following iteration equation:

$$a_i^{n+1} = \frac{1}{2}\left(a_{i+1}^n + a_{i-1}^n\right) - \frac{c\tau}{2h}\left(a_{i+1}^n - a_{i-1}^n\right) \qquad (6\text{-}43)$$

Notice that the Lax method simply replaces the a_i^n term in the FTCS method with the average value of the left and right neighbors.

It is a simple matter to modify the `aftcs` program to implement the Lax scheme; we replace lines 24–27 with

```
% Lax scheme
a_new(2:(N-1))  = .5*(a(3:N)+a(1:(N-2)))  + ...
                     coeff*(a(3:N)-a(1:(N-2)));
a_new(1)  = .5*(a(2)+a(N))  + coeff*(a(2)-a(N));
a_new(N)  = .5*(a(1)+a(N-1))  + coeff*(a(1)-a(N-1));
```

The Lax method is stable if

$$\frac{c\tau}{h} \le 1 \qquad (6\text{-}44)$$

The maximum usable value for τ is thus

$$\tau_{\max} = \frac{h}{c} = t_w \qquad (6\text{-}45)$$

This criterion is known as the *Courant–Friedrichs–Lewy* (CFL) condition. It commonly appears as a stability criterion for numerical schemes that solve hyperbolic equations. Notice that if we use a finer grid (smaller h), we are forced to use a smaller τ. The CFL stability condition is derived in Chapter 8.

Using Lax's method with a grid of $N = 50$ points and a time step of $\tau = t_w = 0.02$, we find that the pulse exactly preserves its shape. The mesh plot of the solution is shown in Figure 6.11.

Lax's method has an interesting property. For values of τ above $\tau_{\max}$ you have problems because the method is numerically unstable. However, for τ signifi-

Amplitude versus x and t

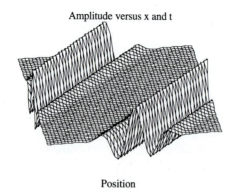

Time

Position

Figure 6.11 Mesh plot of the solution of the advection equation obtained using the Lax method. Parameters used are $N = 50$ grid points and time step $\tau = t_w = 0.02$.

cantly less than τ_{max}, the numerical solution is also wrong. If τ is too small, we find that the pulse dies out as it moves (see Figure 6.12). We get the best results when $\tau = \tau_{max}$. This example should dismiss a popular misconception about numerical methods: The smaller the time step, the better the solution. While it is usually true that the truncation error for many schemes is proportional to τ, this is not a universal property for all methods.

Amplitude versus x and t

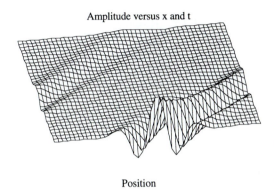

Time

Position

Figure 6.12 Mesh plot of the solution of the advection equation obtained using the Lax method. Parameters used are $N = 50$ grid points and time step $\tau = 0.015$. Notice how the pulse moves to the right but its amplitude dies out since $\tau < t_w = 0.02$.

The averaging term in the Lax method serves to stabilize the numerical solution by introducing an *artificial diffusion* (or artificial viscosity). The magnitude of this artificial diffusion is inversely proportional to the time step τ. When the time step is too large ($\tau > \tau_{max}$), the artificial diffusion is too weak to stabilize the solution. When the time step is too small ($\tau < \tau_{max}$), it is too strong and it damps out the true solution. Many PDE schemes besides the Lax method incorporate some form of numerical diffusion.

Lax–Wendroff Scheme

Let's look at one more scheme for solving hyperbolic PDEs. The Lax–Wendroff scheme is a second-order finite difference scheme; the idea is that we want to take the Taylor expansion

$$a(x,\, t + \tau) = a(x,\, t) + \tau \left(\frac{\partial a}{\partial t} \right) + \frac{\tau^2}{2} \left(\frac{\partial^2 a}{\partial t^2} \right) + O(\tau^3) \qquad (6\text{-}46)$$

and keep the terms through τ^2. The term that is linear in τ is easy to represent using the original equation, which we now write in the more general form

$$\frac{\partial a}{\partial t} = - \frac{\partial}{\partial x} F(a) \qquad (6\text{-}47)$$

where the flux $F(a) = ca$ for the advection equation.

To obtain an expression for the second-order term, we differentiate the equation above:

$$\frac{\partial^2 a}{\partial t^2} = -\frac{\partial}{\partial t}\frac{\partial}{\partial x}F(a) = -\frac{\partial}{\partial x}\frac{\partial F}{\partial t} \tag{6-48}$$

Yet we may write

$$\frac{\partial F}{\partial t} = \frac{dF}{da}\frac{\partial a}{\partial t} = F'(a)\frac{\partial a}{\partial t} = -F'(a)\frac{\partial F}{\partial x} \tag{6-49}$$

where $F'(a) = c$ for the advection equation.

Inserting (6-49) into equation (6-48),

$$\frac{\partial^2 a}{\partial t^2} = \frac{\partial}{\partial x}F'(a)\frac{\partial F}{\partial x} \tag{6-50}$$

Putting it all together in our Taylor expansion, we get

$$a(x, t + \tau) \approx a(x, t) - \tau\left(\frac{\partial}{\partial x}F(a)\right) + \frac{\tau^2}{2}\left(\frac{\partial}{\partial x}F'(a)\frac{\partial}{\partial x}F(a)\right) \tag{6-51}$$

After discretizing the derivatives, we obtain the *Lax–Wendroff scheme*,

$$a_i^{n+1} = a_i^n - \tau\frac{F_{i+1} - F_{i-1}}{2h} + \frac{\tau^2}{2}\frac{1}{h}\left(F'_{i+1/2}\frac{F_{i+1} - F_i}{h} - F'_{i-1/2}\frac{F_i - F_{i-1}}{h}\right) \tag{6-52}$$

where $F_i \equiv F(a_i^n)$ and $F'_{i\pm1/2} \equiv F'[(a_{i\pm1}^n + a_i^n)/2]$.

For the advection equation this equation simplifies considerably, since $F_i = ca_i^n$ and $F'_{i\pm1/2} = c$, and equation (6-52) reduces to

$$a_i^{n+1} = a_i^n - \frac{c\tau}{2h}(a_{i+1}^n - a_{i-1}^n) + \frac{c^2\tau^2}{2h^2}(a_{i+1}^n + a_{i-1}^n - 2a_i^n) \tag{6-53}$$

Notice that the last term is a discretized second derivative in $a(x, t)$. This term gives us an artificial diffusion that stabilizes the numerical solution. The CFL condition (6-45) is also the criterion for stability for the Lax–Wendroff scheme.

To implement the Lax–Wendroff scheme in our `aftcs` program, we simply insert the following code just after line 9:

```
coefflw = 2*coeff^2;    % Coefficient used by L-W scheme
```

and replace lines 24–27 with

```
% Lax-Wendroff scheme
   a_new(2:(N-1)) = a(2:(N-1)) + ...
       coeff*(a(3:N)-a(1:(N-2))) + ...
       coefflw*(a(3:N)+a(1:(N-2))-2*a(2:(N-1)));
   a_new(1) = a(1) + coeff*(a(2)-a(N)) + ...
       coefflw*(a(2)+a(N)-2*a(1));        % Periodic
                                          % boundary
   a_new(N) = a(N) + coeff*(a(1)-a(N-1)) + ...
       coefflw*(a(1)+a(N-1)-2*a(N));      % conditions for
                                          % i=1 and N
```

Notice that if $\tau = \tau_{max} = h/c$, then $coeff = -1/2$ and $coefflw = 1/2$. It is easy to check that in this case the Lax–Wendroff scheme is identical to the Lax scheme. This is good news since we know that Lax is exact when $\tau = \tau_{max}$.

The Lax scheme is flawed because when $\tau < \tau_{max}$ the solution is rapidly damped out by the artificial viscosity (see Figure 6.12). The Lax–Wendroff scheme also has artificial viscosity to control instability, but it does not increase as rapidly with decreasing τ. Figure 6.13 illustrates the use of the Lax–Wendroff scheme for the same parameters as those in Figure 6.12. Comparing the two results, you see that the Lax–Wendroff scheme is still useful even when $\tau < \tau_{max}$.

I want to emphasize again that the one-dimensional advection equation makes a useful test case but is of little interest since the analytical solution is trivial. In the next section we consider a much more interesting hyperbolic equation that (1) is nonlinear, (2) has solutions that develop discontinuities (shocks) even if the initial conditions are smooth and continuous, and (3) is a model for the flow of automobile traffic.

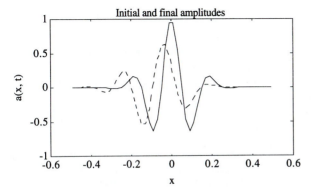

Figure 6.13 Initial (solid) and final (dashed) amplitudes for the advection equation solved using the Lax–Wendroff method. Parameters used are $N = 50$ grid points and time step $\tau = 0.015$. Notice how the amplitude decreases since $\tau < t_w = 0.02$.

EXERCISES

8. By direct substitution into the advection equation, (6-32), check that equation (6-36) is a solution.

9. (a) Modify aftcs to use the FTCS scheme with the Dirichlet boundary conditions

$$a(x = -L/2, t) = \sin(\omega t)$$

$$a(x = L/2, t) = 0$$

Have the program run long enough for the wave generated at $x = -L/2$ to reach the opposite side of the system. Using $N = 50$ grid points and a frequency of $\omega = 10\pi$, what do you observe for time steps of $\tau = 0.015$, 0.02, and 0.03? How do your results change when you vary the frequency?

(b) Repeat part (a) but modify the program to use the Lax scheme.

(c) Repeat part (a) but modify the program to use the Lax–Wendroff scheme.

10. The advection and diffusion of a passive scalar in a one-dimensional flow is commonly described by the transport equation,

$$\frac{\partial T}{\partial t} = -c \frac{\partial T}{\partial x} + \kappa \frac{\partial^2 T}{\partial x^2}$$

Find the solution of this PDE for the initial condition $T(x, t = 0) = \delta(x)$ and periodic boundary conditions at $x = \pm L/2$.

11. Write a program that uses the FTCS scheme to solve the one-dimensional transport equation in the previous problem.

(a) Empirically show that the numerical solution is stable if

$$\left(\frac{c\tau}{h}\right)^2 \le \frac{2\kappa\tau}{h^2} \le 1$$

In this case, physical diffusion can serve to stabilize the numerical scheme.

(b) Compare the results from your program with the solution for the delta function initial condition of the previous exercise.

12. The "upwind" scheme for solving the advection equation uses a left derivative for the $\partial/\partial x$ term,

$$\frac{a_i^{n+1} - a_i^n}{\tau} = -c \frac{a_i^n - a_{i-1}^n}{h}$$

Modify the aftcs program to use this scheme and compare it with the others discussed in this section. For what values of τ is it stable?

13. The leap-frog scheme for solving the advection equation uses centered derivatives for both terms,

$$\frac{a_i^{n+1} - a_i^{n-1}}{2\tau} = -c \frac{a_{i+1}^n - a_{i-1}^n}{2h}$$

Notice that this is a three time-level scheme, that is, it uses a_i^{n+1}, a_i^n, and a_i^{n-1}. To get it started we need to use one of the other schemes (e.g., Lax). Modify the aftcs program to use this scheme and compare it with the others discussed in this section. For what values of τ is it stable?

6.4 PHYSICS OF TRAFFIC FLOW*

Fluid Mechanics

In fluid mechanics the equations of motion are obtained by constructing equations of the form

$$\frac{\partial \mathbf{p}}{\partial t} = -\nabla \cdot \mathbf{F}(\mathbf{p}) \tag{6-54}$$

or in one dimension,

$$\frac{\partial \mathbf{p}}{\partial t} = -\frac{\partial}{\partial x} \mathbf{F}(\mathbf{p}) \tag{6-55}$$

Here **p** is the vector of the conserved quantities

$$\mathbf{p} = \begin{bmatrix} \text{mass density} \\ \text{momentum density} \\ \text{energy density} \end{bmatrix} \tag{6-56}$$

and

$$\mathbf{F}(\mathbf{p}) = \begin{bmatrix} \text{mass flux} \\ \text{momentum flux} \\ \text{energy flux} \end{bmatrix} \tag{6-57}$$

is the vector function of the corresponding fluxes.

While the equations for the momentum and energy are somewhat complicated, the equation for the mass density, ρ, is quite simple. The mass flux equals the mass density times the fluid velocity, v, so

$$\frac{\partial \rho(x, t)}{\partial t} = -\frac{\partial}{\partial x} \{\rho(x, t)v(x, t)\} \tag{6-58}$$

This equation is known as the *equation of continuity*. The equation for the momentum density may be rewritten as an equation for the velocity. This velocity equation involves the energy density (the coupling is in the pressure term), so we must solve the entire set of equations simultaneously.

The full set of hydrodynamics equations is called the *Navier–Stokes equations*.[4] For a variety of reasons, these equations are usually not solved in their full form but rather with a number of approximations. Of course the approximations

---

[4] Sometimes the velocity equation by itself is also referred to as the Navier–Stokes equation.

used depend on the problem at hand. For example, air is incompressible, to a good approximation, in many subsonic flows.

Traffic Flow

One of the simplest, nontrivial flows that may be studied involves fluids for which the velocity is only a function of density,

$$v(x, t) = v(\rho) \qquad (6\text{-}59)$$

For example, suppose that the velocity of the fluid decreased linearly with increasing density as

$$v(\rho) = v_m(1 - \rho/\rho_m) \qquad (6\text{-}60)$$

where $v_m > 0$ is the maximum velocity and $\rho_m > 0$ is the maximum density. What type of fluid behaves this way? One you are probably very familiar with is automobile traffic. The maximum velocity is the speed limit; if the density is near zero (few cars on the road), then the traffic moves at this speed. The maximum density, ρ_m, is achieved when the traffic is bumper-to-bumper. While on real highways the flow may not exactly obey equation (6-60), it turns out to be a good first approximation.[5]

Our equation for the evolution of the density may be written as

$$\frac{\partial \rho}{\partial t} = -\frac{\partial}{\partial x}\{(\alpha + \tfrac{1}{2}\beta\rho)\rho\} \qquad (6\text{-}61)$$

where $\alpha = v_m$ and $\beta = -2v_m/\rho_m$. This equation is called the generalized inviscid Burger's equation. We obtain the standard inviscid Burger's equation when $\alpha = 0$ and $\beta = 1$. This equation has been studied extensively since it is the simplest nonlinear PDE with wave solutions.[6] Equations of this type appear frequently in nonlinear acoustics and shock wave theory.

Returning to our traffic model, we want to develop a method to solve the nonlinear PDE,

$$\frac{\partial \rho}{\partial t} = -\frac{\partial}{\partial x}\{\rho v(\rho)\} \qquad (6\text{-}62)$$

Rewrite this equation as

$$\frac{\partial \rho}{\partial t} = -\left(\frac{d}{d\rho}\,\rho v(\rho)\right)\frac{\partial \rho}{\partial x} \qquad (6\text{-}63)$$

---

[5] R. Haberman, *Mathematical Models* (Englewood Cliffs, N.J.: Prentice Hall, 1977), section 62.

[6] E. Benton and G. Platzman, "A Table of Solutions of the One-Dimensional Burger's Equation," *Q. Appl. Math.*, **30** 195-212 (1972).

or

$$\frac{\partial \rho}{\partial t} = -c(\rho)\frac{\partial \rho}{\partial x} \tag{6-64}$$

where $c(\rho) \equiv d(\rho v)/d\rho$. Using our linear function for $v(\rho)$ as given by equation (6-60), we have

$$c(\rho) = v_m(1 - 2\rho/\rho_m) \tag{6-65}$$

Notice that $c(\rho)$ is also linear in ρ and takes the values $c(0) = v_m$ and $c(\rho_m) = -v_m$. The function $c(\rho)$ is not the speed of the traffic but rather is the speed at which disturbances (or waves) in the flow will travel. Since $c(\rho)$ may be both positive or negative, the waves may move in either direction. Note, however, that $c(\rho) \le v(\rho)$, so the waves may never move faster than the cars.

Method of Characteristics

For $c(\rho) = $ constant, we have the advection equation for which we already know the solution. We may build an analytical solution to (6-64) from our knowledge of the solution of the advection equation by using the *method of characteristics*.[7] The reader who is not interested in learning this method may skim through the introduction to the stoplight problem (Figures 6.17 and 6.18) and skip to its solution, equation (6-71).

For a moment let's return to our solution of the advection equation. We know that, with time, the initial condition, $\rho(x, t = 0) = \rho_0(x)$, is translated with speed c. Consider the sketch of the xt plane shown in Figure 6.14. Suppose that we draw a line with slope $dt/dx = 1/c$ from a point x_1 on the x-axis. This line will be a contour of constant ρ in the xt plane. This is because the solution of the advection equation is just the initial condition displaced by a distance $\Delta x = c\,\Delta t$.

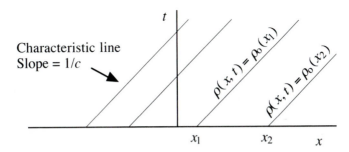

Figure 6.14   Sketch of the characteristic lines for the advection equation.

[7] M. B. Abbot, *An Introduction to the Method of Characteristics* (New York: American Elsevier, 1966).

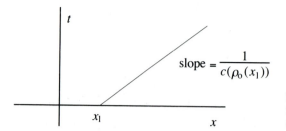

Figure 6.15 Sketch of a single characteristic line for the nonlinear traffic equation.

Now let's return to the nonlinear problem, equation (6-64). In Figure 6.15 we draw the characteristic line from the point x_1; this line has slope $dt/dx = 1/c(\rho_0(x_1))$. Even in the nonlinear problem, *the density is constant along this line.* Here is the proof: For any function of two variables, the chain rule tells us that

$$\frac{d}{dt} f(x(t),\, t) = \frac{\partial}{\partial t} f(x(t),\, t) + \frac{dx}{dt} \frac{\partial}{\partial x} f(x(t),\, t) \tag{6-66}$$

Suppose that we vary x with t such that we move along the characteristic line. This means that $dt/dx = 1/c(\rho_0)$ or $dx/dt = c(\rho_0)$. Using the previous equation with $f(x,\, t) = \rho(x,\, t)$,

$$\frac{d}{dt} \rho(x(t),\, t) = \frac{\partial}{\partial t} \rho(x(t),\, t) + c(\rho_0(x)) \frac{\partial}{\partial x} \rho(x(t),\, t) \tag{6-67}$$

Yet from our original PDE, the right-hand side is zero, so

$$\frac{d}{dt} \rho(x(t),\, t) = 0 \qquad \text{(on the characteristic line)} \tag{6-68}$$

which completes the proof.

To use the method of characteristics to construct our solution, we draw a characteristic line from each point on the x-axis (see Figure 6.16). You should

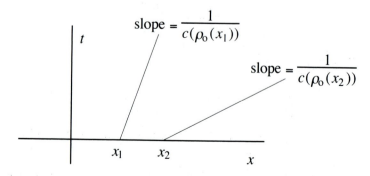

Figure 6.16 Sketch of various characteristic lines for the nonlinear traffic equation.

think of these lines as forming a contour map of $\rho(x, t)$ since each line is a line of constant density.

Traffic at a Stoplight

Now to solve an actual traffic problem. The simplest problem we can solve is the initial distribution

$$\rho(x, t = 0) = \rho_0(x) = \begin{cases} \rho_m & x < 0 \\ 0 & x > 0 \end{cases} \tag{6-69}$$

that is, a step function (see Figure 6.17).

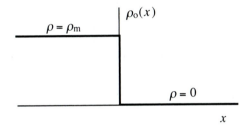

$\rho_0(x)$

$\rho = \rho_m$

$\rho = 0$

x

Figure 6.17 Initial density profile for traffic at a stoplight.

As a traffic problem, this could represent cars at a stoplight. Behind the light (which is at $x = 0$) the traffic is at its maximum density (bumper-to-bumper); there is no traffic on the other side of the light. At time $t = 0$ the light turns green and the cars are free to move. Intuitively, we know that not all cars start moving when the light turns green. The density decreases as the cars separate, but this effect propagates back into the stream of traffic with a finite wave speed (see Figure 6.18). In fluid dynamics this is a *rarefaction wave* problem.

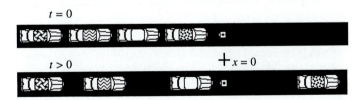

$t = 0$

$t > 0$ $+ x = 0$

Figure 6.18 Traffic moving after a stoplight turns green. Notice that in the second frame the last car toward the rear has not moved.

Let's start by drawing the characteristic lines on the positive x-axis. These lines will have slope $1/c(0) = 1/v_m$. If we shade the region of constant density, we have the sketch shown in Figure 6.19. The first car through the light will move at

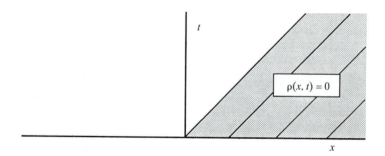

Figure 6.19 Partial construction of $\rho(x, t)$ in the xt plane using characteristic lines. In the shaded region the density $\rho(x, t)$ is zero. The left boundary of this region is given by the position of the lead car.

the maximum velocity since there are no cars in front of it. The border of the $\rho(x, t) = 0$ region is just the location of the lead car.

Next we add the characteristic lines for the points on the negative x-axis, as shown in Figure 6.20. Notice that most cars do not begin to move until long after the light has turned green. This is because the disturbance (or wave) can only move with velocity $c(\rho_m)$. For our linear relation between v and ρ, we have $c(\rho_m) = -v_m$.

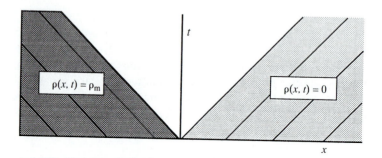

Figure 6.20 Partial construction of $\rho(x, t)$ in the xt plane using characteristic lines. In the shaded region on the left the density $\rho(x, t)$ is maximum (bumper-to-bumper).

To obtain all the characteristic lines we must remember that our initial condition is discontinuous. Suppose that we modified $\rho_0(x)$ so that it varied continuously from ρ_m to zero in a neighborhood of radius ε about $x = 0$. The slopes of the characteristic lines in this neighborhood would vary continuously from $1/v_m$ to $-1/v_m$ (see Figure 6.21).

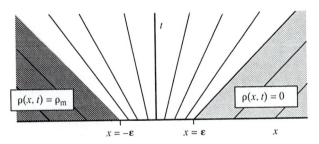

Figure 6.21 Characteristic lines for a continuous initial density profile. This density profile goes to a step function as $\varepsilon \to 0$.

Taking the limit $\varepsilon \to 0$, we have our final picture of the characteristic lines as shown in Figure 6.22. The solution may be written as

$$\rho(x, t) = \begin{cases} \rho_m & \text{for} & x \le -v_m t \\ c^{-1}(x/t) & \text{for} & -v_m t < x < v_m t \\ 0 & \text{for} & x \ge v_m t \end{cases} \tag{6-70}$$

where $c^{-1}(c(\rho)) = \rho$, that is, c^{-1} is the inverse function of $c(\rho)$. Using equation (6-65) for $c(\rho)$ we have

$$\rho(x, t) = \begin{cases} \rho_m & \text{for} & x \le -v_m t \\ \dfrac{\rho_m}{2}\left(1 - \dfrac{x}{v_m t}\right) & \text{for} & -v_m t < x < v_m t \\ 0 & \text{for} & x \ge v_m t \end{cases} \tag{6-71}$$

Notice that in the region $-v_m t < x < v_m t$, the density varies linearly with position (see Figure 6.23).

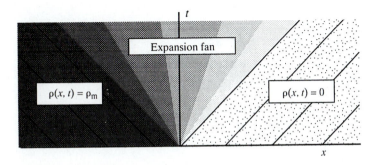

Figure 6.22 Construction of $\rho(x, t)$ in the xt plane using characteristic lines.

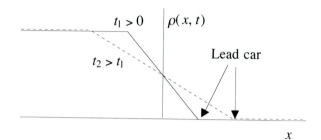

Figure 6.23 Traffic density, $\rho(x, t)$, as a function of position for various times.

Computer Simulation

Now that we have an analytic solution for a simple traffic problem, let's see how well our numerical methods can do. Before writing a program for traffic problems, let's review our three schemes for solving hyperbolic PDEs and how to implement them for the traffic problem. The equation of continuity is

$$\frac{\partial}{\partial t}\,\rho(x, t) = -\frac{\partial}{\partial x}\,F(\rho) \tag{6-72}$$

where the flow is $F(\rho) = \rho(x, t)v(\rho(x, t))$ and the velocity, $v(\rho)$, is given by equation (6-60). The FTCS scheme for solving this equation is

$$\rho_i^{n+1} = \rho_i^n - \frac{\tau}{2h}\,(F_{i+1}^n - F_{i-1}^n) \tag{6-73}$$

where $F_i^n \equiv F(\rho_i^n)$. The Lax scheme uses the equation

$$\rho_i^{n+1} = \tfrac{1}{2}(\rho_{i+1}^n + \rho_{i-1}^n) - \frac{\tau}{2h}\,(F_{i+1}^n - F_{i-1}^n) \tag{6-74}$$

Finally, the Lax–Wendroff scheme uses

$$\rho_i^{n+1} = \rho_i^n - \frac{\tau}{2h}\,(F_{i+1}^n - F_{i-1}^n)$$
$$+ \frac{\tau^2}{2}\frac{1}{h}\left(c_{i+1/2}\,\frac{F_{i+1}^n - F_i^n}{h} - c_{i-1/2}\,\frac{F_i^n - F_{i-1}^n}{h}\right) \tag{6-75}$$

where

$$c_{i\pm1/2} \equiv c(\rho_{i\pm1/2}^n); \qquad \rho_{i\pm1/2}^n \equiv \frac{\rho_{i\pm1}^n + \rho_i^n}{2} \tag{6-76}$$

Notice how the last term is built: We would like to be able to evaluate the function $c(\rho)$ at values between grid points, that is, at $i + \tfrac{1}{2}$ and $i - \tfrac{1}{2}$. Since we only know the value of ρ at grid points, we estimate its value between grid points by using a simple average. We use this estimated value for $\rho_{i\pm1/2}$ to evaluate $c_{i\pm1/2}$.

The program called `traffic` that uses the FTCS scheme is given in Listing 6.3. Notice that the periodic boundary conditions are implemented in a slightly different fashion than before. On lines 29 and 30, the variables ip (for i *plus* 1) and im (for i *minus* 1) are defined; these point to the right and left neighbors. If i is an interior point, then ip(i) = i + 1 and im(i) = i − 1. However, for the boundary points, ip(1) = 2, im(1) = N, and ip(N) = 1, im(N) = N − 1.

LISTING 6.3 Program `traffic`. Solves the equation of continuity for traffic flow.

```
1    % traffic - Program to solve the generalized Burger
2    % equation for the traffic at a stop light problem
3    clear;  help traffic; % Clear memory and print header
4    N = input('Enter the number of grid points - ');
5    L = 400;     % System size (meters)
6    h = L/N;     % Grid spacing for periodic boundary conditions
7    v_max = 25;    % Maximum car speed (m/s)
8    fprintf('Suggested timestep is %g\n',h/v_max);
9    tau = input('Enter time step (tau) - ');
10   coeff = tau/(2*h);    % Coefficient used by the schemes
11   %%%% Initial condition is a square pulse  %%%%
12   %%%% from x = -L/4 to x = 0               %%%%
13   rho_max = 1;                 % Maximum density
14   Flow_max = 1/4*rho_max*v_max;  % Maximum Flow
15   rho = zeros(1,N);
16   for i=N/4:(N/2-1)
17     rho(i) = rho_max;    % Max density in the square pulse
18   end
19   rho(N/2) = rho_max/2;  % Try running without this line
20   %%%% Set loop and plot variables %%%%
21   iplot = 1;
22   xplot = ((1:N)-1/2)*h - L/2;  % Record x scale for plot
23   rplot(:,1) = rho(:);          % Record the initial state
24   tplot(1) = 0;
25   nstep = ceil((L/4)/(v_max*tau)); % Stop after last car moves
26   %nstep = 5*nstep;  % Go further for last plots %*******%
27   fprintf('Number of iterations = %gn',nstep);
28   %%%% MAIN LOOP %%%%
29   ip(1:N) = (1:N)+1;  ip(N) = 1;   % ip = i+1 with periodic
                                      %      boundary conditions
30   im(1:N) = (1:N)-1;  im(1) = N;   % im = i-1 with periodic
                                      %      boundary conditions
31   for istep=1:nstep
32     Flow = rho .* (v_max*(1 - rho/rho_max));
33     %%% FTCS method %%%
34     rho_new(1:N) = rho(1:N) - coeff*(Flow(ip)-Flow(im));
35     %%% ----------- %%%
```

```
36    rho = rho_new;
37    iplot = iplot+1;
38    rplot(:,iplot) = rho(:);     % Record density for plot
39    tplot(iplot) = tau*istep;   % Record time for plot
40    plot(xplot,rho,'-',xplot,Flow/Flow_max,'--');
41    title('Density (solid) and flow (dashed) versus position');
42  end
43  mesh(rplot,[-90 30])    % Wire-mesh plot of density
44  xlabel('Position'); ylabel('Time');
45  title('Density versus x and t');
46  pause;       % Pause between plots; strike any key to continue
47  % Use rot90 function to graph t vs x since
48  % contour(rplot) graphs x vs t.
49  levels = 0:.1:1;      % Contour levels in plot
50  cs = contour(rot90(rplot),levels,xplot,tplot);
51  xlabel('Position');   ylabel('Time');
52  clabel(cs,0:0.1:1);        % Put labels on contour levels
53  title('Density contours');
```

To use the Lax method, we replace lines 33–34 with

```
%%% Lax method %%%
rho_new(1:N) = .5*(rho(ip)+rho(im)) ...
               - coeff*(Flow(ip)-Flow(im));
```

For the Lax–Wendroff method, we add the line

```
coefflw = tau^2/(2*h^2);   % Coefficient used by Lax-Wendroff
```

outside the main loop and replace lines 33–34 with

```
%%% Lax-Wendroff method %%%
cp = v_max*(1 - (rho(ip)+rho(1:N))/rho_max);
cm = v_max*(1 - (rho(1:N)+rho(im))/rho_max);
rho_new(1:N) = rho(1:N) - coeff*(Flow(ip)-Flow(im)) ...
               + coefflw*(cp.*(Flow(ip)-Flow(1:N)) ...
               - cm.*(Flow(1:N)-Flow(im)));
```

As an initial condition we take a square pulse of the form

$$\rho(x, t = 0) = \rho_0(x) = \begin{cases} \rho_m & -L/4 < x < 0 \\ 0 & \text{otherwise} \end{cases} \tag{6-77}$$

This initial value problem is similar to the stoplight problem considered above [see equation (6-69)] except that the line of cars is of finite length. From the solution to the stoplight problem, equation (6-71), we expect the right side of the pulse to expand with the density varying linearly from ρ_m to zero.

The left edge of the pulse should not move until the density there drops below ρ_m. The last car only begins moving when the traffic is no longer bumper-to-bumper. Figure 6.24 shows the characteristic lines for this problem; the discontinuity at $-L/4$ is a *shock front*. Even if the initial condition is smoothed, this shock will develop. The characteristic line solution is valid as long as we terminate the characteristic lines at the shock.

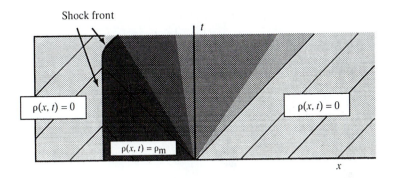

Figure 6.24 Characteristic lines for the finite pulse [see equation (6-77)].

When the last car begins to move, the shock front also moves. Using the condition that the flux, $F(\rho)$, is a continuous function we may compute the motion of the shock. Our formulation using characteristic lines may then be extended to complete the solution. Unfortunately, the complete treatment of shock fronts is beyond the scope of this book.

Numerical Results

Running the `traffic` program using the FTCS method we obtain the results shown in Figure 6.25. Notice that while the FTCS method appears stable in this case, the solution is not at all satisfactory. The right edge of the pulse is curved, yet it should expand as a straight line (see Figure 6.23).

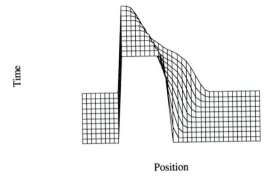

Figure 6.25 Density versus position and time. Results are from the `traffic` program using the FTCS method with 40 grid points and a time step of 0.4.

Using the `traffic` program with the Lax method we get the results shown in Figures 6.26 and 6.27. While the right edge is straighter, its slope is too large. Also the left edge is not maintained constant.

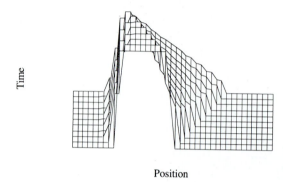

Position

Figure 6.26 Density versus position and time. Results are from the `traffic` program using the Lax method with 40 grid points and a time step of 0.4.

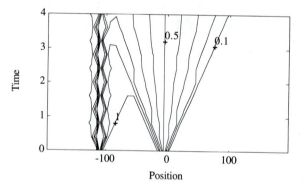

Position

Figure 6.27 Density versus position and time. Results are from the `traffic` program using the Lax method with 40 grid points and a time step of 0.4.

The Lax–Wendroff scheme does an excellent job, as shown in Figures 6.28 and 6.29. The latter is a contour plot of the density in the *xt* plane; compare this result with the characteristic lines shown in Figure 6.22.

If we increase the value of `nstep` in `traffic` we can observe the evolution of the pulse after the last car starts to move (see Figures 6.30 and 6.31). The shock

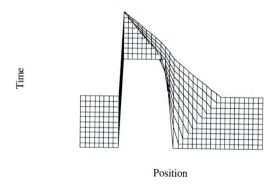

Position

Figure 6.28 Density versus position for various times. Results are from the `traffic` program using the Lax–Wendroff method with 40 grid points and a time step of 0.4.

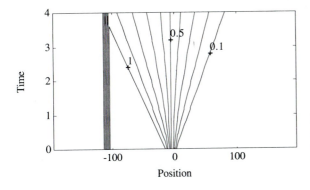

Figure 6.29 Density contours in the *xt* plane. Results are from the `traffic` program using the Lax–Wendroff method with 40 grid points and a time step of 0.4.

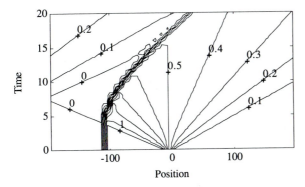

Figure 6.30 Density contours in the *xt* plane as obtained using the Lax–Wendroff method. Parameters are the same as in Figure 6.29 except that the simulation is run five times longer (`nstep` is five times larger).

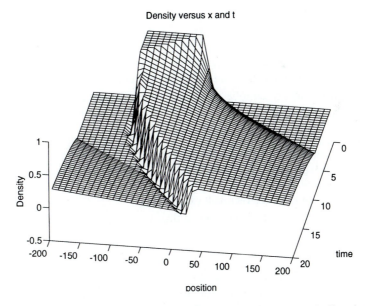

Figure 6.31 Density versus *x* and *t* as obtained using the Lax–Wendroff method. Note: This mesh plot has time moving toward the foreground. The contour plot for this run is shown in Figure 6.30. This plot was produced by MATLAB Version 4.0.

at the left edge of the pulse moves and its strength begins to decrease. Eventually the density becomes uniform everywhere.

Shock waves in real traffic are very dangerous. Drivers have finite reaction times, so sudden changes in traffic density can cause accidents. In our traffic model, the local density determines the traffic velocity. Fortunately, under normal visibility conditions, drivers adjust their speed by judging the global traffic conditions (i.e., they look at more than just the car in front of them). This fact introduces a diffusion term into the model that smooths the discontinuous shock fronts.

EXERCISES

14. The flow of traffic is $F(x, t) = \rho(x, t) \, v(\rho)$. For the stoplight problem, obtain an expression for $F(x, t)$ using the solution (6-71) and the definition for $v(\rho)$, equation (6-60). Sketch $F(x, t)$ versus x for $t > 0$ and show that it is maximum at $x = 0$ (i.e., at the light).

15. Call $x_c(t)$ the position of a given car; then

$$\frac{dx_c}{dt} = v(\rho(x_c(t), t))$$

(a) Using the solution to the stoplight problem, equation (6-71), show that

$$x_c(t) = \begin{cases} x_c(0) & t < -x_c(0)/v_m \\ v_m t - 2\sqrt{-x_c(0)v_m t} & t > -x_c(0)/v_m \end{cases}$$

(b) Plot the trajectories $x_c(t)$ in the xt plane. Typically, how much time does it take the nth car to reach the intersection? (Assume that cars are 4 m long and $v_m = 25$ m/s.)

16. After a time t, the total amount of traffic that has passed through the light is $N(t) = \int_0^\infty \rho(x, t) \, dx$. Show that $N(t) = \int_0^t F(x = 0, t) \, dt$ where $F(x, t) = \rho(x, t)v(x, t)$ is the flow.

17. Modify the `traffic` program so that it uses a Gaussian pulse of width $L/4$ as an initial distribution for the density. Center the pulse at $x = 0$ with $\rho(0, 0) = \rho_m$. Show how the density evolves with time. Explain why one side of the pulse expands while the other contracts [remember that the wave speed, $c(\rho)$, is not a constant]. Describe what a driver on this racetrack will experience.

18. Modify the `traffic` program so that it uses the initial condition

$$\rho(x, t = 0) = \frac{\rho_m}{2} [1 + \cos(4\pi x/L)]$$

(a) Plot the density versus position for a variety of times and show that the cosine wave turns into a sawtooth wave. In nonlinear acoustics this is referred to as an N-wave. If you have ever been to a very loud rock concert, you may have heard one of these.

(b) Modify your program to compute the spatial power spectrum of the density (see Section 5.2). Remove the zero wave number component. Initially, the spectrum will contain a single peak, but in time other peaks appear. Use a wire-mesh plot to graph the spectrum versus wave number and time.

19. Suppose that we have a uniform density of traffic with a small congested area. Modify the `traffic` program so that it uses the initial condition

$$\rho(x, t = 0) = \rho_0[1 + \alpha \exp(-x^2/2\sigma^2)]$$

where $\alpha = \frac{1}{5}$, $\sigma = L/8$, and ρ_0 are constants.
(a) Show that for light traffic (e.g., $\rho_0 = \rho_m/4$) the perturbation moves forward. What is its speed?
(b) Show that for heavy traffic (e.g., $\rho_0 = 3\rho_m/4$) the perturbation moves backwards. Interpret this result physically.
(c) Show that for $\rho_0 = \rho_m/2$ the perturbation is almost stationary; it drifts and distorts slightly.

20. Call $x_s(t)$ the position of the shock wave (see Figures 6.30 and 31). Numerically show that the velocity of the shock is given by

$$\frac{dx_s}{dt} = \frac{F(\rho_+) - F(\rho_-)}{\rho_+ - \rho_-}$$

where $F(x, t) = \rho(x, t)v(x, t)$ is the flow and $\rho_\pm \equiv \rho(x_s \pm h)$, that is, the density on each side of the shock front.

BEYOND THIS CHAPTER

The most complete reference on the analytical treatment of PDEs is still Courant and Hilbert.[8] On the other hand, the standard mathematical physics texts also present the important material in a more easily digestible format.[9] The physics behind most of the PDEs we'll consider is discussed in Morse and Feshbach.[10] For a presentation of the hydrodynamic equations suitable for a physicist, see Tritton.[11]

The FTCS method introduced in Section 6.2 is just one scheme for solving the diffusion equation. Fletcher[12] catalogues and compares many schemes for solving both the one-dimensional and the multidimensional diffusion equation. Numerical instability is the FTCS method's Achilles heel, but in Chapter 8 we introduce implicit methods that are unconditionally stable.

[8] R. Courant and D. Hilbert, *Methods of Mathematical Physics*, vol. II (New York: Interscience, 1962).

[9] For example, J. Mathews and R. Walker, *Mathematical Methods of Physics* (Menlo Park, Calif.: W. A. Benjamin, 1970); G. Arfken, *Mathematical Methods for Physicists* (New York: Academic Press, 1970).

[10] P. Morse and H. Feshback, *Methods of Theoretical Physics*, vol. 1 (New York: McGraw-Hill, 1953), chapter 2.

[11] D. J. Tritton, *Physical Fluid Dynamics*, 2d ed. (Oxford: Clarendon Press, 1988), chapter 4.

[12] C. A. J. Fletcher, *Computational Techniques for Fluid Dynamics*, vol. I (Berlin: Springer-Verlag, 1988), chapters 7 and 8.

In Section 6.4 the method of characteristics is used to obtain an analytical solution to the generalized Burger's equation. The method of characteristics may also be implemented as a numerical scheme for solving hyperbolic equations. For the wave equation we have two sets of characteristic lines (left- and right-moving waves). For more complicated problems (e.g., Euler equations in fluid mechanics) these characteristic lines are computed numerically as trajectories of a nonlinear ODE.[13]

One of the principal difficulties with numerically solving hyperbolic equations is the formation of shocks. At a shock the solution is discontinuous and our PDE description breaks down. One way to treat the problem is to use an uneven grid and concentrate grid points at the location of the shock. Shock-capturing methods automatically adjust the grid spacing to accomplish this. See Anderson et al.[14] and Fletcher[15] for an extensive discussion of finite difference methods for solving hyperbolic equations.

APPENDIX 6A: FORTRAN LISTINGS

LISTING 6A.1   Program `dftcs`. Solves the diffusion equation using the FTCS scheme.

```
      program dftcs
! Program to solve the diffusion equation using the FTCS scheme
      parameter( Nmax = 100 , Pmax = 100 )
      real tt(Nmax),tt_new(Nmax),ttplot(Nmax,Pmax)
      real L,kappa,tplot(Pmax),xplot(Pmax)
      integer plot_step

      write (*,*) 'Enter time step, number of grid points'
      read (*,*) tau,N
      L = 1.            ! System extends from x=-L/2 to x=L/2
      h = L/(N-1)       ! Grid spacing
      kappa = 1.        ! Diffusion coefficient

      coeff = kappa*tau/h**2  ! Coefficient used by FTCS
      if( coeff .lt. 0.5 ) then
        write (*,*) 'Solution is expected to be stable'
      else
        write (*,*) 'Solution is expected to be unstable'
      end if
```

[13] M. Holt, *Numerical Methods in Fluid Dynamics* (Berlin: Springer-Verlag, 1977), chapter 4.

[14] D. Anderson, J. Tannehill, and R. Pletcher, *Computational Fluid Mechanics and Heat Transfer* (New York: Hemisphere, 1984).

[15] C. A. J. Fletcher, *Computational Techniques for Fluid Dynamics*, vol. I (Berlin: Springer-Verlag, 1988), chapters 9 and 10.

```fortran
      do i=1,N
        tt(i) = 0.          ! Initial temperature
        tt_new(i) = 0.      ! New value of temperature (used later)
      end do
      tt(N/2) = 1./h        ! Discretized delta function at x=0
      do i=1,N
        xplot(i) = (i-0.5)*h - L/2.  ! Record x for plotting
      end do
      iplot = 1             ! Plot counter
      nstep = 250           ! Total number of time steps
      nplots = 25           ! Total number of plots
      plot_step = nstep/nplots  ! Number of steps between plots

!!!!! MAIN LOOP !!!!!
      do istep=1,nstep
        do i=2,N-1              !!!!! FTCS scheme !!!!!
          tt_new(i)=tt(i)+coeff*(tt(i+1)+tt(i-1)-2*tt(i))
        end do
        do i=1,N
          tt(i) = tt_new(i)   ! Replace old values with new
        end do
        if ( mod(istep,plot_step) .lt. 1 ) then
          do i=1,N
            ttplot(i,iplot) = tt(i)   ! Record temperature for
                                      ! plots
          end do
          tplot(iplot) = istep*tau ! Record time for plots
          iplot = iplot+1
          write (*,*) istep,' out of ',nstep,' steps completed'
        end if
      end do
! Print out the plotting variables -
!    tplot,xplot,ttplot
!
      open(11,file='tplot.dat')
      open(12,file='xplot.dat')
      open(13,file='ttplot.dat')
      do j=1,iplot-1
        write (11,1001) tplot(j)
      end do
      do i=1,N
        write (12,1001) xplot(i)
        do j=1,iplot-1
          write (13,1002) ttplot(i,j)
        end do
        write(13,1003)
      end do
      stop
```

```
1001 format(e14.6)
1002 format(e12.6,' ',$)   ! The $ suppresses the carriage return
1003 format(/)             ! New line
     end
```

LISTING 6A.2 Program `aftcs`. Solves the advection equation using the FTCS scheme.

```
      program aftcs
! Program to solve the advection equation using the FTCS scheme
      parameter( Nmax = 100, Pmax = 100 )
      real a(Nmax),a_new(Nmax),x(Nmax)
      real tplot(Pmax),aplot(Nmax,Pmax)
      real L,k_wave
      integer plot_step

      pi = 4*atan(1.)
      write (*,*) 'Enter time step, number of grid points'
      read (*,*) tau, N
      L = 1.          ! System size
      h = L/N         ! Grid spacing
      c = 1.          ! Wave speed
      coeff = -c*tau/(2.*h)   ! Coefficient used by FTCS scheme
!!!!! Initial condition is a Gaussian-cosine pulse !!!!!
      sigma = 0.1                ! Width of Gaussian pulse
      k_wave = pi/sigma          ! Wavenumber of the cosine wave
      do i=1,N
        x(i) = (i-1)*h - L/2.   ! Record x for plots
        a(i) = cos(k_wave*x(i)) * exp(-x(i)**2/(2*sigma**2))
      end do

!!!!! Set up plot variables !!!!!
      iplot = 1                  ! Plot counter
      do i=1,N
        aplot(i,1) = a(i)        ! Record initial wave amplitude
      end do
      tplot(1) = 0.              ! Record initial time

!!!!! MAIN LOOP !!!!!
      nstep = int(L/(c*tau))+1       ! Run until wave circles
                                     ! system
      plot_step = int(nstep/50)+1
      do istep=1,nstep
        do i=2,N-1
          a_new(i) = a(i) + coeff*(a(i+1) - a(i-1))   ! FTCS
        end do
        a_new(1) = a(1) + coeff*(a(2)-a(N))        ! Periodic
```

```
        a_new(N) = a(N) + coeff*(a(1)-a(N-1))   ! conditions
        do i=1,N
          a(i) = a_new(i)     ! Replace old amplitude with new
        end do

        if ( mod(istep,plot_step) .lt. 1 ) then
          iplot = iplot+1
          do i=1,N
            aplot(i,iplot) = a(i)   ! Record amplitude for plot
          end do
          tplot(iplot) = tau*istep  ! Record time for plot
          write (*,*) istep,' out of ',nstep,' steps completed'
        end if
      end do

! Print out the plotting variables -
!    tplot,xplot,aplot
!
      open(11,file='x.dat')
      open(12,file='tplot.dat')
      open(13,file='aplot.dat')
      do j=1,iplot
        write (12,1001) tplot(j)
      end do
      do i=1,N
       write (11,1001) x(i)
       do j=1,iplot
        write (13,1002) aplot(i,j)
       end do
       write (13,1003)
      end do
      stop
1001  format(e12.6)
1002  format(e12.6,' ',$)    ! The $ suppresses the carriage
                             ! return
1003  format(/)              ! New line
      end
```

LISTING 6A.3 Program traffic. Solves the equation of continuity for traffic flow.

```
      program traffic
! Program to solve the generalized Burger equation for the
! stop light problem
      parameter( Nmax = 100, Pmax = 100 )
      real rho(Nmax),rho_new(Nmax),Flow(Nmax)
      real xplot(Nmax),tplot(Pmax),rplot(Nmax,Pmax)
      real L
```

```
write (*,*) 'Enter the number of grid points'
read (*,*) N
L = 400.          ! System size (meters)
h = L/N           ! Grid spacing for periodic boundary cond.
v_max = 25.       ! Maximum car speed (m/s)
write (*,*) 'Suggested time step is ',h/v_max
write (*,*) 'Enter time step (tau)'
read (*,*) tau
coeff = tau/(2.*h)
rho_max = 1.                         ! Maximum density
Flow_max = 1/4*rho_max*v_max    ! Maximum flow

!!!!! Initial condition is a square pulse !!!!!
!!!!! from x = -L/4 to x = 0                   !!!!!
do i=1,N
   if( i .ge. N/4 .and. i .le. N/2-1 ) then
     rho(i) = rho_max
   else
     rho(i) = 0
   end if
end do
rho(N/2) = rho_max/2.    ! Try running without this line
iplot = 1
do i=1,N
   xplot(i) = (i-0.5)*h - L/2.
   rplot(i,1) = rho(i)
end do
tplot(1) = 0
! Program stops when last car starts to move
nstep = int((L/4)/(v_max*tau)) + 1
write (*,*) 'Number of iterations = ',nstep

!!!!! MAIN LOOP !!!!!
do istep=1,nstep

   do i=1,N
     Flow(i) = rho(i) * v_max*(1. - rho(i)/rho_max)
   end do
!!!!! FTCS scheme !!!!!
   do i=1,N
     ip = mod(i+N,N)+1      ! i+1 with periodic boundary
     im = mod(i+N-2,N)+1    ! i-1 with periodic boundary
     rho_new(i) = rho(i) - coeff*(Flow(ip) - Flow(im))
   end do
   do i=1,N
     rho(i) = rho_new(i)    ! Reset values of density
   end do
   iplot=iplot+1
```

```fortran
      do i=1,N
        rplot(i,iplot) = rho(i)   ! Record density for plots
      end do
      tplot(iplot) = tau*istep    ! Record time for plots
    end do
! Print out the plotting variables -
!   tplot,xplot,rplot
!
    open(11,file='xplot.dat')
    open(12,file='tplot.dat')
    open(13,file='rplot.dat')
    do i=1,N
      write (11,1001) xplot(i)
    end do
    do j=1,iplot
      write (12,1001) tplot(j)
    end do
    do i=1,N
     do j=1,iplot
      write (13,1002) rplot(i,j)
     end do
     write (13,1003)
    end do
    stop
1001  format(e12.6)
1002  format(e12.6,' ',$)     ! The $ suppresses the carriage
                              ! return
1003  format(/)              ! New line
    end
```

Chapter 7
Partial Differential Equations II:
Relaxation and Spectral Methods

Chapter 6 covered methods for solving parabolic and hyperbolic equations. Now we consider the third type of partial differential equation, elliptic equations. For this kind of PDE we have to solve a boundary value problem and the solution is a static field, such as the electric potential described by Laplace's equation. Despite the dissimilarities, we find that numerical relaxation algorithms link these problems. With spectral methods we explore a completely different way of formulating the numerical solution, using a set of basis functions instead of a spatial grid.

7.1 RELAXATION METHODS

Separation of Variables

Our paradigm for an elliptic PDE is Laplace's equation. In two dimensions it may be written as

$$\frac{\partial^2 \Phi(x, y)}{\partial x^2} + \frac{\partial^2 \Phi(x, y)}{\partial y^2} = 0 \tag{7-1}$$

where Φ is the electrostatic potential. Before going into the numerical methods, let's solve this equation analytically for a simple problem. Take a rectangular geometry with boundaries at $x = 0$, $x = L_x$, $y = 0$, and $y = L_y$. In this case, we can solve Laplace's equation by *separation of variables*.[1]

Our first step is to write $\Phi(x, y)$ as the product

$$\Phi(x, y) = X(x)Y(y) \tag{7-2}$$

Inserting this substitution into Laplace's equation and dividing by Φ, we have

$$\frac{1}{X(x)}\frac{d^2X}{dx^2} + \frac{1}{Y(y)}\frac{d^2Y}{dy^2} = 0 \tag{7-3}$$

Since this equation must hold for all x and y, each term must equal a constant; we write separate equations for X and Y,

$$\frac{1}{X(x)}\frac{d^2X}{dx^2} = -k^2; \qquad \frac{1}{Y(y)}\frac{d^2Y}{dy^2} = k^2 \tag{7-4}$$

where k is a complex constant. These equations are ordinary differential equations since X (or Y) only depends on x (or y).

Two notes are needed for those of you who are less familiar with separation of variables. First, we write the constant as k^2 because it simplifies the notation. Second, that one equation has k^2 and the other $-k^2$ does not matter since k is complex. There is no violation of the original symmetry in Laplace's equation.

The solutions of these ODEs for X and Y are well known:

$$\begin{aligned} X(x) &= C_s \sin(kx) \;\; + C_c \cos(kx) \\ Y(y) &= C_s' \sinh(ky) + C_c' \cosh(ky) \end{aligned} \tag{7-5}$$

where C_s, C_c, C_s', and C_c' are constants. Again, these solutions are the same since k may be complex. In a few lines you will see why I chose to write the solution in this form.

To proceed further, we need to specify boundary conditions. We'll use

$$\Phi(x = 0, y) \;\; = \;\; \Phi(x = L_x, y) = \Phi(x, y = 0) = 0 \tag{7-6a}$$

$$\Phi(x, y = L_y) = \Phi_0 \tag{7-6b}$$

where Φ_0 is a constant. The potential is zero on three of the sides of the rectangle and Φ_0 on the fourth. The boundary condition at $x = 0$ is satisfied if $C_c = 0$; the $y = 0$ boundary condition is satisfied if $C_c' = 0$. The boundary condition at $x = L_x$ is satisfied if $k = n\pi/L_x$ where n is an integer.

---

[1] J. D. Jackson, *Classical Electrodynamics*, 2d ed. (New York: Wiley, 1975), section 2.9.

Using the superposition principle, our general solution takes the form

$$\Phi(x, y) = \sum_{n=1}^{\infty} c_n \sin\left(\frac{n\pi x}{L_x}\right)\sinh\left(\frac{n\pi y}{L_x}\right) \tag{7-7}$$

To find the values of the c_n coefficients we impose the fourth boundary condition, $\Phi(x, y = L_y) = \Phi_o$; hence

$$\Phi_o = \sum_{n=1}^{\infty} c_n \sin\left(\frac{n\pi x}{L_x}\right)\sinh\left(\frac{n\pi L_y}{L_x}\right) \tag{7-8}$$

Multiply both sides by $\sin(m\pi x/L_x)$ and integrate from $x = 0$ to $x = L_x$,

$$\int_0^{L_x} dx \, \Phi_o \sin\left(\frac{m\pi x}{L_x}\right) = \sum_{n=1}^{\infty} c_n \sinh\left(\frac{n\pi L_y}{L_x}\right)\int_0^{L_x} dx \, \sin\left(\frac{m\pi x}{L_x}\right)\sin\left(\frac{n\pi x}{L_x}\right) \tag{7-9}$$

The integral on the left-hand side is easily solved,

$$\int_0^{L_x} dx \, \sin\left(\frac{m\pi x}{L_x}\right) = \begin{cases} \dfrac{2L_x}{\pi m} & m \text{ odd} \\[2mm] 0 & m \text{ even} \end{cases} \tag{7-10}$$

The sum on the right-hand side may be simplified since that integral is

$$\int_0^{L_x} dx \, \sin\left(\frac{m\pi x}{L_x}\right)\sin\left(\frac{n\pi x}{L_x}\right) = \frac{L_x}{2}\,\delta_{n,m} \tag{7-11}$$

which collapses the sum to a single term.
    We then have

$$\Phi_o \frac{2L_x}{\pi m} = c_m \sinh\left(\frac{m\pi L_y}{L_x}\right)\frac{L_x}{2} \qquad (m \text{ odd}) \tag{7-12}$$

or

$$c_m = \frac{4\Phi_o}{\pi m \sinh(m\pi L_y/L_x)} \qquad (m \text{ odd}) \tag{7-13}$$

Our final expression for the solution is the infinite sum

$$\Phi(x, y) = \Phi_o \sum_{n=1,3,5,\ldots}^{\infty} \frac{4}{\pi n} \sin\left(\frac{n\pi x}{L_x}\right)\frac{\sinh(n\pi y/L_x)}{\sinh(n\pi L_y/L_x)} \tag{7-14}$$

When we graph this solution we discover that a large number of terms is needed to represent the solution near the $y = L_y$ boundary accurately (see Figure 7.1). Near this boundary one observes *Gibbs' phenomenon*, which commonly occurs when a Fourier series is used to represent a discontinuous function.[2]

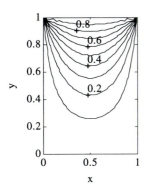 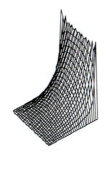

Figure 7.1 Contour and mesh plots of the potential versus x and y as given by the separation of variables solution, equation (7-14), using terms up through $n = 21$.

Jacobi Method

We now develop a numerical method for solving Laplace's equation. To begin, let's go back for a moment to the diffusion equation. Consider the Fourier equation for the two-dimensional diffusion of temperature,

$$\frac{\partial T(x, y, t)}{\partial t} = \kappa \left(\frac{\partial^2 T}{\partial x^2} + \frac{\partial^2 T}{\partial y^2} \right) \tag{7-15}$$

where κ is the thermal diffusion coefficient. We know from physical intuition that given any initial temperature profile plus stationary boundary conditions, the solution will relax to some steady state, call it $T_s(x, y)$. In other words,

$$\lim_{t \to \infty} T(x, y, t) = T_s(x, y) \tag{7-16}$$

When the temperature profile is at the steady state, then it does not change in time, that is, $\partial T_s / \partial t = 0$. This means that the steady state obeys the equation

$$\frac{\partial^2 T_s}{\partial x^2} + \frac{\partial^2 T_s}{\partial y^2} = 0 \tag{7-17}$$

Does this look familiar? Of course, this is just Laplace's equation.

---

[2] J. Mathews and R. L. Walker, *Mathematical Methods of Physics* (Menlo Park, Calif.: W. A. Benjamin, 1970), chapter 4.

The idea is that the solution of Laplace's equation is just the solution of the diffusion equation in the limit $t \to \infty$. Algorithms based on this physical principal are called *relaxation methods*. We already know how to solve the diffusion equation using the FTCS scheme. We start from the two-dimensional diffusion equation

$$\frac{\partial \Phi(x, y, t)}{\partial t} = \mu \left(\frac{\partial^2 \Phi}{\partial x^2} + \frac{\partial^2 \Phi}{\partial y^2} \right) \tag{7-18}$$

The value of the constant μ is unimportant since it drops out later. Using the FTCS scheme in two dimensions,

$$\Phi_{i,j}^{n+1} = \Phi_{i,j}^n + \frac{\mu \tau}{h_x^2} \{ \Phi_{i+1,j}^n + \Phi_{i-1,j}^n - 2\Phi_{i,j}^n \}$$

$$+ \frac{\mu \tau}{h_y^2} \{ \Phi_{i,j+1}^n + \Phi_{i,j-1}^n - 2\Phi_{i,j}^n \} \tag{7-19}$$

where $\Phi_{i,j}^n \equiv \Phi(x_i, y_j, t_n)$, $x_i \equiv (i - 1)h_x$, $y_j \equiv (j - 1)h_y$, and $t_n = (n - 1)\tau$. Remember that we are solving an electrostatics problem so the potential doesn't actually depend on time. We introduce an artificial time-dependence only to assist in the construction of the algorithm. A better way to interpret $\Phi_{i,j}^n$ is to call it the nth guess for the potential with (7-19) serving as a formula for improving this guess.

In Section 6.2 we saw that the FTCS scheme can be numerically unstable. In one-dimensional systems the scheme is stable if $\mu \tau / h^2 \le \frac{1}{2}$; for two-dimensional systems the scheme is stable if

$$\frac{\mu \tau}{h_x^2} + \frac{\mu \tau}{h_y^2} \le \frac{1}{2} \tag{7-20}$$

(see Exercise 8.5). To simplify the analysis, we take $h_x = h_y = h$ and use a square geometry. In that case, the condition for stability is $\mu \tau / h^2 \le \frac{1}{4}$.

Since we are only interested in the steady state solution ($n \to \infty$) we want to use the largest possible time step. Setting $\mu \tau / h^2 = \frac{1}{4}$, (7-19) becomes

$$\Phi_{i,j}^{n+1} = \tfrac{1}{4} \{ \Phi_{i+1,j}^n + \Phi_{i-1,j}^n + \Phi_{i,j+1}^n + \Phi_{i,j-1}^n \} \tag{7-21}$$

Notice that the diffusion constant, μ, has dropped out and that the $\Phi_{i,j}^n$ terms cancel out on the right-hand side. This equation has a good parentage; our first example of a relaxation scheme is called the *Jacobi method*. It is easy to see that the method involves replacing the value of the potential at a point with the average value of the four nearest neighbors. This result may be thought of as a discrete version of the mean-value theorem for electrostatic potential. Of course, equation (7-21) is only used for interior points and not when (i, j) is a boundary point.

Jacobi Method Program

A MATLAB program, called jacobi, that solves Laplace's equation using Jacobi's method is given in Listing 7.1. A few notes about the program: The MATLAB flops (floating point operations) function is used to measure the computational effort. Efficiency will be important since computation time increases quickly with the number of grid points. Jacobi's method is implemented on lines 22–32. Notice that the for loops on lines 24–25 are only over the interior points. The variable temp is used to see how much the solution changes with each iteration (line 29). When the average fractional change, as computed on line 35, is small enough, we halt the computation.

LISTING 7.1 Program jacobi. Solves Laplace's equation using the Jacobi method.

```
1   % jacobi - Program to solve the Laplace equation using
2   % Jacobi's method on a square grid
3   clear; help jacobi;  % Clear memory and print header
4   N = input('input number of grid points on a side - ');
5   L = 1;         % System size (length)
6   h = L/(N-1);   % Grid spacing
7   phi0 = 1;      % Potential at y=L
8   fprintf('Potential at y=L equals %g \n',phi0);
9   fprintf('Potential is zero on all other boundaries\n');
10  change_want = 1e-4;   % Stop when the change is given
                                   fraction
11  %%%% Set initial conditions and boundary conditions %%%%
12  coeff = pi/L;
13  x = (0:N-1)*h;     % x coordinate
14  y = (0:N-1)*h;     % y coordinate
15  % Initial guess is first term in separation of variables
16  phi = phi0 * 4/(pi*sinh(pi)) * sin(coeff*x')*sinh(coeff*y);
17  phi(:,N) = phi0*ones(N,1); % Set the boundary condition
18  %%%% MAIN LOOP %%%%
19  flops(0);  % Reset the flops counter to zero;
20  newphi = phi;
21  max_iter = N^2;  % Set max to avoid excessively long runs
22  for iter=1:max_iter
23    temp = 0;
24    for i=2:(N-1)       % Loop over interior points only
25     for j=2:(N-1)
26       %% Jacobi method %%
27       newphi(i,j) = .25*(phi(i+1,j)+phi(i-1,j)+ . . .
28                            phi(i,j-1)+phi(i,j+1));
```

```
29          temp = temp + abs(1-phi(i,j)/newphi(i,j));
30        end
31      end
32      phi = newphi;     % Reset phi
33      % Break out of the main loop if the fractional change
34      % in an iteration is less than change_want
35      change(iter) = temp/(N-2)^2;
36      fprintf('After %g iterations, fractional change = %g\n',. .
37                           iter,change(iter));
38      if( change(iter) < change_want )
39        disp('Desired accuracy achieved; breaking out of loop');
40        break;
41      end
42    end
43    fprintf('Number of flops = %g\n',flops);
44    subplot(121)
45      cs = contour(rot90(phi),9,x,y);   % Contour plot with labels
46      xlabel('x'); ylabel('y'); clabel(cs,[.2 .4 .6 .8])
47    subplot(122)
48      mesh(phi);     % Wire-mesh plot of potential
49    subplot(111)
50    title('Potential');
51    pause;  % Pause between plots; strike any key to continue
52    semilogy(change);
53    xlabel('iteration');   ylabel('fractional change');
```

The potential, as computed by the jacobi program, is illustrated in Figure 7.2 by a contour map and a wire-mesh figure. Notice that we have no Gibbs' phenomenon (compare with the separation of variables solution, Figure 7.1).

Relaxation algorithms require an initial guess to start the iteration process. The jacobi program uses the first term in the separation of variables solution

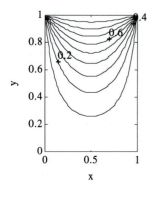

Figure 7.2 Contour and mesh plots of the potential from jacobi. Grid size is $N = 30$. Compare with the separation of variables solution, Figure 7.1.

(7-14). The efficiency of the algorithm is greatly influenced by the accuracy of this initial guess. The best way to appreciate this point is to run the program using a poor initial guess (e.g., $\Phi = 0$ in the interior).

Gauss–Seidel and Simultaneous Overrelaxation

A simple modification of the Jacobi method improves its rate of convergence. Suppose that we use the updated values of $\Phi_{i,j}$ as they become available. The iteration equation is then

$$\Phi_{i,j}^{n+1} = \tfrac{1}{4}\{\Phi_{i+1,j}^{n} + \Phi_{i-1,j}^{n+1} + \Phi_{i,j+1}^{n} + \Phi_{i,j-1}^{n+1}\} \qquad (7\text{-}22)$$

The idea is that the updated values of Φ at two of the nearest neighbors have already been computed, so why not use them. With this modification the method is called the *Gauss–Seidel* method. We may easily modify our jacobi program to implement Gauss–Seidel by replacing lines 26–29 with

```
%% G-S method %%
newphi = .25*(phi(i+1,j)+phi(i-1,j)+phi(i,j-1)+phi(i,j+1));
temp = temp + abs(1-phi(i,j)/newphi);
phi(i,j) = newphi;
```

Furthermore, we eliminate lines 20 and 32. Notice the savings in memory now that newphi is a scalar instead of an $N \times N$ matrix.

We can significantly improve our algorithm by overcorrecting the value of Φ at each iteration of the Gauss–Seidel method. This is achieved by using the iteration equation

$$\Phi_{i,j}^{n+1} = (1 - \omega)\Phi_{i,j}^{n} + \frac{\omega}{4}\{\Phi_{i+1,j}^{n} + \Phi_{i-1,j}^{n+1} + \Phi_{i,j+1}^{n} + \Phi_{i,j-1}^{n+1}\} \qquad (7\text{-}23)$$

where the constant ω is called the overrelaxation parameter. This method is called *simultaneous overrelaxation* (SOR). We may very easily modify the jacobi program to implement SOR by replacing lines 26–29 with,

```
%% SOR method %%
newphi = .25*omega*(phi(i+1,j)+phi(i-1,j)+ . . .
            phi(i,j-1)+phi(i,j+1))  +  (1-omega)*phi(i,j);
temp = temp + abs(1-phi(i,j)/newphi);
phi(i,j) = newphi;
```

and eliminating lines 20 and 32. The variable omega could be input by the user anywhere outside the main loop.

The trick to using SOR effectively is to select a good value for ω. Notice that for $\omega = 1$ we have Gauss–Seidel. For $\omega < 1$ we have underrelaxation and the convergence is slowed. For $\omega > 2$ the SOR method is *unstable*. There is an ideal value for ω between 1 and 2 that gives the best acceleration.

In some geometries this optimal value is known. For example, in an $N_x \times N_y$ rectangular grid,

$$\omega_{\text{opt}} = \frac{2}{1 + \sqrt{1 - r^2}} \tag{7-24}$$

where

$$r = \frac{1}{2}\left(\cos \frac{\pi}{N_x} + \cos \frac{\pi}{ny}\right) \tag{7-25}$$

If $N_x = N_y = N$ (square geometry), this simplifies to

$$\omega_{\text{opt}} = \frac{2}{1 + \sin(\pi/N)} \tag{7-26}$$

In real-life problems the optimal value for ω is obtained by empirical trial and error. Sophisticated programs will automatically adjust ω according to how well the solution is converging.

Poisson Equation

The methods developed so far are easy to generalize to solve the Poisson equation. In MKS units,

$$\frac{\partial^2 \Phi(x, y)}{\partial x^2} + \frac{\partial^2 \Phi(x, y)}{\partial y^2} = -\frac{1}{\varepsilon_o} \rho(x, y) \tag{7-27}$$

where $\rho(x, y)$ is the charge density and ε_o is the permittivity of free space. In discretized form, we have

$$\frac{1}{h_x^2}\{\Phi_{i+1,j} + \Phi_{i-1,j} - 2\Phi_{i,j}\} + \frac{1}{h_y^2}\{\Phi_{i,j+1} + \Phi_{i,j-1} - 2\Phi_{i,j}\} = -\frac{1}{\varepsilon_o} \rho_{i,j} \tag{7-28}$$

Using the analysis presented earlier, we construct the Jacobi relaxation scheme for the Poisson equation as

$$\Phi_{i,j}^{n+1} = \frac{1}{4}\left\{\Phi_{i+1,j}^n + \Phi_{i-1,j}^n + \Phi_{i,j+1}^n + \Phi_{i,j-1}^n + \frac{1}{\varepsilon_o} h^2 \rho_{i,j}\right\} \tag{7-29}$$

where to simplify the formulation we take $h_x = h_y = h$. The other two schemes considered in this section may also be generalized by the simple addition of the charge density term.

EXERCISES

1. Write a program to evaluate the potential $\Phi(x, y)$ numerically as given by equation (7-14), on a 30×30 grid. Take $\Phi_o = 1$ and graph your solution by a mesh or contour plot (see Figures 7.1 and 7.2). Produce a sequence of plots using an increasing number of terms. Estimate how many terms in the infinite sum are needed to obtain about 1% accuracy in the solution.

2. (a) Find the solution to the more general boundary value problem,

$$\begin{aligned} \Phi(x = 0, y) &= \Phi_1 & \Phi(x = L_x, y) &= \Phi_2 \\ \Phi(x, y = 0) &= \Phi_3 & \Phi(x, y = L_y) &= \Phi_4 \end{aligned}$$

where $\Phi_1, \ldots, \Phi_4$ are constants.

   (b) Write a program to graph your solution by a wire-mesh or contour plot. Plot the potential for $\Phi_1 = \Phi_3 = 1$, $\Phi_2 = \Phi_4 = 0$ and for $\Phi_1 = \Phi_2 = 1$, $\Phi_3 = \Phi_4 = 0$.

3. (a) Using separation of variables, find the solution to the three-dimensional cubic boundary value problem,[3]

$$\begin{aligned} \Phi(x = 0, y, z) &= \Phi(x = L, y, z) = 0 \\ \Phi(x, y = 0, z) &= \Phi(x, y = L, z) = 0 \\ \Phi(x, y, z = 0) &= \Phi(x, y, z = L) = \Phi_o \end{aligned}$$

   (b) Write a program to graph your solution for a given height z. Produce wire-mesh or contour plots of $\Phi(x, y, z)$ for $z = L/4$ and $L/2$.

4. A major issue with relaxation methods is their computational speed.
   (a) Try running the jacobi program for different-sized systems ($N_x = N_y = 10$ to 30). Graph the number of iterations performed versus system size. Fit the data to a power law and approximate the exponent.
   (b) Repeat part (a) using a bad initial guess. Set the potential initially to zero everywhere in the interior.
   (c) Modify the program to use SOR and repeat parts (a) and (b). Compare the Jacobi and SOR methods (use the optimum value for ω).

5. Formulate the Jacobi method without assuming that $h_x = h_y$. Modify the jacobi program to implement this modification. Keep $L_x = L_y$ and the boundary conditions (7-6) and try grids of 16×16, 32×8, and 8×32. Do you find any significant differences?

6. The jacobi program uses a good initial guess for the potential $\Phi(x, y)$. To illustrate its importance, run the program with a variety of initial guesses including some poor ones (e.g., $\Phi = 0$ in the interior). Also try an initial guess that uses the first few terms of the separation of variables solution (7-14). Compare and comment on your results.

---

[3] J. D. Jackson, *Classical Electrodynamics*, 2d ed. (New York: Wiley, 1975), problem 2.13.

7. Modify the jacobi program to plot the electric field, $\mathbf{E} = -\nabla\Phi$. You may want to use the gradient (which computes the gradient) and quiver (which produces a field plot such as in Figure 7.8) functions available in MATLAB. Try both proportional and equal length field "arrows."

8. Write a program that uses the SOR method to simulate a Faraday cage (Figure 7.3). Use a square geometry with $N_x = N_y = 30$. Set the left and right walls to $\Phi = 0$ and $\Phi = 100$, respectively. Fix the potential at the top and bottom walls but have it vary linearly across the system.

   (a) The Faraday cage is represented by the following eight points: $(i, j) = (10, 10)$, $(15, 10)$, $(20, 10)$, $(10, 15)$, $(10, 20)$, $(15, 20)$, $(20, 15)$, and $(20, 20)$. The potential at these points is fixed at zero. Plot the potential $\Phi_{i,15}$ versus i (i.e., a horizontal cross section through the center) both with and without the cage.

   (b) Try a cage that has only the four corner points $(10, 10)$, $(10, 20)$, $(20, 10)$, $(20, 20)$ and compare with the results from part (a).

   (c) Try a cage that has only the four side points $(10, 15)$, $(15, 10)$, $(20, 15)$, $(15, 20)$ and compare with the results from part (a).

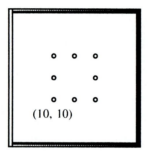

Figure 7.3   Faraday cage for Exercise 7.8.

9. Write a program that uses the SOR method to solve the electrostatics problems shown in Figure 7.4. For each box, the thin lines indicate a boundary where the potential is fixed at zero; a thick line indicates the potential is fixed at one.

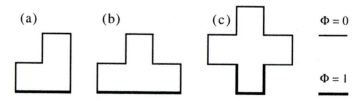

Figure 7.4   Geometries for Exercise 7.9.

10. (a) Write a program that solves the two-dimensional Poisson equation in a square geometry with the Dirichlet boundary conditions $\Phi = 0$ at the boundary. Map the potential for a single charge at the center of the system. Compare with the potential for a charge in free space. Remember that in two dimensions this charge is a line charge and not a point charge (see Figure 7.5).

   (b) Modify your program to use periodic boundary conditions. Compare with the results from part (a). (Hint: Think.)

11. (a) In two dimensions, the Laplace equation may be discretized as

$$\frac{1}{h_x^2}\{\Phi_{i+1,j} + \Phi_{i-1,j} - 2\Phi_{i,j}\} + \frac{1}{h_y^2}\{\Phi_{i,j+1} + \Phi_{i,j-1} - 2\Phi_{i,j}\} = 0$$

Take a very small system that contains only four interior points (2×2 interior grid). Show that the discretized Laplace equation can be reduced to a system of four simultaneous equations.

(b) Write a program that solves this problem using Gaussian elimination. This solution is important since it can be used as an initial guess in multigrid programs (see Beyond This Chapter).

7.2 SPECTRAL METHODS*

Fourier Galerkin Method

At the end of the Section 7.1, we saw that the relaxation methods could be used to solve the Poisson equation

$$\nabla^2\Phi(\mathbf{r}) = -\frac{1}{\varepsilon_o}\rho(\mathbf{r}) \tag{7-30}$$

where $\Phi(\mathbf{r})$ is the electrostatic potential at position $\mathbf{r}$, ρ is the charge density, and ε_o is the permittivity of free space. We now develop a very different approach for solving equation (7-30); for simplicity we'll work in a two-dimensional, square geometry with $0 \le x \le L$ and $0 \le y \le L$. The algorithms in this section are easily extended to rectangular geometries; in Chapter 9 we consider spherical and cylindrical geometries.

Relaxation methods discretize space and assemble a set of equations for $\Phi_{i,j}$. Let's construct a numerical scheme that represents the potential in a different way. In Section 7.1, using separation of variables, we constructed our analytical solution as an infinite sum of trigonometric functions [see equation (7-14)]. Suppose that we build our approximate numerical solution using a finite sum of functions,

$$\Phi(x, y) = a_1f_1(x, y) + a_2f_2(x, y) + \ldots + a_Kf_K(x, y) + T(x, y)$$

$$= \sum_{k=1}^{K} a_kf_k(x, y) + T(x, y) \tag{7-31}$$

$$= \Phi_a(x, y) + T(x, y)$$

where $\Phi_a(x, y)$ is our approximate solution and $T(x, y)$ is the error term. We'll demand that the trial functions be orthogonal,

$$\int_0^L dx \int_0^L dy \, f_k(x, y) f_{k'}(x, y) = A_k \delta_{k,k'} \tag{7-32}$$

This orthogonality condition is not absolutely necessary, but imposing it simplifies the formulation of the algorithm.

Inserting (7-31) into the Poisson equation gives

$$\nabla^2 \left(\sum_k a_k f_k(x, y) \right) + \frac{1}{\varepsilon_o} \rho(x, y) = R(x, y) \tag{7-33}$$

where $R(x, y) = -\nabla^2 T(x, y)$ is the residual. In general, for a partial differential equation of the form $D(\Phi) = 0$ where D is a linear differential operator, the residual is $R = D(\Phi_a)$. Separation of variables follows a similar procedure except that we have an infinite sum with no error term and thus zero residual.

Our next step is to obtain an expression for the coefficients, a_k. There are a variety of approaches depending on how we choose to minimize the error. The *Galerkin method* imposes the condition

$$\int_0^L dx \int_0^L dy \, f_k(x, y) R(x, y) = 0 \tag{7-34}$$

for all k. In other words, we select the coefficients such that the residual is orthogonal to all the trial functions.

Our choice of trial functions is usually motivated by the geometry and the boundary conditions. For a change of pace, let's solve the Poisson equation with the Neumann boundary conditions, $\nabla \Phi \cdot \hat{n} = 0$, where $\hat{n}$ is the unit normal at the boundary. In our square geometry, this condition may be written as

$$\frac{\partial \Phi}{\partial x}\bigg|_{x=0} = \frac{\partial \Phi}{\partial x}\bigg|_{x=L} = \frac{\partial \Phi}{\partial y}\bigg|_{y=0} = \frac{\partial \Phi}{\partial y}\bigg|_{y=L} = 0 \tag{7-35}$$

With these boundary conditions, the normal component of the electric field is zero at the boundary.

Given these boundary conditions, a natural set of trial functions is

$$f_{m,n}(x, y) = \cos\left[\frac{(m-1)\pi x}{L}\right] \cos\left[\frac{(n-1)\pi y}{L}\right] \tag{7-36}$$

with $m, n = 1, \ldots, M$. It is easy to check that

$$\int_0^L dx \int_0^L dy \, f_{m,n}(x, y) f_{m',n'}(x, y) = \frac{L^2}{4} (1 + \delta_{m,1})(1 + \delta_{n,1}) \delta_{m,m'} \delta_{n,n'} \tag{7-37}$$

so these trial functions are orthogonal.

Inserting (7-36) into (7-33) gives

$$-\sum_{m=1}^{M}\sum_{n=1}^{M} a_{m,n} \left[(m-1)^2 + (n-1)^2\right] \frac{\pi^2}{L^2} f_{m,n}(x, y) + \frac{1}{\varepsilon_o} \rho(x, y) = R(x, y) \qquad (7\text{-}38)$$

To solve for the coefficients, apply

$$\int_0^L dx \int_0^L dy \, f_{m',n'}(x, y) \qquad (7\text{-}39)$$

to both sides of this equation. Using the Galerkin condition (7-34) and orthogonality (7-37), we obtain

$$a_{m,n} = \frac{4}{\pi^2 \varepsilon_o} \frac{1}{[(m-1)^2 + (n-1)^2]} \frac{1}{(1 + \delta_{m,1})(1 + \delta_{n,1})}$$
$$\times \int_0^L dx \int_0^L dy \, \rho(x, y)\cos\left[\frac{(m-1)\pi x}{L}\right]\cos\left[\frac{(n-1)\pi y}{L}\right] \qquad (7\text{-}40)$$

Finally, having computed the coefficients, our approximate solution is

$$\Phi_a(x, y) = \sum_{m=1}^{M}\sum_{n=1}^{M} a_{m,n} \cos\left[\frac{(m-1)\pi x}{L}\right]\cos\left[\frac{(n-1)\pi y}{L}\right] \qquad (7\text{-}41)$$

Because our trial functions form a complete basis, the solution would be exact if we used an infinite number of terms.

To obtain the coefficients we need to be able to evaluate the integrals in (7-40). If the charge distribution consists of a finite number of line charges, then the integrals are trivial to evaluate since $\rho(\mathbf{r})$ is a sum of Dirac delta functions. Generally, the charge density is not such a simple function and the integrals in (7-40) have to be evaluated numerically (see Chapter 9).

Dipole Example

As a specific example, consider the charge distribution for a two-dimensional dipole

$$\rho(\mathbf{r}) = \lambda\{\delta[\mathbf{r} - (\mathbf{r}_c + \tfrac{1}{2}\mathbf{d})] - \delta[\mathbf{r} - (\mathbf{r}_c - \tfrac{1}{2}\mathbf{d})]\} \qquad (7\text{-}42)$$

where λ is the charge per unit length, $\mathbf{r}_c$ is the location of the dipole, and $\mathbf{d}$ is the separation vector (see Figure 7.5). For the purpose of comparison, consider the case where the size of the square box is infinite, that is, a dipole in free space. The potential for the free dipole is

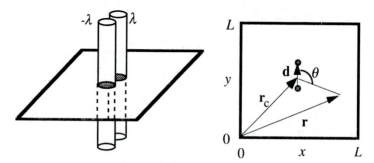

Figure 7.5 Schematic illustration of a two-dimensional dipole. The two line charges, centered at $\mathbf{r}_c$, are separated by a distance d. Their charge densities are $\pm\lambda$ coulombs per unit length.

$$\Phi(\mathbf{r}) = \frac{-\lambda}{2\pi\varepsilon_0}\{\ln[|\mathbf{r} - (\mathbf{r}_c + \tfrac{1}{2}\mathbf{d})|] - \ln[|\mathbf{r} - (\mathbf{r}_c - \tfrac{1}{2}\mathbf{d})|]\} \tag{7-43}$$

If the observation point, $\mathbf{r}$, is far from the dipole, that is, if $|\mathbf{r} - \mathbf{r}_c| \gg |\mathbf{d}|$, then

$$\Phi(\mathbf{r}) \approx \frac{\lambda}{2\pi\varepsilon_0}\frac{|\mathbf{d}|}{|\mathbf{r} - \mathbf{r}_c|}\cos\theta \tag{7-44}$$

where θ is the angle between $\mathbf{d}$ and $\mathbf{r} - \mathbf{r}_c$.

LISTING 7.2 Program `galrkn`. Solves the Poisson equation using the Galerkin method.

```
1   % galrkn - Program to solve the Poisson equation in
2   % 2-dimensions using Galerkin method (Neumann boundary cond.)
3   clear; help galrkn;  % Clear memory and print header
4   eps0 = 8.8542e-12;    % Permittivity (C^2/(N m^2))
5   N = input(' Enter number of terms to use - ');
6   L = 1;      % System size
7   fprintf('System is %g by %g square \n',L,L);
8   M=2;        % Number of charges (M=2 is dipole)
9   % Initialize position and charge of line charges
10  d = 0.1*L;  % Dipole separation
11  xq(1) = L/2;  yq(1) = L/2+d/2;  q(1) = 1;
12  xq(2) = L/2;  yq(2) = L/2-d/2;  q(2) = -q(1);
13  disp('Evaluating the coefficients a(i,j)');
14  a = zeros(N);
15  delt = ones(N,1);    % delt(i) = 1 + delta(i,1)
16  delt(1) = 2;
17  for k=1:M      % Sum over charges
18    tempx = cos((0:N-1)*pi*xq(k)/L);
19    tempy = cos((0:N-1)*pi*yq(k)/L);
20    for i=1:N
```

```
21    for j=1:N
22      a(i,j) = a(i,j) + q(k)*tempx(i)*tempy(j). . .
23                 /( ((i-1)^2+(j-1)^2 + eps)*delt(i)*delt(j) );
24    end
25   end
26  end
27  a = 4/(eps0*pi^2) * a;   % Throw in the factor out in front
28  r = input('Enter radius at which to evaluate phi - ');
29  disp('Computing the potential');
30  Nplot = 50;       % Number of points in the plot
31  phi = zeros(Nplot,1);
32  theta = pi * (0:Nplot-1)/(Nplot-1);
33  for k=1:Nplot
34    x = L/2 + r*sin(theta(k));    % Coordinates at which to
35    y = L/2 + r*cos(theta(k));    % evaluate potential
36    for i=1:N
37      tempx=cos((i-1)*pi*x/L);
38      for j=1:N
39        phi(k) = phi(k) + a(i,j)*tempx*cos((j-1)*pi*y/L);
40      end
41    end
42    % Plot potential and compare with free dipole
43    r_rc = [r*sin(theta(k))  r*cos(theta(k))];
44    free(k) = -q(1)/(2*pi*eps0)*(log(norm(r_rc - [0  d/2])) -
                . . .
45                     log(norm(r_rc + [0  d/2])));
46    fprintf('Finished %g of %g \n',k,Nplot);
47  end
48  plot(theta,phi,'-',theta,free,'--')
49  xlabel('Angle (radians)');   ylabel('Potential (volts)');
50  title(['Radius = ',num2str(r),' (dashed is free dipole)']);
```

The program `galrkn` computes the potential of the dipole-in-a-box using the Galerkin method (see Listing 7.2). Equation (7-40) is used to compute the coefficients `a(i,j)` (lines 13–27). The potential is evaluated [using (7-41)] on a ring of points equidistant from the center of the dipole (lines 28–47). The potential at this radius is plotted along with the potential of the free dipole, equation (7-43).

The output from `galrkn` is shown in Figures 7.6 and 7.7 for $|\mathbf{r} - \mathbf{r}_c| = 0.1$ and 0.4, respectively. The width of the box is $L = 1.0$; the dipole parameters are $\lambda = 1$, $\mathbf{r}_c = (L/2)\hat{\mathbf{x}} + (L/2)\hat{\mathbf{y}}$, and $\mathbf{d} = (L/10)\hat{\mathbf{y}}$. Notice the strong influence of the boundary in Figure 7.7.

Galerkin Method and Separation of Variables

At this point the reader may (incorrectly) have the impression that the Galerkin method is nothing more than using the separation of variables solution retaining only a finite number of terms. For our simple example, with the Poisson equation,

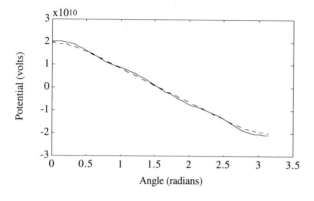

Figure 7.6 Potential as a function of angle for the dipole-in-a-box (see Figure 7.5) as obtained by `galrkn` using $N = 25$ terms. For comparison, the potential for the free dipole is plotted with a dashed line. Width of the box is $L = 1$; radial distance is 0.1.

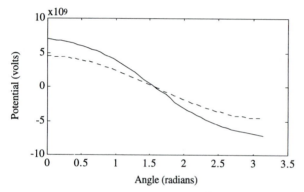

Figure 7.7 Potential as a function of angle for the dipole-in-a-box (see Figure 7.5) as obtained by `galrkn` using $N = 25$ terms. For comparison, the potential for the free dipole is plotted with a dashed line. Width of the box is $L = 1$; radial distance is 0.4.

it turns out that way. However, looking ahead to more complicated partial differential equations we see that the Galerkin method is more versatile. It gives us great latitude as to our choice of trial functions. With separation of variables, we must first find the general solution of a PDE and then build our particular solution by imposing boundary conditions on these functions. With Galerkin, the trial functions do not have to be the eigenfunctions of the PDE we are solving. Rather, any convenient set of functions that is orthogonal and matches the boundary conditions may be used.

A more appropriate way of viewing the Galerkin method is as a spectral transform approach. In our example we represent our solution by its Fourier series that, due to the boundary conditions, is a cosine series. The coefficients of this series are obtained after we compute the Fourier coefficients of the charge density. This line of thought leads us naturally to our next numerical scheme.

Multiple Fourier Transform (MFT) Method

One advantage of the Galerkin method is that it allows us to evaluate the potential at only selected points. Contrast this with relaxation methods that require us to compute the potential everywhere inside the system. On the other hand, if we do need to map the potential over the entire system, the Galerkin method, as constructed above, is relatively inefficient.

To understand why the Galerkin method is slow, consider the following problem. Suppose that we partition the space with an $N \times N$ grid with grid points at locations (x_i, y_j). Define the discretized potential, $\Phi_{i,j} = \Phi(x_i, y_j)$, and charge density, $\rho_{i,j} = \rho(x_i, y_j)$. A simple way of computing the integrals in equation (7-40) is to estimate them as sums,

$$a_{m,n} = \frac{4}{\pi^2 \varepsilon_o} \frac{1}{[(m-1)^2 + (n-1)^2]} \frac{1}{(1 + \delta_{m,1})(1 + \delta_{n,1})}$$

$$\times \sum_{i=1}^{N} \sum_{j=1}^{N} \rho(x_i, y_j) \cos\left[\frac{(m-1)\pi x_i}{L}\right] \cos\left[\frac{(n-1)\pi y_j}{L}\right] h^2 \qquad (7\text{-}45)$$

for $m, n = 1, \ldots, M$, the grid spacing between points is $h = L/N$. Typically $M \approx N$, so computing all the coefficients, using (7-45), requires a calculational effort of $O(N^2 M^2) \approx O(N^4)$. The computation of the potential at all N^2 grid points using (7-41) also requires an effort of $O(N^4)$. In comparison, for simultaneous overrelaxation (SOR) the computation time goes as $O(N^3)$.

If you think about it, in (7-45) we are taking the two-dimensional (cosine) Fourier transform of the density. After getting the coefficients, $a_{i,j}$, we take the inverse transform to obtain the potential. Yet we know that the fast Fourier transform (FFT) algorithm can do this type of operation very efficiently (see Section 5.2). Because the transform is now two-dimensional does not significantly complicate the problem. The discrete Fourier transform is a linear operation, so we may apply it separately in each direction (i.e., it doesn't matter which sum we do first).

The FFT algorithm may be adapted to perform sine or cosine transforms. As we have seen, the cosine transform is useful for Neumann boundary conditions. The sine transform is used with the Dirichlet boundary condition $\Phi = 0$. For simplicity, we'll change the problem and use periodic boundary conditions allowing us to use the standard FFT routines.

Discretized on an $N \times N$ square grid, the Poisson equation, (7-27), may be written as

$$\frac{1}{h^2} \{\Phi_{i+1,j} + \Phi_{i-1,j} - 2\Phi_{i,j}\} + \frac{1}{h^2} \{\Phi_{i,j+1} + \Phi_{i,j-1} - 2\Phi_{i,j}\} = -\frac{1}{\varepsilon_o} \rho_{i,j} \qquad (7\text{-}46)$$

We define the two-dimensional Fourier transforms of the potential and the charge density as

$$F_{m+1,n+1} = \sum_{i=0}^{N-1} \sum_{j=0}^{N-1} \Phi_{i+1,j+1} e^{-\alpha i m} e^{-\alpha j n} \qquad (7\text{-}47)$$

$$R_{m+1,n+1} = \sum_{i=0}^{N-1} \sum_{j=0}^{N-1} \rho_{i+1,j+1} e^{-\alpha im} e^{-\alpha jn} \tag{7-48}$$

where $\alpha \equiv 2\pi\sqrt{-1}/N$. The inverse transforms are

$$\Phi_{i+1,j+1} = \frac{1}{N^2} \sum_{m=0}^{N-1} \sum_{n=0}^{N-1} F_{m+1,n+1} e^{\alpha im} e^{\alpha jn} \tag{7-49}$$

$$\rho_{i+1,j+1} = \frac{1}{N^2} \sum_{m=0}^{N-1} \sum_{n=0}^{N-1} R_{m+1,n+1} e^{\alpha im} e^{\alpha jn} \tag{7-50}$$

Notice that we have as many Fourier coefficients as grid points, that is, $M = N$. Transforming (7-46),

$$\{e^{-\alpha(m-1)} + e^{\alpha(m-1)} + e^{-\alpha(n-1)} + e^{\alpha(n-1)} - 4\}F_{m,n} = -\frac{1}{\varepsilon_o} h^2 R_{m,n} \tag{7-51}$$

Solving for the matrix **F**, we have

$$F_{m,n} = P_{m,n} R_{m,n} \tag{7-52}$$

where

$$P_{m,n} = \frac{-h^2/2\varepsilon_o}{\cos[2\pi(m-1)/N] + \cos[2\pi(n-1)/N] - 2} \tag{7-53}$$

Taking the inverse transform of **F** gives us the potential. This algorithm is called the *multiple Fourier transform* (MFT) method.

Multiple Fourier Transform (MFT) Program

The program `fftpoi` uses the MFT method to solve the dipole-in-a-box problem with periodic boundary conditions (see Listing 7.3). A few notes about the program: The user is prompted for the number of charges, their location, and charge density. The charges are placed at the nearest grid points (lines 13–21). The matrix **P** is computed on lines 22–32. The matrix **R** is computed on line 34; **F** and Φ are computed on the next two lines. The MATLAB functions `fft2` and `ifft2` are used to compute the two-dimensional transforms. The function `fft2 (X)` applies the one-dimensional FFT to each row of X and then to each column; `ifft2 (X)` is the inverse transform.

LISTING 7.3 Program `fftpoi`. Solves the Poisson equation using the multiple Fourier transform method.

```
1   % fftpoi - Program to solve the Poisson equation using
2   % MFT method (Periodic boundary conditions)
```

```
3   clear; help fftpoi;   % Clear memory and print header
4   eps0 = 8.8542e-12;    % Permittivity (C^2/(N m^2))
5   N = 32;   % Number of grid points on a side (square grid)
6   L = 1;    % System size
7   h = L/N;   % Grid spacing for periodic boundary cond.
8   x = ((1:N)-1/2)*h;   % Coordinates of grid points
9   y = x;                % Square grid
10  fprintf('System is a square of length %g \n',L);
11  % Set up charge density rho(i,j)
12  rho = zeros(N,N);   % Initialize charge density to zero
13  M = input('Enter number of line charges - ');
14  for i=1:M
15    fprintf('\n For charge #%g \n',i);
16    r = input('Enter position [x y] - ');
17    ii=round(r(1)/h + 1/2);   % Place charge at nearest
18    jj=round(r(2)/h + 1/2);   % grid point
19    q = input('Enter charge density - ');
20    rho(ii,jj) = rho(ii,jj) + q;
21  end
22  disp('Computing matrix P')
23  coeff = 2*pi/N;
24  cx = cos(coeff*(0:N-1));
25  cy = cx;
26  for i=1:N
27    for j=1:N
28      P(i,j) = 1/(cx(i)+cy(j)-2);
29    end
30  end
31  P(1,1) = 0;              % Clean up divide by zero
32  P = -h^2/(2*eps0) * P;   % Throw in the factor in front
33  disp('Computing potential and E field');
34  rhoT = fft2(rho);   % Transform rho into frequency domain
35  phiT = rhoT .* P;   % Computing phi in the frequency domain
36  phi = ifft2(phiT);   % Inv. transf. phi into the real domain
37  [Ex Ey] = gradient(rot90(phi)); % Compute E field using
38  temp = sqrt(Ex.^2 + Ey.^2);      %     E = - grad phi
39  Ex = -Ex ./ temp;                % Normalize components so
40  Ey = -Ey ./ temp;                % vectors are equal length
41  % Plot potential and E field
42  axis('square');        % Use a square aspect ratio
43  subplot(121)
44    contour(rot90(phi),15,x,y);   % Contour plot of potential
45    title('Potential'); xlabel('x'); ylabel('y');
46  subplot(122)
47    [xmax ymax] = size(Ex);
48    axis([0 xmax+1 0 ymax+1]);
49    quiver(Ex,Ey)            % Plot E field with vectors
50    title('E-field (Direction)'); xlabel('i'); ylabel('j');
```

```
51  subplot(111)
52  axis; axis('normal');   % Reset auto-scaling and normal axes
```

Given the potential the electric field is $\mathbf{E} = -\nabla\Phi$. The MATLAB gradient function is used to take the gradient of the potential (see line 37). The MATLAB quiver function produces a vector arrow plot given Ex and Ey, the x- and y-components of the electric field (line 49). For better visualization, we normalize $\mathbf{E}$ so that its magnitude is unity; the quiver plot shows the direction of the field (see Figure 7.8). Try plotting the field without normalizing to see why this is useful.

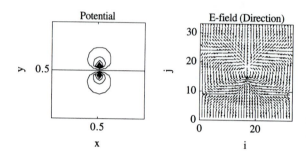

Figure 7.8 Potential and field lines from fftpoi. Charges of the dipole are located at $[x \quad y] = [0.5 \quad 0.55]$ and $[0.5 \quad 0.45]$.

EXERCISES

12. Derive equation (7-44) given (7-43).

13. (a) Using the method of images, find an expression for the potential for the dipole-in-a-box problem.

    (b) Write a program to evaluate this solution at a ring of points centered on the dipole. Compare your output with that of galrkn (see Figures 7.6 and 7.7).

14. (a) Write a version of galrkn that uses the Dirichlet boundary conditions

$$\Phi(x = 0, y) = \Phi(x = L, y) = \Phi(x, y = 0) = \Phi(x, y = L) = 0$$

    Compare the results with those from the original version.

    (b) Using the method of images, find an expression for the potential of a dipole-in-a-box with Dirichlet boundary conditions. Write a program to evaluate this solution at a ring of points centered on the dipole. Check your answers with those obtained in part (a).

15. Consider the trivial ODE $df/dt = f$ with $f(0) = 1$. Suppose we construct an approximate solution as

$$f_a(t) = 1 + \sum_{k=1}^{K} a_k t^k$$

in the interval $t \in [0,1]$. Using the Galerkin method, find the coefficients, a_k, for $K = 3$. Notice that our basis functions are not orthogonal but the integrals are easy to evaluate. Compare with the Taylor expansion of the true solution.

16. The `fftpoi` program uses a rather coarse method for placing the charges on the grid: It assigns a charge to the nearest grid point. Modify the program so that it proportionally assigns a fraction of the charge to each of the nearest four grid points (see Figure 7.9). Compare with the unmodified version by plotting $\Phi(x = L/2, y)$ for a dipole with charges at $[x\ y] = [0.5\quad 0.505]$ and $[0.5\quad 0.495]$. Take $L = 1$.

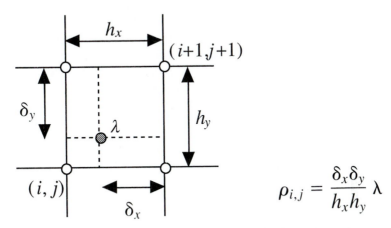

$$\rho_{i,j} = \frac{\delta_x \delta_y}{h_x h_y}\,\lambda$$

Figure 7.9 *Proportional partitioning of charge density on the mesh.*

17. In optical diffraction, the Fraunhofer irradiance is the two-dimensional Fourier transform of the aperture function.[4] An example of a square aperture and its irradiance are shown in Figure 7.10.[5] Write a program, using `fft2`, that computes the irradiance for a given aperture. You will probably want to use the MATLAB `fftshift(X)` function that swaps quadrants one and three of the matrix X with quadrants two and four (i.e., the four corners are mapped to the center).

Aperture   Irradiance

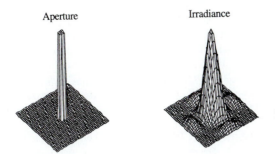

Figure 7.10 Aperture function for a square aperture and its corresponding Fraunhofer irradiance pattern.

[4] F. L. Pedrotti and L. S. Pedrotti, *Introduction to Optics*, 2d ed. (Englewood Cliffs, N.J.: Prentice Hall, 1993), chapter 24.

[5] For a complete discussion and more examples, see R. G. Wilson, S. M. McCreary, and F. L. Thompson, "Optical transformations in three-space: Simulations with a PC," *Am. J. Phys.*, **60**, 49–56 (1992).

BEYOND THIS CHAPTER

The relaxation methods can be accelerated by using multiple grids.[6] Recall the physical analogy between relaxation and the diffusion equation. The long wavelength modes decay the slowest; this is why it is useful to use the first term of the separation of variables solution as an initial guess. In a multigrid algorithm, one starts the relaxation process on the finest grid. When the convergence begins to slow, we transfer (by averaging) to a coarser grid. When we have converged on the coarse grid, we transfer back down (by interpolation) to the finer grid. Multigrid schemes can solve the Laplace equation on an $N \times N$ grid in a time $O(N^2)$ as compared with $O(N^4)$ for Jacobi and $O(N^3)$ for SOR.

The Laplace (or Poisson) equation in one dimension is an ordinary differential equation. It should be clear that the methods described in this section may be applied to this type of boundary value ODE problem. See Ascher et al.[7]

Exercise 6.11 illustrates that the discretized Laplace equation may be written as a large system of linear equations. I say large because the number of unknowns (and equations) equals the number of interior grid points. This approach for solving elliptic PDEs is most useful when the grid size is small (e.g., coarsest grid in a multigrid scheme). However, you can get a lot of extra mileage by making use of the sparseness of the matrix (see Section 8.3). A rapid, iterative algorithm for solving such sparse matrices is the *conjugate-gradient method*.[8]

Spectral methods have been used extensively to compute the fields in plasma simulations.[9] They are also use in computational fluid dynamics,[10] especially in turbulence studies.[11] Haltiner and Williams discuss spectral methods in the context of geophysical problems.[12] Sadiku[13] and Booton[14] review a wide variety of numerical techniques as applied to electromagnetic problems.

[6] W. Hackbusch, *Multi-Grid Methods and Applications* (Berlin: Springer-Verlag, 1985); P. Wesseling, *An Introduction to Multigrid Methods* (Chichester: Wiley, 1992).

[7] U. Ascher, R. Mattheij, and R. Russell, *Numerical Solution of Boundary Value Problems for Ordinary Differential Equations* (Englewood Cliffs, N.J.: Prentice Hall, 1988).

[8] G. H. Golum and C. F. Van Loan, *Matrix Computations*, 2d ed. (Baltimore: Johns Hopkins University Press, 1989), section 10.2.

[9] C. K. Birdsall and A. B. Langdon, *Plasma Physics via Computer Simulation* (New York: McGraw-Hill, 1985); R. W. Hockney and J. W. Eastwood, *Computer Simulation Using Particles* (Bristol: Adam Hilger, 1988); T. Tajima, *Computational Plasma Physics: With Applications to Fusion and Astrophysics* (Redwood Calif.: Addison-Wesley, 1989).

[10] C. A. J. Fletcher, *Computational Galerkin Methods* (New York: Springer-Verlag, 1984).

[11] C. Canuto, M. Y. Hussaini, A. Quarteroni, and T. A. Zang, *Spectral Methods in Fluid Dynamics* (Berlin: Springer-Verlag, 1988).

[12] G. J. Haltiner and R. T. Williams, *Numerical Prediction and Dynamic Meteorology*, 2d ed. (New York: Wiley, 1980).

[13] M. N. O. Sadiku, *Numerical Techniques in Electromagnetics* (Boca Raton, Fla.: CRC Press, 1992).

[14] R. C. Booton, *Computational Methods for Electromagnetics and Microwaves* (New York: Wiley, 1992).

APPENDIX 7A: FORTRAN LISTINGS

LISTING 7A.1 Program `jacobi`. Solves Laplace's equation using the Jacobi method.

```
      program jacobi
! Program to solve the Laplace equation using Jacobi's method
! on a square grid
      parameter ( Nmax = 100 )
      real phi(Nmax,Nmax),newphi(Nmax,Nmax),change(Nmax*Nmax)
      real x(Nmax),y(Nmax),L

      pi = 4.*atan(1.)
      write (*,*) 'Enter number of grid points on a side'
      read (*,*) N
      L = 1.               ! System size (length)
      h = L/(N-1)          ! Grid spacing
      phi0 = 1.            ! Potential at y=L
      write (*,*) 'Potential at y=L equals ',phi0
      write (*,*) 'Potential is zero on all other boundaries'
      change_want = 1e-4   ! Stop when change is given fraction
!!!!! Set initial conditions and boundary conditions !!!!!
      coeff = pi/L
      do i=1,N
        x(i) = (i-1)*h     ! x and y coordinates for
        y(i) = x(i)        ! square grid
      end do
      do i=1,N             ! Initialize phi with first term of
       do j=1,N            ! separation of variables solution
         phi(i,j) = phi0 * 4./(pi*sinh(pi)) *
     &                 sin(coeff*x(i))*sinh(coeff*y(j))
         newphi(i,j) = 0
       end do
      end do
      do i=1,N
        phi(i,N) = phi0        ! Set the boundary condition
        newphi(i,N) = phi0
      end do

!!!!! MAIN LOOP !!!!!
      max_iter = N**2          ! Set max to avoid long runs
      do iter=1,max_iter
        temp = 0
        do i=2,(N-1)           ! Jacobi method
         do j=2,(N-1)
```

```fortran
          newphi(i,j) = 0.25 * (phi(i+1,j) + phi(i-1,j) +
     &                                  phi(i,j+1) +phi(i,j-1))
          ! Measure average fractional change in solution
          temp = temp + abs(1.-phi(i,j)/newphi(i,j))
         end do
        end do
        do i=1,N
         do j=1,N
           phi(i,j) = newphi(i,j)  ! Reset values of potential
         end do
        end do
        change(iter) = temp/(N-2)**2
        write (*,*) 'After ',iter,' iterations, change =',
     &                                  change(iter)
        if ( change(iter) .lt. change_want ) then
          write (*,*) 'Desired accuracy; breaking out of loop'
          goto 100  ! Break out of main loop
        end if
       end do
100    continue ! Break to this point when desired accuracy
               ! achieved

! Print out the plotting variables -
!   phi,change,x,y
!
       open(11,file='x.dat')
       open(12,file='y.dat')
       open(13,file='phi.dat')
       open(14,file='change.dat')
       do i=1,N
        write (11,1001) x(i)
        write (12,1001) y(i)
        do j=1,N
          write (13,1002) phi(i,j)
        end do
        write (13,1003)
       end do
       do i=1,iter
          write (14,1001) change(i)
       end do
       stop
1001   format(e12.6)
1002   format(e12.6,' ',$)       ! The $ suppresses the carriage
                                 ! return
1003   format(/)                 ! New line
       end
```

LISTING 7A.2 Program `galrkn`. Solves the Poisson equation using the Galerkin method.

```fortran
      program galrkn
! Program to solve the Poisson equation in 2-dimensions using
! Galerkin method (Neumann boundary conditions)
      parameter ( Nmax = 100, Mmax = 10, Pmax = 200 )
      real L,xq(Mmax),yq(Mmax),q(Mmax),a(Nmax,Nmax)
      real theta(Pmax),phi(Pmax),free(Pmax),delt(Nmax)

      pi = 4.*atan(1.)
      eps = 1e-16             ! Epsilon (used to avoid division by
                              ! zero)
      eps0 = 8.8542e-12       ! Permittivity (C^2/(N m^2)
      write (*,*) 'Enter number of terms to use in sum'
      read (*,*) N
      L = 1.                  ! System size
      write (*,*) 'System is ',L,' by ',L,' square'
      M = 2                   ! Number of charges (M=2 is dipole)
      d = 0.1*L               ! Dipole separation
      xq(1) = L/2.            ! x-coordinate of first line charge
      yq(1) = L/2.+d/2.       ! y-coordinate of first line charge
      q(1)  = 1.              ! charge density of first line charge
      xq(2) = L/2.            ! x-coordinate of second line charge
      yq(2) = L/2.-d/2.       ! y-coordinate of second line charge
      q(2)  = -q(1)           ! charge density of second line charge
      write (*,*) 'Evaluating the coefficients a(i,j)'
      do i=1,N
       delt(i) = 1.
       do j=1,N
        a(i,j) = 0.           ! Initialize matrix a to zero
       end do
      end do
      delt(1) = 2.            ! Delt(i) = 1 + delta(i,1)
!! Construct the matrix a !!
      do k=1,M
       do i=1,N
         tempx = 4./(eps0*pi**2)*q(k)*cos((i-1)*pi*xq(k)/L)
         do j=1,N
           a(i,j) = a(i,j) + tempx*cos((j-1)*pi*yq(k)/L)
     &            /( ((i-1)**2 + (j-1)**2 + eps)*delt(i)*delt(j) )
         end do
       end do
      end do

      write (*,*) 'Enter radius at which to evaluate phi'
      read (*,*) r
      write (*,*) 'Computing the potential'
```

```fortran
      Nplot = 200
      do i=1,Nplot
        phi(i) = 0.
      end do
      do k=1,Nplot
        theta(k) = pi * (k-1)/float(Nplot-1) ! Angle theta in
                                             ! radians
        x = L/2 + r*sin(theta(k))    ! x coordinate given r and
                                     ! theta
        y = L/2 + r*cos(theta(k))    ! y coordinate given r and
                                     ! theta
        do i=1,N
          tempx = cos((i-1)*pi*x/L)
          do j=1,N
            phi(k) = phi(k) + a(i,j)*tempx*cos((j-1)*pi*y/L)
          end do
        end do
        write (*,*) 'Finished ',k,' of ',Nplot
      end do
      do k=1,Nplot
        rcx = r*sin(theta(k))   ! x-component of vector r sub c
        rcy = r*cos(theta(k))   ! y-component of vector r sub c
        temp1 = rcx**2 + (rcy-d/2)**2
        temp2 = rcx**2 + (rcy+d/2)**2
        ! Compute potential for a dipole in free space
        free(k) = -q(1)/(2*pi*eps0)
     &                 *(alog(sqrt(temp1))-alog(sqrt(temp2)))
      end do
! Print out the plotting variables -
!   tplot,xplot,aplot
!
      open(11,file='theta.dat')
      open(12,file='phi.dat')
      open(13,file='free.dat')
      open(14,file='r.dat')
      do k=1,Nplot
        write (11,1001) theta(k)
        write (12,1001) phi(k)
        write (13,1001) free(k)
      end do
      write (14,1001) r
      stop
1001  format(e12.6)
      end
```

LISTING 7A.3 Program `fftpoi`. Solves the Poisson equation using the multiple Fourier transform method. Uses `fft2` (Listing 7A.4) and `ifft2` (Listing 7A.5).

```fortran
      program fftpoi
! Program to solve the Poisson equation using the MFT method
! Periodic boundary conditions
      parameter( Nmax = 32 )
      real L,x(Nmax),y(Nmax),rho(Nmax,Nmax),phi(Nmax,Nmax)
      real cx(Nmax),cy(Nmax),P(Nmax,Nmax)
      complex rhoT(Nmax,Nmax),phiT(Nmax,Nmax)

      pi = 4*atan(1.) ! = 3.14159...
      eps = 1.e-15       ! A tiny number
      eps0 = 8.8542e-12      ! Permittivity of free space
      npow = 5           ! Must use a power of 2 for FFT routine
      N = 2**npow        ! Number of grid points on a side
      L = 1.             ! System size
      h = L/N            ! Grid spacing (periodic boundaries)
      do i=1,N
        x(i) = (i-0.5)*h ! x-components of grid points
        y(i) = x(i)         ! y-components of grid points
      end do
      write (*,*) 'System is a square of length ',L

! Set up the charge density rho(i,j)
      do i=1,N
       do j=1,N
        rho(i,j) = 0.
       end do
      end do

      write (*,*) 'Enter number of charges'
      read (*,*) M
      do i=1,M
        write (*,*) 'For # ',i,' enter x,y and charge'
        read (*,*) rx,ry,q
        ii = int(rx/h+0.5) + 1      ! Place charge at nearest
        jj = int(ry/h+0.5) + 1      ! grid point
        rho(ii,jj) = rho(ii,jj) + q
      end do

      write (*,*) 'Computing matrix P'
      coeff = 2*pi/N
      do i=1,N
        cx(i) = cos(coeff*(i-1))   ! Used to compute matrix P
        cy(i) = cx(i)                ! Used to compute matrix P
      end do
```

```
      temp = -h**2/(2*eps0)
      do i=1,N     ! Compute the matrix P
       do j=1,N
         P(i,j) = temp/((cx(i)+cy(j)-2)+eps)
       end do
      end do
      P(1,1) = 0. ! Clean up the divide by zero
      write (*,*) 'Computing potential'
      do i=1,N
       do j=1,N
         rhoT(i,j) = rho(i,j)   ! Copy values into rhoT
       end do
      end do
      call fft2(rhoT,npow,Nmax) ! Take 2D transform
      do i=1,N
       do j=1,N
         phiT(i,j) = rhoT(i,j)*P(i,j)   ! Compute phi in the
       end do                           ! frequency domain
      end do
      call ifft2(phiT,npow,Nmax)     ! Inv. transform into
      do i=1,N                       ! the real (space) domain
       do j=1,N
         phi(i,j) = real(phiT(i,j))     ! Copy into phi
       end do
      end do
! Print out the plotting variables -
!    x,y,rho,phi
!
      open(11,file='x.dat')
      open(12,file='y.dat')
      open(13,file='rho.dat')
      open(14,file='phi.dat')
      do i=1,N
        write (11,1001) x(i)
        write (12,1001) y(i)
        do j=1,N
          write (13,1002) rho(i,j)
          write (14,1002) phi(i,j)
        end do
        write (13,1003)
        write (14,1003)
      end do
      stop
1001 format(e12.6)
1002 format(e12.6,' ',$)    ! The $ suppresses the carriage
                            ! return
1003 format(/)              ! New line
      end
```

LISTING 7A.4 Function `fft2`. Computes two-dimensional discrete Fourier transform. Uses `fft` (Listing 5A.8).

```
      subroutine fft2(A,npow,NA)
! Routine to compute the two dimensional Fourier transform
! Uses the fft routine
      parameter( Nmax = 1024 )   ! Maximum number of data points
      complex A(NA,NA), Atemp(Nmax)

      N = 2**npow     ! Number of data points
      do j=1,N
        do i=1,N
          Atemp(i) = A(i,j)    ! Copy out a column
        end do
        call fft(Atemp,npow)   ! Take its transform
        do i=1,N
          A(i,j) = Atemp(i)    ! Copy it back in
        end do
      end do
      do i=1,N
        do j=1,N
          Atemp(j) = A(i,j)    ! Copy out a row
        end do
        call fft(Atemp,npow)   ! Take its transform
        do j=1,N
          A(i,j) = Atemp(j)    ! Copy it back in
        end do
      end do

      return
      end
```

LISTING 7A.5 Function `ifft2`. Computes two-dimensional inverse discrete Fourier transform. Uses `fft` (Listing 5A.8).

```
      subroutine ifft2(A,npow,NA)
! Routine to compute the two dimensional inverse Fourier
! transform
! Uses the fft routine
      parameter( Nmax = 4096 )   ! Maximum number of data points
      complex A(NA,NA), Atemp(Nmax)

      N = 2**npow     ! Number of data points
      do j=1,N
        do i=1,N
          Atemp(i) = conjg(A(i,j))   ! Copy out a column
        end do
```

```
      call fft(Atemp,npow)   ! Take its transform
      do i=1,N
        A(i,j) = Atemp(i)    ! Copy it back in
      end do
    end do

    do i=1,N
      do j=1,N
        Atemp(j) = A(i,j)    ! Copy out a row
      end do
      call fft(Atemp,npow)   ! Take its transform
      do j=1,N
        A(i,j) = conjg(Atemp(j))/N**2   ! Copy it back in
      end do
    end do

    return
    end
```

Chapter 8
Partial Differential Equations III:
Stability and Implicit Methods

Chapters 6 and 7 covered various methods for solving partial differential equations. Along the way, we discovered that some methods were numerically unstable when the time step was too large. This chapter presents two techniques for testing numerical stability. Section 8.2 discusses some algorithms that are unconditionally stable, at a price. The price is having to invert large matrices. Section 8.3 covers some specialized routines handling these matrices.

8.1 STABILITY ANALYSIS

Von Neumann Stability

Consider the `aftcs` program we used in Chapter 6 for solving the advection equation using the FTCS scheme. The method is numerically unstable, as shown in Figure 8.1. The solution looks like a standing wave that rapidly grows in amplitude. In fact, the amplitude of the solution at the last time step is many orders of magnitude larger than the initial condition; this is why the mesh plot looks flat for the earlier times. We get a very similar picture if we employ a large time step ($\tau > h/|c|$) when solving the advection equation using any of the other methods in Chapter 6.

From the above observations it is reasonable to use a trial solution that has the form

$$a(x, t) = z(t)\,e^{ikx} \tag{8-1}$$

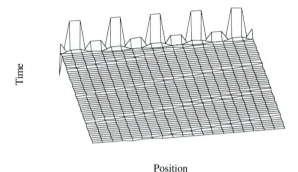

Time

Position

Figure 8.1 Amplitude versus x and t. Mesh plot of the solution of the advection equation obtained by the aftcs program using $N = 20$ mesh points and time step $\tau = 0.05$. The number of iteration steps was set to nstep = $10L/(c\tau)$ so that the wave circles the system ten times.

This solution is a wave with wave number k and (complex) amplitude $z(t)$. In discretized form, we have

$$a_j^n = \xi^n e^{ikhj} \tag{8-2}$$

The notation here is standard but a bit confusing: The superscript n on a_j^n is an index that labels the time step. On the other hand, ξ^n means ξ raised to the nth power. In other words, $z(t + \tau) = \xi z(t)$. The coefficient ξ is called the *amplification factor*.

The strategy of the analysis is to insert the trial solution, (8-2), into the numerical scheme. If the equations are linear, it is not difficult to solve for the amplification factor in terms of the grid spacing, h, and the time step, τ. A scheme is unstable if the magnitude of the amplification factor exceeds unity, that is, if $|\xi| > 1$. This approach is called *von Neumann stability analysis*.

Stability of FTCS for the Advection Equation

To illustrate the von Neumann analysis, we'll work through some examples. Recall from Chapter 6 that for the advection equation, the FTCS scheme may be written as

$$a_j^{n+1} = a_j^n - \frac{c\tau}{2h}\left(a_{j+1}^n - a_{j-1}^n\right) \tag{8-3}$$

where a is the wave amplitude, τ is the time step, and h is the spatial grid spacing. Inserting our trial solution, (8-2), we get

$$\xi^{n+1}e^{ikjh} = \xi^n e^{ikjh} - \frac{c\tau}{2h}\left(\xi^n e^{ik(j+1)h} - \xi^n e^{ik(j-1)h}\right)$$

$$= \xi^n e^{ikjh}\left(1 - \frac{c\tau}{2h}\left(e^{ikh} - e^{-ikh}\right)\right) \tag{8-4}$$

Dividing both sides by $\xi^n e^{ikjh}$, we find that the amplification factor is

$$\xi = 1 - \frac{c\tau}{2h}(e^{ikh} - e^{-ikh})$$

$$= 1 - i\frac{c\tau}{h}\sin(kh) \qquad (8\text{-}5)$$

The magnitude of the amplification factor is

$$|\xi| = \sqrt{1 + \left(\frac{c\tau}{h}\right)^2 \sin^2(kh)} \qquad (8\text{-}6)$$

Clearly, the magnitude of ξ is always greater than one. The solution is unstable since its amplitude grows with each time step.

Since $|\xi| \geq 1$ for all k, all the modes become numerically unstable. However, some grow faster than others. We find the fastest-growing mode, k_{max}, by solving $\sin^2(k_{max}\, h) = 1$. Since $k = 2\pi/\lambda$, where λ is the wavelength, then $\lambda_{max} = 4h$. Compare this result with the mesh plot in Figure 8.1.

Stability of the Lax Scheme

As a second example, let's apply the von Neumann analysis to the Lax method for solving the advection equation. Recall that the Lax scheme may be written as

$$a_j^{n+1} = \tfrac{1}{2}(a_{j+1}^n + a_{j-1}^n) - \frac{c\tau}{2h}(a_{j+1}^n - a_{j-1}^n) \qquad (8\text{-}7)$$

As before, we insert the trial solution, $a_j^n = \xi^n e^{ikhj}$, and get

$$\xi^{n+1}e^{ikjh} = \tfrac{1}{2}(\xi^n e^{ik(j+1)h} + \xi^n e^{ik(j-1)h}) - \frac{c\tau}{2h}(\xi^n e^{ik(j+1)h} - \xi^n e^{ik(j-1)h})$$

$$= \xi^n e^{ikjh}\left[\tfrac{1}{2}(e^{ikh} + e^{-ikh}) - \frac{c\tau}{2h}(e^{ikh} - e^{-ikh})\right] \qquad (8\text{-}8)$$

The amplification factor is thus

$$\xi = \cos(kh) - i\frac{c\tau}{h}\sin(kh) \qquad (8\text{-}9)$$

and its magnitude is

$$|\xi| = \sqrt{\cos^2(kh) + \left(\frac{c\tau}{h}\right)^2 \sin^2(kh)} \qquad (8\text{-}10)$$

Thus, $|\xi| \le 1$ if and only if $|c\tau/h| \le 1$, which is the Courant–Friedrichs–Lewy (CFL) stability criterion discussed in Chapter 6.

Matrix Stability

The von Neumann approach is not the only way to investigate the stability of a scheme but, being the easiest to do, it is the most popular. One of its shortcomings is that it neglects the possible influence of the boundary conditions. To include their effect, we introduce *matrix stability analysis*.[1]

For linear problems, most schemes may be written in the form, $\mathbf{x}^{n+1} = \mathbf{A}\mathbf{x}^n$, where $\mathbf{x}$ is the solution at time $t = (n - 1)\tau$ and τ is the time step. Consider the FTCS scheme for solving the thermal diffusion equation,

$$T_j^{n+1} = T_j^n + \frac{\tau}{2t_\sigma} (T_{j+1}^n + T_{j-1}^n - 2T_j^n) \tag{8-11}$$

where T is the temperature, h is the spatial grid spacing, κ is the diffusion coefficient, and $t_\sigma = h^2/2\kappa$. For Dirichlet boundary conditions (i.e., values of T_1^n and T_N^n fixed), we may write the FTCS scheme as

$$\mathbf{T}^{n+1} = \mathbf{T}^n + \frac{\tau}{2t_\sigma} \mathbf{D}\mathbf{T}^n$$

$$= \left(\mathbf{I} + \frac{\tau}{2t_\sigma} \mathbf{D}\right) \mathbf{T}^n = \mathbf{A}\mathbf{T}^n \tag{8-12}$$

where

$$\mathbf{T}^n = \begin{bmatrix} T_1^n \\ T_2^n \\ T_3^n \\ \vdots \\ T_N^n \end{bmatrix}; \quad \mathbf{D} = \begin{bmatrix} 0 & 0 & 0 & \cdots & 0 \\ 1 & -2 & 1 & \cdots & 0 \\ 0 & 1 & -2 & \cdots & 0 \\ \vdots & \vdots & \vdots & \ddots & \vdots \\ 0 & 0 & 0 & \cdots & 0 \end{bmatrix} \tag{8-13}$$

and $\mathbf{I}$ is the identity matrix. Notice that the matrix $\mathbf{D}$ has the following structure: Because of the Dirichlet boundary conditions the elements of the first and last rows are all zero, guaranteeing that the values of the end points (T_1 and T_N) remain unchanged. All the other rows have a -2 on the main diagonal and a 1 at the first off-diagonal elements.

---

[1] G. D. Smith, *Numerical Solution of Partial Differential Equations: Finite Difference Methods*, 3d ed. (Oxford: Oxford University Press, 1985).

To determine the stability of $\mathbf{T}^{n+1} = \mathbf{A}\mathbf{T}^n$, we consider the eigenvalue problem for the matrix $\mathbf{A}$,

$$\mathbf{A}\mathbf{v}_k = \lambda_k \mathbf{v}_k \tag{8-14}$$

where $\mathbf{v}_k$ is the eigenvector corresponding to the eigenvalue λ_k. Assuming that the eigenvectors form a complete basis, we may write our initial condition as

$$\mathbf{T}^1 = \sum_{k=1}^{N} c_k \mathbf{v}_k \tag{8-15}$$

From equation (8-12),

$$\mathbf{T}^{n+1} = \mathbf{A}\mathbf{T}^n = \mathbf{A}(\mathbf{A}\mathbf{T}^{n-1}) = \mathbf{A}^n \mathbf{T}^1 \tag{8-16}$$

In other words, the solution at time step $n + 1$ may be obtained by repeatedly multiplying the initial condition n times by the matrix $\mathbf{A}$.

Using our decomposition, (8-15),

$$\mathbf{T}^{n+1} = \sum_{k=1}^{N} c_k \mathbf{A}^n \mathbf{v}_k = \sum_{k=1}^{N} c_k (\lambda_k)^n \mathbf{v}_k \tag{8-17}$$

Clearly, if $|\lambda_k| > 1$ for any eigenvalue, then $|\mathbf{T}^n| \to \infty$ as $n \to \infty$.

The spectral radius of the matrix $\mathbf{A}$, $\rho(\mathbf{A})$, is defined as the magnitude of its largest eigenvalue. A scheme is matrix stable if the spectral radius is less than or equal to unity. There are many powerful theorems that allow us to set bounds on the spectral radius (e.g., the Gerschgorin circle theorem).[2]

Alternatively, we may obtain the spectral radius by numerically computing the eigenvalues. In MATLAB the eigenvalues of a matrix may be obtained using the built-in `eig` function. With this function, the spectral radius is ρ(A) = max (abs(eig(A))). Figure 8.2 shows the spectral radius of the matrix $\mathbf{A}$ used by the FTCS scheme in solving the diffusion equation with Dirichlet boundary conditions [equation (8-12)]. Notice that the spectral radius is one if the time step is less than or equal to t_σ. This agrees with our empirical findings from Chapter 6.

Power Method

The ideas developed above for matrix stability may be turned around to construct an algorithm for computing eigenvalues and eigenvectors. Consider the general eigenvalue problem

$$\mathbf{M}\mathbf{v}_k = \lambda_k \mathbf{v}_k \tag{8-18}$$

---

[2] J. Ortega, *Numerical Analysis–A second course* (New York: Academic Press, 1972).

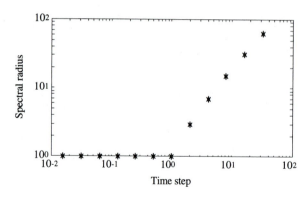

Figure 8.2   Spectral radius $\rho(\mathbf{A})$ as a function of τ for $N = 41$ grid points [see equation (8-12)]. Notice that $\rho(\mathbf{A}) = 1$ for $\tau \le t_\sigma = 1$.

where $\mathbf{v}_k$ is the (normalized) eigenvector corresponding to the (nondegenerate) eigenvalue λ_k. We label the eigenvalues in decreasing order, so $|\lambda_1| > |\lambda_2| > \ldots > |\lambda_N|$. Take any vector $\mathbf{x}$; write it as[3]

$$\mathbf{x} = \sum_{k=1}^{N} c_k \mathbf{v}_k \tag{8-19}$$

We assume that $\mathbf{x}$ is not orthogonal to any of the eigenvectors, so $c_k \ne 0$.

The *power method*[4] is a simple technique for computing λ_1, the largest eigenvalue, and $\mathbf{v}_1$, its corresponding eigenvector. First, we repeatedly multiply the vector $\mathbf{x}$ by the matrix $\mathbf{M}$. Doing so n times, we have

$$\mathbf{M}^n \mathbf{x} = \mathbf{M}^n \left(\sum c_k \mathbf{v}_k \right) = \sum c_k \lambda_k^n \mathbf{v}_k \tag{8-20}$$

Since the λ's are ordered by decreasing magnitude,

$$\mathbf{M}^n \mathbf{x} \approx c_1 \lambda_1^n \mathbf{v}_1 \tag{8-21}$$

as $n \to \infty$. The eigenvectors are normalized, that is, $|\mathbf{v}_k| = 1$, so we obtain the first eigenvector

$$\mathbf{v}_1 \approx \frac{\mathbf{M}^n \mathbf{x}}{|\mathbf{M}^n \mathbf{x}|} \tag{8-22}$$

as $n \to \infty$. To use the power method we iteratively compute $\mathbf{M}^n \mathbf{x}$ until the value of $\mathbf{v}_1$ appears to be converging. Using (8-18) and (8-22) we immediately obtain the largest eigenvalue, λ_1.

---

[3] Here we assume that the eigenvectors form a complete basis. There are exceptional matrices (e.g., Jordan matrices) for which this assumption is not valid. The power method still works, but the derivation is longer.

[4] F. S. Acton, *Numerical Methods that Work* (New York: Harper and Row, 1970), chapter 8.

Using MATLAB interactively, here's a simple example of the power method with $n = 10$:

```
>>M=[2 -1 0;-1 2 -1;0 -1 2];
>>x=[1; 1; 1];
>>v=M^10*x;
>>v=v/norm(v)

  v =

      0.5000
     -0.7071
      0.5000
```

This eigenvalue problem was considered in Section 5.3; the eigenvalues and eigenvectors are given by (5-58) and (5-59). The power method correctly gives the eigenvector $\mathbf{a}_+$ corresponding to the largest eigenvalue $\omega_+ = 2 + \sqrt{2}$.

There are several ways to modify the power method to accelerate its convergence to the largest eigenvalue. The method may also be extended to obtain the next largest eigenvalue by forming the new matrix

$$\tilde{\mathbf{M}} = \mathbf{M} - \frac{\lambda_1}{\mathbf{v}_1^T \mathbf{v}_1} (\mathbf{v}_1 \mathbf{v}_1^T) \tag{8-23}$$

where $\mathbf{v}_1^T$ is the transpose of $\mathbf{v}_1$. This matrix has the same eigenvalues and eigenvectors as $\mathbf{M}$ except that λ_1 is replaced by zero. This method is called *deflation*.

The power method is useful if we require only the first few largest eigenvalues. If we need to compute all the eigenvalues, then there are more efficient methods. Finally, you could use the power method to compute the spectral radius and determine the matrix stability of a PDE scheme. However, you would essentially be running the scheme and seeing if the solution diverged as $t \to \infty$.

EXERCISES

1. Apply the von Neumann stability analysis to the FTCS scheme for the diffusion equation. Confirm that the method is stable only if $\tau \leq h^2/2\kappa$.

2. Another method for solving the advection equation is the Lax–Wendroff scheme,

$$a_j^{n+1} = a_j^n - \frac{c\tau}{h} \left\{ \tfrac{1}{2}(a_{j+1}^n - a_{j-1}^n) - \frac{c\tau}{2h} (a_{j+1}^n + a_{j-1}^n - 2a_j^n) \right\}$$

Apply the von Neumann stability analysis to this scheme. Confirm that the method is stable only if $\tau \leq h/|c|$ (i.e., the CFL criterion).

3. Another method for solving the advection equation is the leap-frog scheme

$$a_j^{n+1} = a_j^{n-1} - \frac{c\tau}{h}(a_{j+1}^n - a_{j-1}^n)$$

Apply the von Neumann stability analysis to this scheme; notice that you will have a quadratic equation for ξ. Confirm that the method is stable only if $\tau \leq h/|c|$ (i.e., the CFL criterion).

4. The *Richardson method* for solving the diffusion equation uses centered derivatives in both space and time:

$$\frac{T_i^{n+1} - T_i^{n-1}}{2\tau} = \kappa \frac{T_{i+1}^n + T_{i-1}^n - 2T_i^n}{h^2}$$

Apply the von Neumann stability analysis to this scheme; notice that you will have a quadratic equation for ξ. Show that this scheme is unconditionally *unstable*.

5. The two dimensional diffusion equation for temperature (Fourier equation) may be solved using the FTCS scheme as,

$$T_{i,j}^{n+1} = T_{i,j}^n + \frac{\kappa\tau}{h_x^2}(T_{i+1,j}^n + T_{i-1,j}^n - 2T_{i,j}^n) + \frac{\kappa\tau}{h_y^2}(T_{i,j+1}^n + T_{i,j-1}^n - 2T_{i,j}^n)$$

where h_x, h_y are the x and y grid spacings, respectively. Apply the von Neumann stability analysis to this scheme and show that it is stable if $\tau \leq \frac{1}{2\kappa}\left(\frac{1}{h_x^2} + \frac{1}{h_y^2}\right)^{-1}$.

6. Write a program that evaluates the spectral radius of the matrix $\mathbf{A}$ defined in equation (8-12). Reproduce the graph in Figure 8.2.

7. The FTCS scheme for the advection equation with periodic boundary conditions may be written as

$$\mathbf{a}^{n+1} = \left(\mathbf{I} - \frac{c\tau}{2h}\mathbf{B}\right)\mathbf{a}^n$$

where

$$\mathbf{a}^n = \begin{bmatrix} a_1^n \\ a_2^n \\ a_3^n \\ \vdots \\ a_N^n \end{bmatrix}; \quad \mathbf{B} = \begin{bmatrix} 0 & 1 & 0 & \cdots & -1 \\ -1 & 0 & 1 & \cdots & 0 \\ 0 & -1 & 0 & \cdots & 0 \\ \vdots & \vdots & \vdots & \ddots & \vdots \\ 1 & 0 & 0 & \cdots & 0 \end{bmatrix}$$

Modify the program from the previous exercise and demonstrate that this scheme is unconditionally unstable.

8. The Lax scheme for the advection equation may be written as

$$\mathbf{a}^{n+1} = \left(\tfrac{1}{2}\mathbf{C} - \frac{c\tau}{2h}\mathbf{B}\right)\mathbf{a}^n$$

where **a** and **B** are defined in the previous problem and

$$
\mathbf{C} =
\begin{bmatrix}
0 & 1 & 0 & \cdots & 1 \\
1 & 0 & 1 & \cdots & 0 \\
0 & 1 & 0 & \cdots & 0 \\
\vdots & \vdots & \vdots & \ddots & \vdots \\
1 & 0 & 0 & \cdots & 0
\end{bmatrix}
$$

Modify the program from Exercise 8.6 and demonstrate that matrix stability for the Lax scheme is given by the CFL condition.

9. A stricter condition for matrix stability is that the norm of **A**, $\|\mathbf{A}\|$, is less than or equal to unity. This condition is stronger since $\rho(\mathbf{A}) \leq \|\mathbf{A}\|$. There are a variety of ways of defining the norm; two of the easiest to compute are the 1-norm,

$$
\|\mathbf{A}\|_1 \equiv \max_{j=1,\ldots N} \left\{ \sum_{i=1}^{N} |A_{ij}| \right\}
$$

and the ∞-norm,

$$
\|\mathbf{A}\|_\infty \equiv \max_{i=1,\ldots N} \left\{ \sum_{j=1}^{N} |A_{ij}| \right\}
$$

Modify the program from Exercise 8.6 to compute these two norms and plot them along with the spectral radius.

10. Write a function that computes the largest eigenvalue of a matrix using the power method. Establish a suitable criterion for when to stop the iteration process.

8.2 IMPLICIT SCHEMES

Schrödinger Equation

As a physicist you need no introduction to the Schrödinger equation.[5] For a particle of mass m in one dimension, it may be written as

$$
i\hbar \frac{\partial}{\partial t} \psi(x, t) = -\frac{\hbar^2}{2m} \frac{\partial^2}{\partial x^2} \psi + V(x)\psi
\tag{8-24}
$$

where $\psi(x, t)$ is the wave function and $V(x)$ is the potential. In operator notation, we may write the Schrödinger equation as

$$
i\hbar \frac{\partial \psi}{\partial t} = H\psi
\tag{8-25}
$$

---

[5] But if you need a refresher, see D. Saxon, *Elementary Quantum Mechanics* (San Francisco: Holden-Day, 1968), or L. Schiff, *Quantum Mechanics* (New York: McGraw-Hill, 1968).

where H is the Hamiltonian operator,

$$H = -\frac{\hbar^2}{2m}\frac{\partial^2}{\partial x^2} + V(x) \qquad (8\text{-}26)$$

As before, we discretize space and time in increments of h and τ, respectively; Planck's constant always appears as $\hbar$ so there should be no confusion with the grid spacing h. In our notation, the discretized wave function is

$$\begin{aligned}\psi_j^n &\equiv \psi(x_j, t_n) \\ &= \psi[x = (j-1)h, t = (n-1)\tau]\end{aligned} \qquad (8\text{-}27)$$

The FTCS scheme discretizes the Schrödinger equation as

$$i\hbar\,\frac{\psi_j^{n+1} - \psi_j^n}{\tau} = -\frac{\hbar^2}{2m}\frac{\psi_{j+1}^n + \psi_{j-1}^n - 2\psi_j^n}{h^2} + V_j\psi_j^n \qquad (8\text{-}28)$$

where $V_j \equiv V(x = (j-1)h)$.

Since the Hamiltonian is a linear operator, we may write the previous equation as

$$i\hbar\,\frac{\psi_j^{n+1} - \psi_j^n}{\tau} = \sum_{k=1}^{N} H_{jk}\psi_k^n \qquad (8\text{-}29)$$

where the matrix $\mathbf{H}$ is the discretized form of the Hamiltonian operator

$$H_{ij} = -\frac{\hbar^2}{2m}\frac{\delta_{i,j+1} + \delta_{i,j-1} - 2\delta_{ij}}{h^2} + V_i\delta_{ij} \qquad (8\text{-}30)$$

Solving (8-29) for ψ_j^{n+1} gives us our numerical scheme; in matrix notation it may be written as

$$\boldsymbol{\psi}^{n+1} = \left(\mathbf{I} - \frac{i\tau}{\hbar}\,\mathbf{H}\right)\boldsymbol{\psi}^n \qquad (8\text{-}31)$$

where $\boldsymbol{\psi}^n$ is a column vector and $\mathbf{I}$ is the identity matrix. Equation (8-31) is the explicit FTCS scheme for solving the Schrödinger equation.

Implicit Schemes

The disadvantage of the FTCS scheme is that it is numerically unstable if the time step, τ, is too large. This stability problem motivates us to consider alternative approaches. For example, suppose that we apply the Hamiltonian to the *future value* of ψ,

$$i\hbar\,\frac{\psi_j^{n+1} - \psi_j^n}{\tau} = \sum_{k=1}^{N} H_{jk}\psi_k^{n+1} \tag{8-32}$$

or

$$\psi^{n+1} = \psi^n - \frac{i\tau}{\hbar}\,\mathbf{H}\psi^{n+1} \tag{8-33}$$

Collecting ψ^{n+1} we have

$$\left(\mathbf{I} + \frac{i\tau}{\hbar}\,\mathbf{H}\right)\psi^{n+1} = \psi^n \tag{8-34}$$

Solving for ψ^{n+1} we have

$$\psi^{n+1} = \left(\mathbf{I} + \frac{i\tau}{\hbar}\,\mathbf{H}\right)^{-1}\psi^n \tag{8-35}$$

This scheme is called the *implicit FTCS method*; compare it with the explicit FTCS scheme, equation (8-31). Note that since $(1 - \varepsilon)^{-1} \to (1 + \varepsilon)$ as $\varepsilon \to 0$, our implicit and explicit schemes are equivalent in the limit $\tau \to 0$.

Our new method requires the evaluation of a matrix inverse; this is a common feature of implicit schemes. Of course we wouldn't even consider doing this extra work without some benefit. The advantage of the implicit FTCS scheme is that it is *unconditionally stable*, as may be shown using von Neumann stability analysis. Unconditional stability is a general feature of implicit schemes.

While the fully implicit scheme is very appealing because of its stability, we also want a method to be accurate. Just because the solution doesn't blow up doesn't mean that it is correct. A more accurate scheme is the *Crank–Nicolson method*. It basically takes the average between the implicit and explicit FTCS schemes

$$i\hbar\,\frac{\psi_j^{n+1} - \psi_j^n}{\tau} = \frac{1}{2}\sum_{k=1}^{N} H_{jk}(\psi_k^{n+1} + \psi_k^n) \tag{8-36}$$

In matrix form it may be written as

$$\psi^{n+1} = \psi^n - \frac{i\tau}{2\hbar}\,\mathbf{H}(\psi^{n+1} + \psi^n) \tag{8-37}$$

or

$$\left(\mathbf{I} + \frac{i\tau}{2\hbar}\,\mathbf{H}\right)\psi^{n+1} = \left(\mathbf{I} - \frac{i\tau}{2\hbar}\,\mathbf{H}\right)\psi^n \tag{8-38}$$

Finally, isolating the ψ^{n+1} term on the left-hand side we have

$$\psi^{n+1} = \left(\mathbf{I} + \frac{i\tau}{2\hbar}\,\mathbf{H}\right)^{-1}\left(\mathbf{I} - \frac{i\tau}{2\hbar}\,\mathbf{H}\right)\psi^n \tag{8-39}$$

As nasty as this looks, the Crank–Nicolson scheme is the best of the three schemes. Being centered in both space and time, it is second-order accurate and unitary.[6] The Crank–Nicolson scheme is also unconditionally stable.

Wave Packet for a Free Particle

Before putting together our program to solve the Schrödinger equation, we need to think about what initial conditions we want to use. A reasonable initial condition would be the wave packet for a particle localized about x_0 with a packet width of σ_0 and an average momentum $p_0 = \hbar k_0$ (k_0 is the average wave number).

We will use a Gaussian wave packet; the initial wave function is

$$\psi(x,\ t = 0) = \frac{1}{\sqrt{\sigma_0\sqrt{\pi}}}\,e^{ik_0 x}e^{-(x-x_0)^2/2\sigma_0^2} \tag{8-40}$$

Notice that this wave function is normalized so that $\int_{-\infty}^{\infty} |\psi|^2\,dx = 1$. The Gaussian wave packet has the special property that the uncertainty product $\Delta x\,\Delta p$ has its minimum theoretical value of $\hbar/2$.

In free space, the wave function evolves as

$$\psi(x,\ t) = \frac{1}{\sqrt{\sigma_0\sqrt{\pi}}}\,\frac{\sigma_0}{\alpha}\,e^{ik_0(x-p_0 t/2m)}e^{-(x-x_0-p_0 t/m)^2/2\alpha^2} \tag{8-41}$$

where $\alpha^2 = \sigma_0^2 + i\hbar t/m$. The probability density $P(x,\ t) = |\psi(x,\ t)|^2$ is

$$P(x,\ t) = \frac{\sigma_0}{|\alpha|^2\sqrt{\pi}}\,\exp\left[-\left(\frac{\sigma_0}{|\alpha|}\right)^4\frac{(x-x_0-p_0 t/m)^2}{\sigma_0^2}\right] \tag{8-42}$$

thus, $P(x,\ t)$ remains a Gaussian in time.

By symmetry, the maximum of the Gaussian equals the expectation value $\langle x \rangle = \int x P(x,\ t)\,dx$. In time, it moves as $\langle x(t) \rangle = x_0 + p_0 t/m$, that is, the packet

---

[6] A. Goldberg, H. Schey, and J. Schwartz, "Computer-Generated Motion Pictures of One-Dimensional Quantum-Mechanical Transmission and Reflection Phenomena," *Am. J. Phys.*, **35**, 177–86 (1967).

moves with a velocity p_o/m. The Gaussian spreads in time; its standard deviation is

$$\sigma(t) = \sigma_0 \sqrt{\left(\frac{|\alpha|}{\sigma_0}\right)^4} = \sigma_0 \sqrt{1 + \frac{\hbar^2 t^2}{m^2 \sigma_0^4}} \tag{8-43}$$

The details of this calculation are in any undergraduate quantum mechanics text.[7]

Crank–Nicolson Program for a Free Particle

The program `schro` solves the Schrödinger equation using the Crank–Nicolson scheme (see Listing 8.1). The initial condition is a Gaussian packet (8-40); it is set up on lines 25–32. The boundary conditions are periodic so when the particle moves out the right side it reappears on the left.

On lines 10–20 the Hamiltonian matrix is defined. The interior rows are defined according to equation (8-30); in this version the potential, $V(x)$, is zero. The first and last rows of the matrix ham are defined so as to give periodic boundary conditions. On lines 22–23 we define the matrix dCN that implements the Crank–Nicolson scheme (8-39). This matrix is used inside the main loop on line 45 to compute the new value of the wave function.

LISTING 8.1 Program `schro`. Computes the evolution of a Gaussian wave packet by solving the Schrödinger equation using the Crank–Nicolson scheme.

```
1   %  schro - Program to solve the Schrodinger equation
2   %  using Crank-Nicolson scheme (Free particle version)
3   clear;  help schro;   % Clear memory and print header
4   i_imag = sqrt(-1);    % Imaginary i
5   N = input('Enter number of grid points - ');
6   L = 100;              % System extends from -L/2 to L/2
7   h = L/(N-1);          % Grid size
8   h_bar = 1; mass = 1;  % Natural units
9   tau = input('input time step - ');
10  %%%%%  Set up the Hamiltonian operator matrix ham(,) %%%%%
11  ham = zeros(N);  % Set all elements to zero
12  coeff = -h_bar^2/(2*mass*h^2);
13  for i=2:(N-1)
14    ham(i,i-1) = coeff;
15    ham(i,i) = -2*coeff;  % Set interior rows
16    ham(i,i+1) = coeff;
17  end
```

---

[7] See, for example, D. Saxon, *Elementary Quantum Mechanics* (San Francisco: Holden-Day, 1968), chapter 4, section 5; L. Schiff, *Quantum Mechanics* (New York: McGraw-Hill, 1968), chapter 3, section 12.

```
18   % Periodic boundary conditions
19   ham(1,N) = coeff;    ham(1,1) = -2*coeff; ham(1,2) = coeff;
20   ham(N,N-1) = coeff; ham(N,N) = -2*coeff; ham(N,1) = coeff;
21   disp('Computing the Crank-Nicolson matrix')
22   dCN = ( inv(eye(N) + .5*i_imag*tau/h_bar*ham) * ...
23                   (eye(N) - .5*i_imag*tau/h_bar*ham) );
24   %%%% Initialize wavefunction %%%%
25   x0 = 0;            % Location of the center of the wavepacket
26   velocity = 0.5;    % Average velocity of the packet
27   k0 = mass*velocity/h_bar;        % Average wavenumber
28   sigma0 = L/10;    % Standard deviation of the wavefunction
29   Norm_psi = 1/(sqrt(sigma0*sqrt(pi)));  % Normalization
30   x = h*(0:N-1) - L/2;
31   psi = Norm_psi * exp(i_imag*k0*x') .* ...
32                        exp(-(x'-x0).^2/(2*sigma0^2));
33   plot(x,real(psi),x,imag(psi));
34   title('Hit any key to continue');
35   xlabel('Position');  ylabel('Real(psi) and imag(psi)');
36   pause;
37   %%%% Set loop and plot variables %%%%
38   max_iter = L/(velocity*tau); % The particle should circle
                                    system
39   plot_iter = max_iter/20;        % Produce 20 curves
40   p_plot(:,1) = psi.*conj(psi);   % Record initial condition
41   iplot = 2;
42   %%%% MAIN LOOP %%%%
43   disp('Entering main loop')
44   for iter=1:max_iter
45     psi = dCN*psi;  % Crank-Nicolson scheme
46     if( rem(iter,plot_iter) < 1 )  % Every plot_iter steps
47       p_plot(:,iplot) = psi.*conj(psi); % record amplitude
48       plot(x,p_plot(:,iplot));
49       xlabel('position'); ylabel('Prob. density');
50       title(sprintf('Finished %g of %g iteration',
                                    iter,max_iter));
51       iplot = iplot+1;
52     end
53   end
54   plot(x,p_plot(:,1:3:iplot-2),x,p_plot(:,iplot-1));
55   xlabel('Position'); ylabel('Probability density');
56   title('Probability density at various times');
```

When the program is run it first displays a plot of $\psi(x,\ t = 0)$; since the function is complex we plot the real and imaginary parts separately. For the wave function shown in Figure 8.3 (for $N = 30$ grid points), the discretization is rather noticeable.

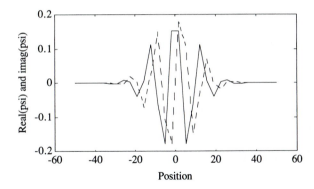

Figure 8.3 Real and imaginary parts of $\psi(x, t = 0)$ as computed by schro for $N = 30$.

The program computes $\psi(x, t)$ up to a time such that the pulse should circle the system once and return to the center (see line 38). As discussed above, the width of the pulse increases with time as given by equation (8-43). The plot of $|\psi(x, t)|^2$ versus x for various values of t is shown in Figure 8.4. Notice that the evolution appears normal except that the pulse only travels about half the expected distance. If we lower the time step, the result is not significantly affected.

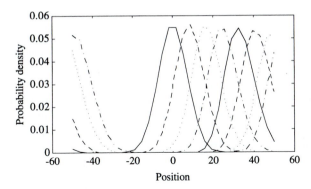

Figure 8.4 Probability density for the particle as a function of position for various times for $N = 30$. Notice that the packet moves to the right and, due to the periodic boundary conditions, reappears from the left. The time step is $\tau = 1.0$.

However, if we increase the number of grid points to $N = 80$, the spatial discretization is less prominent; the plot of the wave function is smoother (see Figure 8.5).[8] The probability density, $|\psi(x, t)|^2$, versus x for various values of t, is shown in Figure 8.6. This result looks much better since the pulse almost returns to the origin. If we further increase N, we get even better results (although this is problematic on some machines because of the amount of memory required by the large arrays).

Our error when $N = 30$ arises from how well we are representing the initial condition. The spatial discretization suppresses the shorter wavelengths. Because of this suppression of the higher wave number modes, the discretized Gaussian

---

[8] Note: The student version of MATLAB cannot handle matrices of this size. However, see Section 8.3.

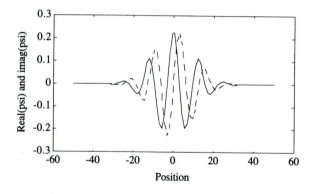

Figure 8.5 Real and imaginary parts of $\psi(x, t = 0)$ as computed by schro for $N = 80$.

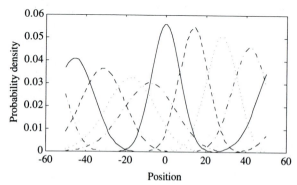

Figure 8.6 Probability density for the particle as a function of position for various times for $N = 80$. Notice that the packet moves to the right and, due to the periodic boundary conditions, reappears from the left. The time step is $\tau = 1.0$.

pulse ψ_i^n has a lower momentum than $\psi(x, t)$. This error arises from the same aliasing problem we encountered in Chapter 5. Our highest wave number is limited by the grid spacing. The solution is to use a finer grid with more points. Unfortunately, we are already running into memory problems with our large matrices. In the next section we reformulate the Crank–Nicolson scheme to avoid these difficulties.

EXERCISES

11. Write a program that uses equation (8-42) to compute and plot $P(x, t) = |\psi(x, t)|^2$ versus x for various values of t. Compare with the results from schro.

12. (a) Write a program that solves the diffusion equation using the implicit FTCS scheme. Repeat Exercise 6.2 and compare the results from your program with those from the explicit FTCS program, dftcs.

    (b) Repeat part (a) using the Crank–Nicolson scheme.

13. (a) Using von Neumann stability analysis, show that the Crank–Nicolson scheme for solving the diffusion equation is unconditionally stable.

(b) Modify your program from Exercise 8.6 to show that the Crank–Nicolson scheme for solving the diffusion equation is matrix stable.

14. (a) Modify the schro program to compute the energy of the particle as

$$\langle E \rangle = \frac{\int dx \psi^*(x, t) H \psi(x, t)}{\int dx \psi^*(x, t) \psi(x, t)} \approx \frac{\sum_{i,j} \psi_i^* H_{ij} \psi_j}{\sum_i \psi_i^* \psi_i}$$

Is energy conserved?

(b) Obtain an expression for the total momentum of the particle. In schro compute the total momentum as a function of time; is it conserved?

15. Modify the schro program to include the delta function potential $V(x) = U\delta(x - L/2)$. Vary the amplitude U and do runs where it is less than, equal to, and more than $E = \hbar^2 k_0^2 / 2m$, the energy of the particle. Show that some of the wave function penetrates the potential even when $E < U$. If memory allows, increase L, the system size, to distinctly separate the reflected and transmitted waves.

16. An important PDE from nonlinear acoustics is *Burger's equation*,

$$\frac{\partial a}{\partial t} = -a\frac{\partial a}{\partial x} + \kappa \frac{\partial^2 a}{\partial x^2}$$

Write a program that solves it by the explicit/implicit scheme

$$\frac{a_j^{n+1} - a_j^n}{\tau} = -a_j^n \frac{a_{j+1}^n - a_{j-1}^n}{2h}$$

$$+ \frac{1}{2}\kappa \left(\frac{a_{j+1}^n + a_{j-1}^n - 2a_j^n}{h^2} + \frac{a_{j+1}^{n+1} + a_{j-1}^{n+1} - 2a_j^{n+1}}{h^2} \right)$$

Take the initial condition: $a(x, 0) = 1$ if $x < 0$ and $a(x, 0) = -1$ if $x > 0$. Use the Dirichlet boundary conditions: $a(\pm L/2, t) = \mp 1$; try $L = 10$, $\kappa = 1$. Compare your results with the exact (for $L \to \infty$) solution,

$$a(x, t) = \kappa \frac{F(x, t) - F(-x, t)}{F(x, t) + F(-x, t)}$$

where

$$F(x, t) \equiv \frac{1}{2} e^{t-x} \left\{ 1 - \text{erf}\left(\frac{x - 2t}{2\sqrt{t}}\right) \right\}$$

and erf(x) is the error function.

17. An important equation from the theory of solitons is the *Korteweg–de Vries* (KdV) equation,[9]

$$\frac{\partial \rho}{\partial t} = -6\rho \frac{\partial \rho}{\partial x} - \frac{\partial^3 \rho}{\partial x^3}$$

[9] P. G. Drazin and R. S. Johnson, *Solitons, An Introduction* (Cambridge: Cambridge University Press, 1989).

Write a program that solves it using the explicit/implicit scheme

$$\frac{\rho_j^{n+1} - \rho_j^n}{\tau} = -6\rho_j^n \frac{\rho_{j+1}^n - \rho_{j-1}^n}{2h}$$

$$-\frac{1}{2}\left(\frac{\rho_{j+2}^n - 2\rho_{j+1}^n + 2\rho_{j-1}^n - \rho_{j-2}^n}{2h^3} + \frac{\rho_{j+2}^{n+1} - 2\rho_{j+1}^{n+1} + 2\rho_{j-1}^{n+1} - \rho_{j-2}^{n+1}}{2h^3}\right)$$

Use Dirichlet boundary conditions, $\rho(x = \pm L/2) = 0$. Test your program for the solitary wave solution of the KdV equations: $\rho(x, t) = 2 \text{ sech}^2(x - 4t)$.

8.3 SPARSE MATRICES*

General Properties

As the complexity of our problems increases we find ourselves working with larger and larger matrices. You may have already run into memory problems, especially when using the student version of MATLAB. However, you probably noticed that the Hamiltonian matrix used in the previous section is very sparse (i.e., almost all the elements of the matrix are zero). In this section we examine how to exploit this feature to allow us to work with larger systems.

Sparse matrices fall into two categories depending on their structure. The more general case is when the nonzero elements are arbitrarily distributed, such as in the matrix sketched below:

$$\begin{bmatrix}
0 & * & 0 & 0 & * & 0 & 0 & 0 \\
* & 0 & 0 & * & * & 0 & 0 & 0 \\
0 & * & 0 & 0 & 0 & 0 & 0 & 0 \\
* & * & * & 0 & 0 & 0 & 0 & * \\
0 & 0 & * & 0 & 0 & * & * & 0 \\
0 & 0 & 0 & 0 & * & 0 & 0 & 0 \\
* & * & 0 & 0 & 0 & 0 & * & * \\
0 & 0 & 0 & * & 0 & 0 & 0 & 0
\end{bmatrix}$$

where the nonzero elements are indicated by asterisks. Such matrices may be stored in a compressed format by recording the values of the nonzero elements and their locations. Typically this is done using a linked list.[10]

---

[10] I. S. Duff, A. M. Erisman, and J. K. Reid, *Direct Methods for Sparse Matrices* (Oxford: Clarendon Press, 1989), chapter 2.

The simpler, and fortunately more common, case is when the matrix has a definite, known structure. Some examples of such matrices are sketched below:

$$
\begin{bmatrix}
* & * & 0 & 0 & 0 & 0 & 0 & 0 \\
0 & * & * & 0 & 0 & 0 & 0 & 0 \\
0 & 0 & * & * & 0 & 0 & 0 & 0 \\
0 & 0 & 0 & * & * & 0 & 0 & 0 \\
0 & 0 & 0 & 0 & * & * & 0 & 0 \\
0 & 0 & 0 & 0 & 0 & * & * & 0 \\
* & * & * & * & * & * & * & * \\
0 & 0 & 0 & 0 & 0 & 0 & 0 & *
\end{bmatrix}
\qquad
\begin{bmatrix}
* & * & 0 & 0 & 0 & 0 & 0 & 0 \\
* & * & 0 & 0 & 0 & 0 & 0 & 0 \\
0 & 0 & 0 & 0 & 0 & 0 & 0 & 0 \\
0 & 0 & 0 & * & * & * & 0 & 0 \\
0 & 0 & 0 & * & * & * & 0 & 0 \\
0 & 0 & 0 & * & * & * & 0 & 0 \\
0 & 0 & 0 & 0 & 0 & 0 & * & * \\
0 & 0 & 0 & 0 & 0 & 0 & * & *
\end{bmatrix}
$$

The first matrix is an example of a banded matrix and the second is a block diagonal matrix.

The solution of sparse matrix problems is so important that a significant industry has arisen in the numerical analysis community.[11] It would be far beyond the scope of this book to go into these specialized methods, but there is one special case that is so simple and so common that I believe it is valuable for you to learn it. It also gives you a bit of a flavor of what is involved when working with sparse matrices.

Tridiagonal Matrices

The special case we consider is the tridiagonal matrix; it has the following structure:

$$
\mathbf{A} =
\begin{bmatrix}
\beta_1 & \gamma_1 & 0 & \cdots & 0 \\
\alpha_1 & \beta_2 & \gamma_2 & \cdots & 0 \\
0 & \alpha_2 & \beta_3 & \cdots & 0 \\
\vdots & \vdots & \vdots & \ddots & \vdots \\
0 & 0 & 0 & \cdots & \beta_N
\end{bmatrix}
\tag{8-44}
$$

Since only the elements on the main diagonal (the β's) and on the first sub and superdiagonals (the α's and γ's) are nonzero, we may store the matrix in a com-

---

[11] S. Pissanetsky, *Sparse Matrix Technology* (London: Academic Press, 1984).

pressed (or packed) form using the matrix

$$
\mathbf{A}_c = \begin{bmatrix} ? & \beta_1 & \gamma_1 \\ \alpha_1 & \beta_2 & \gamma_2 \\ \alpha_2 & \beta_3 & \gamma_3 \\ \vdots & \vdots & \vdots \\ \alpha_{N-1} & \beta_N & ? \end{bmatrix} \tag{8-45}
$$

The two corner elements marked with question marks are unused.

We now reformulate the Gaussian elimination algorithm, but specialized for the case of tridiagonal matrices. The basic method requires no major modification; many operations may be skipped since most elements of the matrix are zero. To solve the linear system $\mathbf{Ax} = \mathbf{b}$, the forward elimination stage requires only that we modify the values on the main diagonal as

$$
\beta_i \leftarrow \beta_i - \frac{\alpha_{i-1}}{\beta_{i-1}} \gamma_{i-1} \qquad i = 2, \ldots, N \tag{8-46}
$$

and the elements of $\mathbf{b}$ as

$$
b_i \leftarrow b_i - \frac{\alpha_{i-1}}{\beta_{i-1}} b_{i-1} \qquad i = 2, \ldots, N \tag{8-47}
$$

For the backsubstitution stage we may easily obtain $\mathbf{x}$ using $x_N = b_N/\beta_N$ and the recursion relation

$$
x_i = \frac{b_i - \gamma_i x_{i+1}}{\beta_i} \qquad i = N - 1, \ldots, 1 \tag{8-48}
$$

This formulation of Gaussian elimination for a tridiagonal matrix is sometimes called the *Thomas algorithm*. The function `tri_ge` (Listing 8.2) performs Gaussian elimination on a packed tridiagonal matrix.

LISTING 8.2 Function `tri_ge`. Gaussian elimination routine for tridiagonal matrices.

```
1   function x = tri_ge(a,b)
2   % Function to solve b = a*x by Gaussian elimination where
3   % the matrix a is a packed tridiagonal matrix
4   % Input -
5   %    a = Packed tridiagonal matrix, N by N unpacked
6   %    b = Column vector of length N
7   % Output -
8   %    x = Solution of b = a*x; Column vector of length N
```

```
9    [N,M] = size(a);
10   [NN,MM] = size(b);
11   if( N ~= NN | MM ~= 1)
12     error('Problem in tri_GE, inputs are incompatible');
13   end
14   %%%% Unpack diagonals of triangular matrix into vectors %%%%
15   alpha(1:N-1) = a(2:N,1);
16   beta(1:N) = a(1:N,2);
17   gamma(1:N-1) = a(1:N-1,3);
18   %%%% Forward elimination %%%%
19   for i=2:N
20     coeff = alpha(i-1)/beta(i-1);
21     beta(i) = beta(i) - coeff*gamma(i-1);
22     b(i) = b(i) - coeff*b(i-1);
23   end
24   %%%% Back Substitution %%%%
25   x(N) = b(N)/beta(N);
26   for i=N-1:-1:1
27     x(i) = (b(i) - gamma(i)*x(i+1))/beta(i);
28   end
29   x = x.';     % Return x as a column vector
30   return;
```

You might expect that we would next assemble a program to compute the inverse of a tridiagonal matrix. There is only one problem: The inverse of a tridiagonal matrix is not necessarily sparse. Go into MATLAB and try a few matrices constructed at random; you will find that the inverse is usually a full matrix.

Crank–Nicolson for Tridiagonal Matrices

Returning to the Schrödinger equation, we want to take the Crank–Nicolson scheme

$$\psi^{n+1} = \left(\mathbf{I} + \frac{i\tau}{2\hbar}\,\mathbf{H}\right)^{-1}\left(\mathbf{I} - \frac{i\tau}{2\hbar}\,\mathbf{H}\right)\psi^n \tag{8-49}$$

and rewrite it in such a way that we do not have to compute a matrix inverse. Rearranging terms gives

$$\psi^{n+1} = \left(\mathbf{I} + \frac{i\tau}{2\hbar}\,\mathbf{H}\right)^{-1}\left[2\mathbf{I} - \left(\mathbf{I} + \frac{i\tau}{2\hbar}\,\mathbf{H}\right)\right]\psi^n \tag{8-50}$$

or

$$\psi^{n+1} = \left[2\left(\mathbf{I} + \frac{i\tau}{2\hbar}\,\mathbf{H}\right)^{-1} - \mathbf{I}\right]\psi^n \tag{8-51}$$

or

$$\psi^{n+1} = (\mathbf{Q}^{-1} - \mathbf{I})\psi^n$$
$$= \mathbf{Q}^{-1}\psi^n - \psi^n \tag{8-52}$$

where $\mathbf{Q} = \frac{1}{2}[\mathbf{I} + (i\tau/2\hbar)\mathbf{H}]$.

The computation of a matrix inverse is avoided by splitting the problem into two stages. First we solve the following linear system for the vector χ,

$$\mathbf{Q}\chi = \psi^n \tag{8-53}$$

and then update our solution as

$$\psi^{n+1} = \chi - \psi^n \tag{8-54}$$

Notice that while we do not have to take the inverse of a matrix, we do have to solve a linear system, equation (8-53), at each time step.

We are now ready to assemble the program. For a change of pace, let's use Dirichlet boundary conditions and set the wave function to zero on the boundaries. This is equivalent to having an infinite potential at the boundaries. Our program will thus be computing $\psi(x, t)$ for a particle in a box.

The implementation of the scheme discussed above is straightforward. Our Hamiltonian matrix, in packed form, may be constructed by replacing lines 10–20 in schro with

```
%%%%  Set up the Hamiltonian operator matrix ham(,) %%%%
coeff = -h_bar^2/(2*mass*h^2);
for i=2:(N-1)
   ham(i,1) = coeff;
   ham(i,2) = -2*coeff;
   ham(i,3) = coeff;
end
% Dirichlet boundary conditions
ham(1,1)=0;  ham(1,2)=0;  ham(1,3)=0;
ham(N,1)=0;  ham(N,2)=0;  ham(N,3)=0;
```

Notice that we implement the Dirichlet boundary condition by setting the first and last rows of ham to zero. The matrix dCN is no longer used (eliminate lines 21–23); instead, we build the matrix $\mathbf{Q}$ using

```
%%%% Set up the matrix Q %%%%
tri_eye = zeros(N,3);
tri_eye(1:N,2) = ones(N,1);    % Identity matrix (packed)
Q = 0.5*( tri_eye + 0.5*i_imag*tau/h_bar * ham );
```

To get a perfectly reflecting boundary, we set $\psi_1 = \psi_N = 0$; add the following lines below line 32:

```
% Set psi to zero at boundary; perfect reflector
psi(1) = 0;   psi(N) = 0;
```

Finally, to implement the Crank–Nicolson scheme, we replace line 45 with the following two lines:

```
chi = tri_ge(Q,psi);   % Solve for vector chi
psi = chi - psi;       % Crank-Nicolson scheme
```

Compare with equations (8-53) and (8-54).

With these modifications we now increase the number of grid points to $N = 101$, set the time step to $\tau = 1$ and let it run (and go get some coffee if you are on a slow machine). The solution obtained is illustrated as a wire-mesh plot in Figure 8.7. Notice the interesting structure in the wave function as it rebounds off the reflecting wall. After it moves away from the wall, the wave packet regains its original Gaussian shape (with the appropriate spreading). For a collection of pictures showing wave packets interacting with potentials, see Saxon.[12]

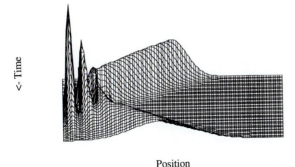

Position

Figure 8.7 Square amplitude of the wavefunction $\psi(x, t)$ as a function of position and time.

<- Time

EXERCISES

18. Solve the following problem by hand using the Thomas algorithm,

$$\begin{bmatrix} 1 & 1 & 0 & 0 \\ 1 & 2 & 0 & 0 \\ 0 & -1 & 3 & 2 \\ 0 & 0 & 1 & 4 \end{bmatrix} \begin{bmatrix} x_1 \\ x_2 \\ x_3 \\ x_4 \end{bmatrix} = \begin{bmatrix} 3 \\ 5 \\ 15 \\ 19 \end{bmatrix}$$

[12] D. Saxon, *Elementary Quantum Mechanics* (San Francisco: Holden-Day, 1968), section 6.10.

What do the matrix and the right-hand side look like at the end of forward elimination?

19. Using the `tri_ge` function for the Crank–Nicolson scheme is inefficient for two reasons. The first part of forward elimination (8-46) is repeated at every time step even though the matrix **Q** is fixed. Second, several terms are recomputed at every time step even though they remain constant. Improve the efficiency by breaking up the `tri_ge` function into two separate functions. The first function will be invoked once outside the main loop and the second function will be invoked at each iteration. You will essentially be implementing LU decomposition. Rewrite `schro` to use your routines and quantitatively measure the improvement (e.g., use the flops command).

20. Modify `schro` to include the potential,

$$V_i = U[\delta(x + L/4) + \delta(x - L/4)]$$

which makes the center of the system a "box." Due to tunneling, the particle is not contained by the box even when its energy is less than U. Compute the probability that the particle is inside the box,

$$P[t = (n - 1)\tau] = \sum_{i=N/4}^{3N/4} (\psi_i^n)^* \psi_i^n$$

and plot $P(t)$ versus t for various values of U.

21. For periodic boundary conditions the Hamiltonian matrix is not tridiagonal. The elements at the opposite corners, ham(1, N) and ham(N, 1), are nonzero. Derive a modified version of the Thomas algorithm that can perform Gaussian elimination on matrices of this type. Write a function that implements this algorithm and demonstrate its use in a modified version of `schro`.

22. (a) Derive a modified version of the Thomas algorithm that applies to pentadiagonal matrices, that is, matrices for which only elements on the five central diagonals are nonzero.

    (b) Write a computer routine that implements your algorithm from part (a).

BEYOND THIS CHAPTER

Our two stability analyses are suitable only for linear partial differential equations. On the other hand, most interesting research problems involve nonlinear equations. We may still use von Neumann or matrix analysis by linearizing our PDE about a reference state. Also, the stability criteria we've seen have a physical basis (e.g., the CFL condition is given by the time it takes a wave to move one grid spacing). Going back to the original physical problem, you can usually find some characteristic time scale to guide your selection of a time step. Finally, there are some specialized techniques (e.g., energy stability analysis) that can sometimes be used with nonlinear PDEs. See Richtmyer and Morton for a more complete discussion.[13]

[13] R. D. Richtmyer and K. W. Morton, *Difference Methods for Initial Value Problems*, 2d ed. (New York: Wiley, 1967).

This chapter covers two techniques for studying stability. A related problem is determining the dissipation and dispersion of a numerical scheme. We saw in Chapter 6 that the Lax scheme had an undesirably large numerical dissipation when the time step was too small. Furthermore, we want our numerical scheme to preserve the same dispersion relation as the original PDE. Both dissipation and dispersion may be studied by a simple extension of von Neumann stability analysis; see Anderson, Tannehill, and Pletcher for details.[14]

Implicit techniques are more difficult to use in higher-dimensional problems. This is because the conventional extension would require us to manipulate huge matrices. A more efficient approach is to use operator splitting to separately perform the implicit step in each direction. This is known as *alternating direction implicit* (ADI).[15]

APPENDIX 8A: FORTRAN LISTINGS

LISTING 8A.1 Program `schro`. Computes the evolution of a Gaussian wave packet by solving the Schrödinger equation using the Crank–Nicholson scheme. Uses `cinv` (Listing 8A.4)

```
      program schro
! Program to solve the Schrodinger equation using
! Crank-Nicolson scheme
! Free particle version
      parameter( Nmax = 100, Pmax = 100 )
      real ham(Nmax,Nmax),L,mass,k0,Norm_psi
      real p_plot(Nmax,Pmax),x(Nmax)
      complex dCN(Nmax,Nmax),ctemp1(Nmax,Nmax),ctemp2(Nmax,Nmax)
      complex i_imag,ctemp
      complex psi(Nmax),psip(Nmax),newpsi(Nmax)
      integer plot_iter,eye

      eye(i,j) = min(i,j)/max(i,j)   ! Kronecker delta function

      pi = 4.*atan(1.)
      i_imag = (0. , 1.)     ! Imaginary i
      write(*,*) 'Enter number of grid points'
      read (*,*) N
      L = 100.                ! System extends from -L/2 to L/2
      h = L/N                 ! Grid size
      h_bar = 1.              ! Natural units
      mass = 1.
```

[14] D. A. Anderson, J. C. Tannehill, and R. H. Pletcher, *Computational Fluid Mechanics and Heat Transfer* (New York: Hemisphere, 1984), chapter 4.

[15] R. Peyret and T. D. Taylor, *Computational Methods for Fluid Flow* (New York: Springer-Verlag, 1983), section 2.8.

```
       write (*,*) 'Enter time step (tau)'
       read (*,*) tau
!!!!! Set up the Hamiltonian operator matrix ham(,) !!!!!
       do i=1,N
        do j=1,N
          ham(i,j) = 0.    ! Most of the matrix is zero
        end do
       end do
       coeff = -h_bar**2/(2.*mass*h**2)
       do i=2,(N-1)
          ham(i,i-1) = coeff      ! First off-diagonals = coeff
          ham(i,i) = -2.*coeff    ! Main diagonal = -2*coeff
          ham(i,i+1) = coeff      ! First off-diagonals = coeff
       end do
!!!!! Periodic boundary conditions !!!!!
       ham(1,N) = coeff           ! Careful to do the corners
       ham(1,1) = -2.*coeff       ! correctly
       ham(1,2) = coeff
       ham(N,N-1) = coeff
       ham(N,N) = -2.*coeff
       ham(N,1) = coeff
       write (*,*) 'Computing the Crank-Nicolson matrix'
       do i=1,N
        do j=1,N
          ctemp2(i,j) = eye(i,j) + 0.5*i_imag*tau/h_bar*ham(i,j)
        end do
       end do
       call cinv(ctemp2,N,Nmax,ctemp1) ! Invert matrix ctemp2
       do i=1,N
        do j=1,N
          ctemp2(i,j) = eye(i,j) - 0.5*i_imag*tau/h_bar*ham(i,j)
        end do
       end do
       do i=1,N            ! Matrix dCN is matrix product
        do j=1,N
          dCN(i,j) = (0. , 0.)
          do k=1,N
            dCN(i,j) = dCN(i,j) + ctemp1(i,k)*ctemp2(k,j)
          end do
        end do
       end do

!!!!! Initialize wavefunction !!!!!
       x0 = 0.                    ! Location of the center of the
                                  ! wavepacket
       velocity = 0.5            ! Average velocity of the packet
       k0 = mass*velocity/h_bar   ! Average wavenumber
```

```fortran
      sigma0 = L/10.           ! Standard deviation of the
                               ! wavefunction
      Norm_psi = 1./sqrt(sigma0*sqrt(pi))   ! Normalization
      do i=1,N
        x(i) = h*(i-0.5) - L/2.
        ctemp = cos(k0*x(i)) + i_imag*sin(k0*x(i))
        psi(i) = Norm_psi * ctemp *
     &             exp( -(x(i)-x0)**2/(2.*sigma0**2) )
        psip(i) = psi(i)     ! Save copy of initial condition
      end do

!!!!! Set loop and plot variables
      max_iter = L/(velocity*tau) !Particle should circle system
      plot_iter = max_iter/20     ! Plot 20 curves
      do i=1,N
        p_plot(i,1) = psi(i)*conjg(psi(i))   ! Probability density
      end do
      iplot = 2

!!!!! MAIN LOOP !!!!!
      write (*,*) 'Entering main loop'
      do iter=1,max_iter
        do i=1,N
          newpsi(i) = 0.    ! Crank-Nicolson scheme
          do j=1,N
            newpsi(i) = newpsi(i) + dCN(i,j)*psi(j)
          end do
        end do
        do i=1,N
          psi(i) = newpsi(i)   ! Update wavefunction
        end do

        if( mod(iter,plot_iter) .lt. 1 ) then
          do i=1,N
            p_plot(i,iplot) = psi(i)*conjg(psi(i))
          end do
          write (*,*) 'Finished ',iter,' of ',max_iter
          iplot = iplot + 1
        end if
      end do

! Print out the plotting variables -
!   x,p_plot,psip
!
      open(11,file='x.dat')
      open(12,file='p_plot.dat')
      open(13,file='psip.dat')
```

```
      do i=1,N
        write (11,1001) x(i)
        do j=1,(iplot-1)
          write (12,1003) p_plot(i,j)
        end do
        write (12,1004)
      end do
      do i=1,N
        write (13,1002) real(psip(i)),imag(psip(i))
      end do
      stop
1001  format(e12.6)
1002  format(e12.6,2x,e12.6)
1003  format(e12.6,' ',$)        ! The $ suppresses carriage return
1004  format(/)
      end
```

LISTING 8A.2 Function `tri_ge`. Gaussian elimination routine for tridiagonal matrices.

```
      subroutine tri_ge(a,b,N,Nmax,x)
! Solve b=a*x by Gaussian elimination where
! the matrix a is a packed tridiagonal matrix
      parameter(Mmax = 500)
      complex a(Nmax,3),b(Nmax),x(Nmax)
      complex alpha(Mmax),beta(Mmax),gamma(Mmax),coeff,
     &btemp(Mmax)

!!!!! Copy b into btemp
      do i=1,N
        btemp(i) = b(i)
      end do
!!!!! Unpack diagonals of tridiagonal matrix into vectors
      do i=1,N-1
        alpha(i) = a(i+1,1)
        beta(i) = a(i,2)
        gamma(i) = a(i,3)
      end do
      beta(N) = a(N,2)
!!!!! Forward elimination !!!!!
      do i=2,N
        coeff = alpha(i-1)/beta(i-1)
        beta(i) = beta(i) - coeff*gamma(i-1)
        btemp(i) = btemp(i) - coeff*btemp(i-1)
      end do
!!!!! Back Substitution !!!!!
      x(N) = btemp(N)/beta(N)
```

```
      do i=(N-1),1,-1
        x(i) = (btemp(i) - gamma(i)*x(i+1))/beta(i)
      end do
      return
      end
```

LISTING 8A.3 Program schrot. Similar to program schro but using the tridiagonal matrix routine, tri_ge (Listing 8A.2).

```
      program schrot
! Program to solve the Schrodinger equation using the sparse
! Crank-Nicolson scheme. Particle-in-a-box version
      parameter( Nmax = 300, Pmax = 100 )
      real ham(Nmax,3),L,mass,k0,Norm_psi
      real p_plot(Nmax,Pmax),x(Nmax),etemp(3)
      complex Q(Nmax,3),chi(Nmax)
      complex i_imag,ctemp
      complex psi(Nmax),psip(Nmax)
      integer plot_iter

      pi = 4.*atan(1.)
      i_imag = (0. , 1.)      ! Imaginary i
      write(*,*) 'Enter number of grid points'
      read (*,*) N
      L = 100.                ! System extends from -L/2 to L/2
      h = L/(N-1)             ! Grid size
      h_bar = 1.              ! Natural units
      mass = 1.
      write (*,*) 'Enter time step (tau)'
      read (*,*) tau

!!!!! Set up the Hamiltonian operator matrix ham(,) !!!!!
      coeff = -h_bar**2/(2.*mass*h**2)
      do i=2,(N-1)
        ham(i,1) = coeff
        ham(i,2) = -2.*coeff
        ham(i,3) = coeff
      end do
!!!!! Dirichlet boundary conditions !!!!!
      ham(1,1) = 0.
      ham(1,2) = 0.
      ham(1,3) = 0.
      ham(N,1) = 0.
      ham(N,2) = 0.
      ham(N,3) = 0.
!!!!! Set up the matrix Q(,) !!!!!
      etemp(1) = 0.
```

```
      etemp(2) = 1.   ! etemp() is used in place of eye_tri
      etemp(3) = 0.
      do i=1,N
        do j=1,3
          Q(i,j) = 0.5*(etemp(j)+0.5*i_imag*tau/h_bar*ham(i,j))
        end do
      end do
!!!!! Initialize wavefunction !!!!!
      x0 = 0.                   ! Location of the center of the
                                ! wavepacket
      velocity = 0.5           ! Average velocity of the packet
      k0 = mass*velocity/h_bar   ! Average wavenumber
      sigma0 = L/10.            ! Standard deviation of the
                                ! wavefunction
      Norm_psi = 1./sqrt(sigma0*sqrt(pi))   ! Normalization
      do i=1,N
        x(i) = h*(i-1) - L/2.
        if( i .eq. 1 .or. i .eq. N ) then
          psi(i) = 0.
        else
          ctemp = cos(k0*x(i)) + i_imag*sin(k0*x(i))
          psi(i) = Norm_psi * ctemp *
     &            exp( -(x(i)-x0)**2/(2.*sigma0**2) )
        endif
        psip(i) = psi(i)        ! Save a copy of the initial
                                ! condition for plot
      end do
!!!!! Set loop and plot variables
      max_iter = L/(velocity*tau) ! Particle should circle system
      plot_iter = max_iter/50    ! Plot 50 curves
      do i=1,N
        p_plot(i,1) = psi(i)*conjg(psi(i))   ! Probability density
      end do
      iplot = 2

!!!!! MAIN LOOP !!!!!
      write (*,*) 'Entering main loop'
      do iter=1,max_iter
        call tri_ge(Q,psi,N,Nmax,chi)         ! Crank-Nicolson
        do i=1,N
          psi(i) = chi(i) - psi(i)
        end do

        if( mod(iter,plot_iter) .lt. 1 ) then
          do i=1,N
            p_plot(i,iplot) = psi(i)*conjg(psi(i))
          end do
```

```
            write (*,*) 'Finished ',iter,' of ',max_iter
            iplot = iplot + 1
          end if
        end do

! Print out the plotting variables -
!   x,p_plot,psip
!
        open(11,file='x.dat')
        open(12,file='p_plot.dat')
        open(13,file='psip.dat')
        do i=1,N
          write (11,1001) x(i)
          do j=1,(iplot-1)
            write (12,1003) p_plot(i,j)
          end do
          write (12,1004)
        end do
        do i=1,N
          write (13,1002) real(psip(i)),imag(psip(i))
        end do
        stop
1001    format(e12.6)
1002    format(e12.6,2x,e12.6)
1003    format(e12.6,' ',$)         ! The $ suppresses carriage return
1004    format(/)
        end
```

LISTING 8A.4 Subroutine `cinv`. Computes the inverse of a complex matrix using Gaussian elimination with pivoting.

```
        subroutine cinv(aa,n,np,ainv)
! Compute inverse of n by n complex matrix aa
! Physical size of aa is np by np (np >= n); inverse is ainv
        parameter( nmax = 100 )
        complex aa(np,np),ainv(np,np)
        complex a(nmax,nmax), b(nmax,nmax)
        complex sum, coeff
        integer index(nmax)
        real scale(nmax)

        if( np .gt. nmax ) then
          print *, 'ERROR - Matrix is too large for cinv routine'
          stop
        end if
```

```
        do i=1,n
          do j=1,n
            a(i,j) = aa(i,j)    ! Copy matrix
            b(i,j) = 0.         ! Create identity matrix
          end do
          b(i,i) = 1.
        end do

!!!!! Forward elimination !!!!!

        do i=1,n
          index(i) = i
          scalemax = 0.
          do j=1,N
            scalemax = amax1(scalemax,cabs(a(i,j)))
          end do
          scale(i) = scalemax
        end do

        do k=1,N-1
          ratiomax = 0.
          do i=k,n
            ratio = cabs(a(index(i),k))/scale(index(i))
            if( ratio .gt. ratiomax ) then
              j=i
              ratiomax = ratio
            end if
          end do
          indexk = index(j)
          index(j) = index(k)
          index(k) = indexk
          do i=k+1,n
            coeff = a(index(i),k)/a(indexk,k)
            do j=k+1,n
              a(index(i),j) = a(index(i),j) - coeff*a(indexk,j)
            end do
            a(index(i),k) = coeff
            do jj=1,n
              b(index(i),jj)
     &         =b(index(i),jj)-a(index(i),k)*b(indexk,jj)
            end do
          end do
        end do

!!!!! Back substitution !!!!!

        do jj=1,n
          ainv(n,jj) = b(index(n),jj)/a(index(n),n)
```

```fortran
   do i=n-1,1,-1
     sum = b(index(i),jj)
     do j=i+1,n
       sum = sum - a(index(i),j)*ainv(j,jj)
     end do
     ainv(i,jj) = sum/a(index(i),i)
   end do
end do

return
end
```

Chapter 9
Special Functions and Quadrature

You can go quite far with the elementary transcendental functions (exponential, sine, etc.). However, eventually you will find it useful to add more functions to your toolbox. This chapter discusses two important special functions: Legendre polynomials and Bessel functions. The second topic of this chapter is quadrature, a fancy term for evaluating integrals numerically. Two general-purpose methods are covered: Romberg and Gaussian integration.

9.1 SPECIAL FUNCTIONS

Eigenfunctions

In Chapter 7 we solved the Laplace equation in rectangular coordinates using separation of variables. Our PDE was separated into ODEs, all of which were of the form

$$\frac{d^2}{dx^2} f(x) + k^2 f(x) = 0 \tag{9-1}$$

The general solution of this simple equation is a linear combination of trigonometric functions,

$$f(x) = A \cos(kx) + B \sin(kx) \tag{9-2}$$

where the coefficients A and B are determined by the boundary conditions.

This ODE is a special case of the *Sturm–Liouville equation*,

$$Lf_i(x) - \lambda_i \rho(x) f_i(x) = 0 \tag{9-3}$$

where λ_i is the eigenvalue and f_i is its corresponding eigenfunction. The linear differential operator L is

$$L = \frac{d}{dx} p(x) \frac{d}{dx} + q(x) \tag{9-4}$$

and $\rho(x)$ is a weight function.

It is not difficult to show that this operator, with homogeneous boundary conditions[1] in the interval $[a, b]$, is *Hermitian*,

$$\int_a^b f_i^*(x) L f_j(x) \, dx = \int_a^b f_j^*(x) L f_i(x) \, dx \tag{9-5}$$

where the asterisk denotes complex conjugate. This Hermitian property leads to the two important results: (1) the eigenvalues λ_i are real (i.e., $\lambda_i = \lambda_i^*$), and (2) the eigenfunctions are orthogonal,

$$\int_a^b \rho(x) f_i^*(x) f_j(x) \, dx = N_i \delta_{i,j} \tag{9-6}$$

where N_i is the normalization. If this sounds vaguely familiar, you probably recall that in quantum mechanics the Hamiltonian operator (which is Hermitian) has real eigenvalues. These eigenvalues are the energy levels and the eigenfunctions are the wave functions of those states.

When separation of variables is performed in a nonrectangular coordinate system (e.g., spherical, cylindrical) we commonly obtain ODEs of the form given by (9-3). The solutions of these more complicated problems are important enough to be studied as named functions. In this section we cover two of these special functions: Legendre polynomials and Bessel functions. Special functions are also discussed in the standard mathematical physics texts;[2] many useful identities may be found in Abramowitz and Stegun.[3]

---

[1] The boundary condition is homogeneous if at the boundary $f = 0$ (Dirichlet) or $df/dx = 0$ (Neumann) or a linear combination of f and df/dx is zero (mixed).

[2] J. Mathews and R. Walker, *Mathematical Methods of Physics* (Menlo Park, Calif.: W. A. Benjamin, 1970); G. Arfken, *Mathematical Methods for Physicists* (New York: Academic Press, 1970).

[3] M. Abramowitz and I. Stegun, *Handbook of Mathematical Functions* (New York: Dover, 1972).

Legendre Polynomials

Laplace's equation is $\nabla^2 \Phi(\mathbf{r}) = 0$; in spherical coordinates, it may be written as

$$\left(\frac{1}{r} \frac{\partial^2}{\partial r^2} r + \frac{1}{r^2 \sin \theta} \frac{\partial}{\partial \theta} \sin \theta \frac{\partial}{\partial \theta} + \frac{1}{r^2 \sin^2 \theta} \frac{\partial^2}{\partial \phi^2} \right) \Phi(r, \theta, \phi) = 0 \qquad (9\text{-}7)$$

We'll assume that our problem is azimuthally symmetric, so Φ is independent of the angle ϕ. To use separation of variables, we insert

$$\Phi(r, \theta) = U(r)P(\theta) \qquad (9\text{-}8)$$

into (9-7) and obtain the pair of equations

$$\frac{1}{\sin \theta} \frac{d}{d\theta} \left(\sin \theta \frac{dP}{d\theta} \right) + \lambda P = 0 \qquad (9\text{-}9)$$

$$r^2 \frac{d^2}{dr^2} U + 2r \frac{dU}{dr} - \lambda U = 0 \qquad (9\text{-}10)$$

The former equation may be solved by power series expansion in $\cos \theta$. We find that the solution is finite at $\theta = 0$ and $\theta = \pi$ only if $\lambda = n(n + 1)$, where n is an integer.[4]

The solution of the radial equation (9-10) is

$$U(r) = c_1 r^n + c_2 r^{-(n+1)} \qquad (9\text{-}11)$$

where the constants c_1 and c_2 are determined by the boundary conditions.

It is convenient to rewrite (9-9) as

$$\frac{d}{dx} (1 - x^2) \frac{d}{dx} P_n(x) + n(n + 1)P_n(x) = 0 \qquad (9\text{-}12)$$

by using the change of variable $x = \cos \theta$. Note that this is equation (9-3) with $p(x) = 1 - x^2$, $q(x) = 0$, $\lambda_n = -n(n + 1)$, and $\rho(x) = 1$. The power series solution of (9-12) terminates after a finite number of terms, so $P_n(x)$ is a polynomial of order n.

The most compact way to write the *Legendre polynomial* $P_n(x)$ is by the Rodrigues formula

$$P_n(x) = \frac{1}{2^n n!} \left(\frac{d}{dx} \right)^n (x^2 - 1)^n \qquad (9\text{-}13)$$

---

[4] J. Mathews and R. Walker, *Mathematical Methods of Physics* (Menlo Park, Calif.: W. A. Benjamin, 1970), section 1-2.

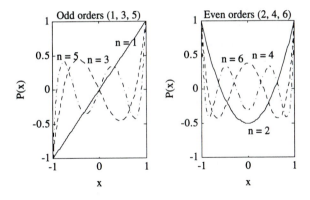

Figure 9.1 Graph of the first few Legendre polynomials, $P_n(x)$. Solid line indicates $n = 1$ and $n = 2$; dashed line is $n = 3$ and $n = 4$; dash-dot line is $n = 5$ and $n = 6$.

The first few polynomials are listed below and plotted in Figure 9.1.

$$P_0(x) = 1 \qquad\qquad P_3(x) = \tfrac{1}{2}(5x^3 - 3x)$$
$$P_1(x) = x \qquad\qquad P_4(x) = \tfrac{1}{8}(35x^4 - 30x^2 + 3) \qquad (9\text{-}14)$$
$$P_2(x) = \tfrac{1}{2}(3x^2 - 1) \qquad P_5(x) = \tfrac{1}{8}(63x^5 - 70x^3 + 15x)$$

The Legendre polynomials are orthogonal in the interval $[-1, 1]$ with the normalization

$$\int_{-1}^{1} P_m(x)P_n(x)\,dx = \frac{2}{2n + 1}\,\delta_{m,n} \qquad (9\text{-}15)$$

Many other Legendre polynomial identities are compiled in Abramowitz and Stegun.[5]

For the purpose of numerical computation, the most useful identity is the recursion relation (see Exercise 9.2),

$$(n + 1)P_{n+1}(x) = (2n + 1)xP_n(x) - nP_{n-1}(x) \qquad (9\text{-}16)$$

Since the first two polynomials are trivial to compute $[P_0(x) = 1; P_1(x) = x]$, we may use this recursion relation to boot-strap up to the desired P_n. The function legndr(n, x) returns a vector containing the values $[P_0(x)\, P_1(x) \ldots P_n(x)]$ (see Listing 9.1). To compute P_n using (9-16) we need to compute all the lower index values, so we might as well return them.

LISTING 9.1 Function legndr. Computes Legendre polynomials.

```
1    function p = legndr (n, x)
2    % Legendre polynomials
```

```
3    % Inputs -
4    %      n - Function returns value of all polynomials up
5    %          to and including order n
6    %      x - Value at which polynomial is evaluated
7    % Outputs -
8    %      p - Vector containing P(x) for order 0,1,...,n
9    p(1)=1;       % P(x) for n=0
10   p(2)=x;       % P(x) for n=1
11   for i=3:n+1   % Use upward recursion to obtain other n's
12     p(i) = ((2*i-3)*x*p(i-1) - (i-2)*p(i-2))/(i-1);
13   end
14   return;
```

Bessel Functions

A ubiquitous PDE in mathematical physics is the Helmholtz equation,

$$\nabla^2\psi(\mathbf{r}) + k^2\psi = 0 \tag{9-17}$$

In cylindrical coordinates it is

$$\left(\frac{1}{\rho}\frac{\partial}{\partial\rho}\,\rho\,\frac{\partial}{\partial\rho} + \frac{1}{\rho^2}\frac{\partial^2}{\partial\phi^2} + \frac{\partial^2}{\partial z^2}\right)\psi(\rho, \phi, z) + k^2\psi = 0 \tag{9-18}$$

Using the separation of variables substitution, $\psi(\rho, \phi, z) = R(\rho)Q(\phi)Z(z)$, we find that the equations for Q and Z have solutions

$$Q(\phi) = c_1 \cos(m\phi) + c_2 \sin(m\phi) \tag{9-19}$$

$$Z(z) = c_3 \cos(\alpha z) + c_4 \sin(\alpha z) \tag{9-20}$$

where $c_1, \ldots, c_4$, m, and α are constants fixed by the boundary conditions. The radial component is given by Bessel's equation

$$\rho^2 \frac{d^2R}{d\rho^2} + \rho\,\frac{dR}{d\rho} + [(k^2 - \alpha^2)\rho^2 - m^2]R = 0 \tag{9-21}$$

If we require that R be regular on the z-axis (i.e., finite for $\rho = 0$), the solution is the *Bessel function* of the first kind

$$R(\rho) = J_m(\sqrt{k^2 - \alpha^2}\rho) \tag{9-22}$$

See Figure 9.2 for a plot of $J_m(x)$ for the first few values of m.

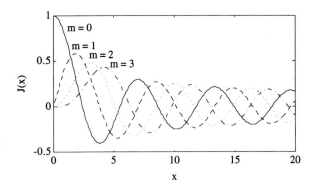

Figure 9.2   Bessel function $J_m(x)$ for $m = 0$ (solid line), $m = 1$ (dashed), $m = 2$ (dotted), and $m = 3$ (dash-dot).

The MATLAB toolbox has an excellent Bessel function routine, but it is instructive to build one ourselves. To evaluate $J_m(x)$ numerically we make use of the recursion relation (see Exercise 9.6)

$$J_{m-1}(x) = (2m/x)J_m(x) - J_{m+1}(x) \qquad (9\text{-}23)$$

There are two reasons why this equation is not as easy to use as the recursion relation for Legendre polynomials. First, the functions $J_0(x)$ and $J_1(x)$ are not as simple to evaluate as the first two Legendre polynomials. There are, however, tabulated polynomial approximations for the low-order Bessel functions.[6]
The more serious difficulty is that when $m > x$ upward recursion is numerically unstable. However, the recursion relation is stable if we iterate *downward*. To illustrate the algorithm, let's evaluate $J_0(0.5)$. The idea is to use

$$J_m(x) \rightarrow \frac{1}{\sqrt{2\pi m}}\left(\frac{ex}{2m}\right)^m \qquad \text{as } m \rightarrow \infty \qquad (9\text{-}24)$$

thus, $J_m(x) \ll 1$ if $m \gg x$. We start the recursion using the (incorrect) values

$$J_5(0.5) = 0; \qquad J_4(0.5) = 1$$

The values at this stage are arbitrary since we are going to renormalize the result in a moment. Using (9-23) we obtain

$$J_3(0.5) = 16; \qquad J_2(0.5) = 191; \qquad J_1(0.5) = 1512; \qquad J_0(0.5) = 5857$$

Finally, we normalize the values by using the identity (see Exercise 9.7),

$$J_0(x) + 2J_2(x) + 2J_4(x) + \ldots = 1 \qquad (9\text{-}25)$$

---

[6] M. Abramowitz and I. Stegun, *Handbook of Mathematical Functions* (New York: Dover, 1972), section 9.4.

Our final result is $J_0(0.5) \cong 5857/6241 = 0.938471$; tables give 0.938470. Of course our estimate for $J_4(0.5) \cong 1/6241 = 1.6023 \times 10^{-4}$ is not as accurate (tables give 1.6074×10^{-4}).

The function bess (m, x) returns a vector of values $[J_0(x)\ J_1(x)\ \ldots\ J_m(x)]$ (see Listing 9.2). For the downward recursion to work accurately, we need to start at a sufficiently large value of m (see line 9). The recursion is performed on lines 11–15; to normalize the values, (9-25) is used on lines 16–22. This routine would be more efficient if it used upward recursion for $m < x$ (see Exercise 9.9).

LISTING 9.2 Function bess. Returns values of Bessel functions $[J_m(x) \ldots J_0(x)]$.

```
1   function jj = bess(m_max,x)
2   % Function to calculate of Bessel function
3   % jj = bess(m_max,x)
4   % Inputs
5   %      m_max - Largest desired order
6   %      x - Function returns J(x) for all orders <= m_max
7   % Output
8   %      jj - Bessel function J(x) for all orders <= m_max
9   m_top = max(m_max,x)+15;    % Top value of m for recursion
10  m_top = 2*ceil( m_top/2 );  % Round up to an even number
11  j(m_top+1) = 0;
12  j(m_top) = 1;
13  for m=m_top-2:-1:0          % Downward recursion
14    j(m+1) = 2*(m+1)/(x+eps)*j(m+2) - j(m+3);
15  end
16  norm = j(1);               % NOTE: Be careful, m=0,1,... but
17  for m=2:2:m_top            % vector goes j(1),j(2),...
18    norm = norm + 2*j(m+1);
19  end
20  for m=0:m_max             % Send back only the values for
21    jj(m+1) = j(m+1)/norm;  % m=0,...,m_max and discard values
22  end                       % for m=m_max+1,...,m_top
```

Zeros of the Bessel Function

The Bessel function $J_m(x)$ is oscillatory in a fashion similar to the trigonometric functions. This fact is evident from the graph of the function (Figure 9.2) and from the asymptotic formula

$$J_m(x) \approx \sqrt{\frac{2}{\pi x}} \cos(x - \tfrac{1}{2}m\pi - \tfrac{1}{4}\pi) \tag{9-26}$$

as $x \to \infty$. Like the trigonometric functions, the Bessel function has an infinite number of zeros. However, the zeros of J_m are *not* evenly spaced.

The sine function satisfies the well-known orthogonality relation

$$\int_0^L dx \; \sin(\alpha_i x/L)\sin(\alpha_j x/L) = \frac{L}{2} \, \delta_{i,j} \qquad (9\text{-}27)$$

where $\alpha_i = \pi i$ is the ith zero of $\sin(x)$. The Bessel function, $J_m(x)$, satisfies a similar orthogonality relation

$$\int_0^R \rho J_m(\zeta_{m,s}\rho/R)J_m(\zeta_{m,t}\rho/R)\, d\rho = \frac{R^2}{2}\, J_{m+1}^2(\zeta_{m,s})\delta_{s,t} \qquad (9\text{-}28)$$

where $\zeta_{m,s}$ is the sth zero of $J_m(x)$.

The function `zeroj` (Listing 9.3) computes $\zeta_{m,s}$ using Newton's method (see Section 4.4). This algorithm requires a good initial guess to converge to the right root. The function `zeroj` uses the asymptotic formula

$$\zeta_{m,s} \approx \beta - \frac{\mu - 1}{8\beta} - \frac{4(\mu - 1)(7\mu - 31)}{3(8\beta)^3} \qquad (9\text{-}29)$$

where $\beta = (s + \frac{1}{2}m - \frac{1}{4})\pi$ and $\mu = 4m^2$. This formula is valid for $s \gg m$. If $m \gg s$, then (9-29) is not an accurate estimate for $\zeta_{m,s}$. However, it lands us close enough to the root that we converge after only a few iterations.

Newton's method needs the derivative of $J_m(x)$; this is easy to obtain from the recurrence relation (see Exercise 9.8)

$$J_m'(x) = -J_{m+1}(x) + \frac{m}{x}\, J_m(x) \qquad (9\text{-}30)$$

This relation is used on line 16 of `zeroj`. For a more advanced algorithm, see Temme.[7]

LISTING 9.3 Function `zeroj`. Returns the sth zero of $J_m(x)$.

```
1    function z = zeroj (m_order, n_zero)
2    % Function which returns the zeros of the Bessel function J(x)
3    %    z = zeroj (m_order, n_zeros)
4    % Inputs
5    %    m_order - Order of the Bessel function
6    %    n_zero  - Number of the zero
7    % Output
```

[7] N. M. Temme, "An Algorithm with ALGOL 60 Program for the Computation of the Zeros of Ordinary Bessel Functions and those of their Derivatives," *J. Comp. Phys.*, **32**, 270 (1979).

```
8    %   z - The "n_zero th" zero of the Bessel function
9    %% Use asymtotic formula for initial guess
10   beta = (n_zero + 0.5*m_order - 0.25)*pi;
11   mu = 4*m_order^2;
12   z = beta-(mu-1)/(8*beta)-4*(mu-1)*(7*mu-31)/(3*(8*beta)^3);
13   for i=1:5
14      jj = bess(m_order+1,z);        % Use recursion
15      deriv = -jj(m_order+2) + ...   % to evaluate derivative
16            m_order/z * jj(m_order+1);
17      z = z - jj(m_order+1)/deriv;   % Newton's root finding
18   end
19   return;
```

   In the next two sections we use special functions to solve some physics problems. You'll discover that we need to solve integrals of special functions, motivating us to find a way of computing them numerically.

EXERCISES

1. One of the most useful identities for Legendre polynomials is

$$\frac{1}{\sqrt{1 - 2hx + h^2}} = \sum_{n=0}^{\infty} h^n P_n(x)$$

where the left-hand side is called the *generating function* for $P_n(x)$. Using this identity, show that

$$\frac{1}{|\mathbf{r} - \mathbf{r'}|} = \sum_{n=0}^{\infty} \frac{r_<^n}{r_>^{n+1}} P_n(\cos \theta)$$

where $\mathbf{r}$ and $\mathbf{r'}$ are three-dimensional vectors, θ is the angle between them, and $r_< = \min(|\mathbf{r}|, |\mathbf{r'}|)$, $r_> = \max(|\mathbf{r}|, |\mathbf{r'}|)$.

2. Using the generating function (see Exercise 9.1), derive the recursion relation (9-16). (Hint: Differentiate with respect to h.)

3. Using the generating function (see Exercise 9.1), show that

$$P_n(0) = \begin{cases} 0 & n \text{ odd} \\ \dfrac{(-1)^{n/2}(n - 1)!!}{2^{n/2}(n/2)!} & n \text{ even} \end{cases}$$

where $n!! = n \times (n - 2) \times \ldots \times 5 \times 3 \times 1$.

4. Show that

$$\int_0^1 P_n(x)\,dx = \begin{cases} 0 & n \text{ even} \\[2mm] \dfrac{(-1)^{(n-1)/2}(n-2)!!}{2^{(n+1)/2}\left(\dfrac{n+1}{2}\right)!} & n \text{ odd} \end{cases}$$

where $n!! = n \times (n-2) \times \ldots \times 5 \times 3 \times 1$.

5. Write a function that finds $\zeta_{n,i}$, the ith root of $P_n(x)$, using Newton's method. The derivative of the Legendre polynomials may be found using $(x^2 - 1)P_n'(x) = nxP_n(x) - nP_{n-1}(x)$. Since a good initial guess is crucial for locating the right root, you may want to use the following facts: (1) $\zeta_{1,1} = 0$; (2) all the roots are real; and (3) the roots are intertwined, so $\zeta_{n+1,i} < \zeta_{n,i} < \zeta_{n+1,i+1}$. Compare your results with the Gaussian nodes listed in Table 9.1.

6. A useful Bessel function identity is

$$\exp\left[\frac{1}{2} x \left(h - \frac{1}{h}\right)\right] = \sum_{m=-\infty}^{\infty} J_m(x)h^m$$

where the left-hand side is called the generating function of $J_m(x)$. Using this identity derive the recursion relation, equation (9-23). (Hint: Take the derivative with respect to h.)

7. Using the generating function (see the previous exercise), derive the normalization identity, equation (9-25). [Hint: $J_{-m}(x) = (-1)^m J_m(x)$.]

8. Using the generating function (see Exercise 9.6), derive the recursion relation, equation (9-30).

9. Modify the function bess to use polynomial approximation with upward recursion when $m < x$. Compute $J_0(x)$ using the old and new routines and plot the absolute difference for $0 \le x \le 100$.

10. The spherical Bessel function of the first kind[8] is defined as

$$j_m(x) = \sqrt{\frac{\pi}{2x}}\, J_{m+\frac{1}{2}}(x)$$

Write a function that computes $j_m(x)$ using the recursion relation

$$j_{m-1}(x) + j_{m+1}(x) = \frac{2m+1}{x} j_m(x)$$

Use upward recursion with the starting values $j_0(x) = \sin x/x$ and $j_1(x) = \sin x/x^2 - \cos x/x$. Show that this scheme works well except when $x \ll m$.

11. The second solution to Bessel's equation is $Y_m(x)$, the Bessel function of the second kind. Write a function to compute $Y_m(x)$ using upward recursion

$$Y_{m+1}(x) = \frac{2m}{x} Y_m(x) - Y_{m-1}(x)$$

[8] M. Abramowitz and I. Stegun, *Handbook of Mathematical Functions* (New York: Dover, 1972), chapter 10.

To obtain the starting values for recursion, use the identities

$$Y_0(x) = \frac{2}{\pi} \{\ln(x/2) + \gamma\} J_0(x) - \frac{4}{\pi} \sum_{k=1}^{\infty} (-1)^k \frac{J_{2k}(x)}{k}$$

and

$$J_1(x) Y_0(x) - J_0(x) Y_1(x) = \frac{2}{\pi x}$$

where $\gamma = 0.577215664 \ldots$. Demonstrate your routine by producing some plots of $Y_m(x)$ for $0 < x < 50$ and various m.

9.2 BASIC NUMERICAL INTEGRATION

Laplace Equation in Spherical Coordinates

As an application of our special functions, we consider the following electrostatics problem. Take a sphere of radius R; the outer surface of the sphere is held at the fixed potential $V(\theta)$. The solution of Laplace's equation in spherical coordinates for azymuthally symmetric problems is[9]

$$\Phi(r, \theta) = \sum_{n=0}^{\infty} [A_n r^n + B_n r^{-(n+1)}] P_n(\cos \theta) \tag{9-31}$$

To find the potential everywhere outside the sphere, we need to obtain the coefficients A_n and B_n that match the boundary conditions.

The implicit boundary condition at infinity requires that the potential goes to zero as $r \rightarrow \infty$. To meet this requirement, the A_n's must all be zero. Matching the potential at the surface of the sphere

$$V(\theta) = \sum_{n=0}^{\infty} B_n R^{-(n+1)} P_n(\cos \theta) \tag{9-32}$$

To solve for B_n we multiply both sides by $P_m(\cos \theta)$ and integrate,

$$\int_0^{\pi} d\theta \sin \theta P_m(\cos \theta) V(\theta) = \int_0^{\pi} d\theta \sin \theta P_m(\cos \theta) \sum_{n=0}^{\infty} B_n a^{-(n+1)} P_n(\cos \theta) \tag{9-33}$$

[9] J. D. Jackson, *Classical Electrodynamics*, 2d ed. (New York: Wiley, 1975), section 3.3.

Using the orthogonality relation, equation (9-15), we may solve for B_n as

$$B_n = \frac{2n + 1}{2} R^{n+1} \int_0^{\pi} d\theta \ \sin \theta V(\theta) P_n(\cos \theta)$$

$$= \frac{2n + 1}{2} R^{n+1} \int_{-1}^{1} dx \ V(x) P_n(x)$$

(9-34)

The problem is now reduced to solving this integral to obtain the B's.

Unfortunately, (9-34) is not always simple to evaluate for an arbitrary potential $V(\theta)$. One example that is not so difficult is the split-hemisphere potential

$$V(\theta) = \begin{cases} -V_0 & \text{for } 0 \le \theta < \pi/2 \\ -V_0 & \text{for } \pi/2 < \theta < \pi \end{cases}$$

(9-35)

where V_0 is a constant. In this case, (see Exercise 9.4) we obtain

$$B_n = V_0 R^{n+1} \left(-\frac{1}{2} \right)^{(n-1)/2} \frac{(2n + 1)(n - 2)!!}{2[\frac{1}{2}(n + 1)]!} \quad (n \text{ odd})$$

(9-36)

and $B_n = 0$ for n even.

Ideally, we would like to compute the coefficients for an arbitrary potential $V(\theta)$. In fact, we may not even know this potential as a function; we may just have a table of values for selected angles. This motivates our study of numerical integration.

Trapezoidal Rule

Consider the integral

$$I = \int_a^b f(x) \, dx$$

(9-37)

Our strategy for estimating I is to evaluate $f(x)$ at a few points and fit a simple curve (e.g., piecewise linear) through these points. First subdivide the interval $[a, b]$ into $N - 1$ subintervals. Define the points x_i as

$$x_1 = a, \quad x_N = b, \quad x_1 < x_2 < \ldots < x_{N-1} < x_N$$

(9-38)

The function is only evaluated at these points, so we use the short-hand notation $f_i \equiv f(x_i)$.

The simplest, most practical quadrature scheme is the *trapezoidal rule*. As illustrated in Figure 9.3, straight lines connect the points and this piecewise linear

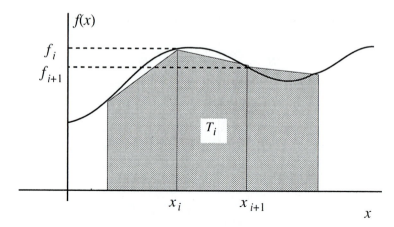

Figure 9.3   Schematic illustrating the trapezoidal rule.

function serves as our fitting curve. The integral of this fitting function is easy to compute since it is the sum of the areas of trapezoids. The area of a single trapezoid is

$$T_i = \tfrac{1}{2}(x_{i+1} - x_i)(f_{i+1} + f_i) \tag{9-39}$$

The true integral is estimated as the sum of the areas of the trapezoids, so

$$I \approx I_T = T_1 + T_2 + \ldots + T_{N-1} \tag{9-40}$$

Notice that the last term in the sum is $N - 1$ since there is one fewer panel than grid point.

The general formula simplifies if we take equally spaced grid points. The spacing is $h = (b - a)/(N - 1)$, so $x_i = a + (i - 1)h$. Our formula for the area of a trapezoid simplifies to

$$T_i = \tfrac{1}{2}h(f_{i+1} + f_i) \tag{9-41}$$

The trapezoidal rule for equally spaced points is

$$I_T(h) = \tfrac{1}{2}hf_1 + hf_2 + hf_3 + \ldots + hf_{N-1} + \tfrac{1}{2}hf_N$$
$$= \tfrac{1}{2}h(f_1 + f_N) + h \sum_{i=2}^{N-1} f_i \tag{9-42}$$

Notice that for all the interior points ($i \neq 1$, $i \neq N$) the coefficient is h, while on the two exterior points ($i = 1$ and $i = N$) it is $h/2$. This is because each interior point appears in two trapezoids; it is on the right of one trapezoid and on the left of the neighboring trapezoid.

A quick example shows you that something as simple as the trapezoidal rule does quite well. Consider the error function

$$\text{erf}(x) = \frac{2}{\sqrt{\pi}} \int_0^x e^{-y^2} \, dy \tag{9-43}$$

For $x = 1$, $\text{erf}(1) \cong 0.842701$. The trapezoidal rule with $N = 5$ gives a value of 0.83837, which is good to about two decimal places. Of course the integrand in this example is very smooth and well behaved.

Most numerical analysis texts give the truncation error for trapezoidal rule as[10]

$$I - I_T(h) = -\tfrac{1}{12}(b - a)h^2 f''(\zeta) \tag{9-44}$$

for some ζ in $[a, b]$. An alternative way of writing the truncation error makes use of the Euler-Maclaurin formula

$$I - I_T(h) = -\tfrac{1}{12}h^2[f'(b) - f'(a)] + O(h^4) \tag{9-45}$$

Two remarks: The error is proportional to h^2 and the latter expression warns you that the trapezoidal rule will have difficulties if the derivative diverges at the end points. For example, the integral $\int_0^b \sqrt{x} \, dx$ is problematic (see Exercise 9.12).

Romberg Integration

A common question is, "How many panels should I use?" One way to decide is to repeat the calculation with a smaller interval. If the answer doesn't change significantly, then we accept it as correct. We might get tricked by pathological functions or in unusual scenarios, but don't be paranoid about this. With the trapezoidal rule, if the number of panels is a power of two, we can halve the interval size without having to recompute all the points (see Figure 9.4).

We define the sequence of interval sizes,

$$h_1 = (b - a), \; h_2 = \tfrac{1}{2}(b - a) \ldots h_n = \frac{1}{2^{n-1}} (b - a) \tag{9-46}$$

For $n = 1$ we have only one panel, so

$$I_T(h_1) = \tfrac{1}{2}(b - a)[f(a) + f(b)] \tag{9-47}$$

[10] For example, W. Cheney and D. Kincaid, *Numerical Mathematics and Computing* (Monterey, Calif.: Brooks Cole, 1985).

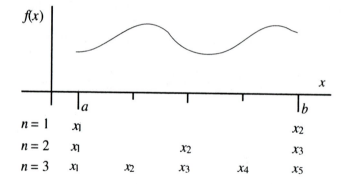

Figure 9.4   Schematic illustrating intervals used by the recursive trapezoidal rule.

There is a simple recursive formula for calculating $I_T(h_{n+1})$ using $I_T(h_n)$:

$$I_T(h_{n+1}) = \tfrac{1}{2} I_T(h_n) + h_{n+1} \sum_{i=1}^{2^{n-1}} f[a + (2i - 1)h_{n+1}] \qquad (9\text{-}48)$$

This formula is best understood by working out a small example by hand.

By using the recursive method described above, we can keep adding panels until the answer appears to converge. However, we can greatly improve this process by using a method called *Romberg integration*. I'll first describe the mechanics of Romberg integration and then show you why it works. The method builds a triangular table of the form:

$$R(1, 1)$$

$$R(2, 1) \quad R(2, 2)$$

$$R(3, 1) \quad R(3, 2) \quad R(3, 3)$$

$$\vdots \qquad \vdots \qquad \vdots \qquad \ddots$$

The formula for the first column is just the recursive trapezoidal rule.

$$R(n, 1) = I_T(h_n) \qquad (9\text{-}49)$$

The successive columns to the right are computed using the Richardson extrapolation formula

$$R(n + 1, m + 1) = R(n + 1, m) + \frac{1}{4^m - 1} [R(n + 1, m) - R(n, m)] \qquad (9\text{-}50)$$

The most accurate estimate for the integral is $R(n, n)$, the value at the bottom right corner of the table.

Romberg Integration Routine

The function `rombf` performs Romberg integration for a given integrand and interval (see Listing 9.4). The integrand, $f(x)$, is defined as a separate function; for example, see the simple routine `my_f` in Listing 9.5. The name of the integrand function is passed to rombf as a text string in the variable `func`. To evaluate $f(x)$ the `rombf` function uses the MATLAB `feval` command (see line 22). The variable `param` serves to pass any parameters that are needed by the function; see Chapter 3 for more on `feval`.

LISTING 9.4 Function `rombf`. Computes integrals using the Romberg algorithm.

```
1   function R = rombf(a,b,N,func,param)
2   %  Function to compute integrals by Romberg algorithm
3   %  R = rombf(a,b,N,func,param)
4   %  Inputs -
5   %    a,b    Lower and upper bound of the integral
6   %    N      Romberg table is N by N
7   %    func   Name of integrand function in a string such as
8   %           func='my_f'.  Calling sequence is my_f(x,param)
9   %    param  Set of parameters to be passed to function
10  %  Output -
11  %    R      Romberg table; Entry R(N,N) is best estimate of
12  %           the value of the integral
13  h = b - a;          % This is the coarsest panel size
14  % Compute the first term R(1,1)
15  R(1,1) = h/2 * (feval(func,a,param) + feval(func,b,param));
16  np = 1;             % np is the number of panels
17  for i=2:N
18    h = h/2;          % Use panels half the previous size
19    np = 2*np;        % Use twice as many panels
20    sum = 0;
21    for k=1:2:np-1    % This for loop goes k=1,3,5,...,np-1
22      sum = sum + feval(func, a + k*h, param);
23    end
24  % Compute the first column entry R(i,1)
25    R(i,1) = 1/2 * R(i-1,1) + h * sum;
26    m = 1;
27  % Compute the other columns R(i,2), ..., R(i,i)
28    for j=2:i
29      m = 4*m;
30      R(i,j) = R(i,j-1) + (R(i,j-1) - R(i-1,j-1))/(m-1);
31    end
32  end
33  return;
```

LISTING 9.5   Function `my_f`. Defines the integrand function for `rombf`.

```
1    function y = my_f (x, param)
2    % Error function integrand
3    y = exp (-x^2);
4    return;
```

To illustrate the use of `rombf`, consider our error function example, equation (9-43). The simple function `my_f` supplies the integrand (see Listing 9.5). Working interactively from the command line we obtain the small Romberg table shown below:

```
>>format long   % Print answer to full precision
>>2/sqrt(pi)*rombf(0,1,3,'my_f',0)

ans =
```

0.77174333225805	0	0
0.82526295559675	0.84310283004298	0
0.83836777744121	0.84273605138936	0.84271159947912

Our best estimate of the integral is given by the bottom right entry in the table, $R(3, 3)$. Comparing with the exact result we find that we have almost five digits of accuracy using only four panels! Again, this is a very smooth integrand; life is not always so kind.

It is useful that the function returns the entire table and not just the last entry, since this gives us an estimate on the error. One should not be too eager to use an excessive amount of computer time to make the table as large as possible. Eventually, round-off error begins to degrade the answer (see Exercise 9.12), so it is better to quit while you're ahead.

Why Romberg Works

To understand why the Romberg scheme works, consider the truncation error, $E_T(h_n) = I - I_T(h_n)$. Using (9-45),

$$E_T(h_n) = -\tfrac{1}{12}h_n^2[f'(b) - f'(a)] + O(h_n^4) \tag{9-51}$$

Since $h_{n+1} = h_n/2$,

$$E_T(h_{n+1}) = -\tfrac{1}{48}h_n^2[f'(b) - f'(a)] + O(h_n^4) \tag{9-52}$$

Consider the second column of the Romberg table. The truncation error for $R(n + 1, 2)$ is

$$I - R(n + 1, 2) = I - \{I_T(h_{n+1}) + \tfrac{1}{3}[I_T(h_{n+1}) - I_T(h_n)]\}$$

$$= E_T(h_{n+1}) + \tfrac{1}{3}[E_T(h_{n+1}) - E_T(h_n)]$$

$$= -[\tfrac{1}{48} + \tfrac{1}{3}(\tfrac{1}{48} - \tfrac{1}{12})]h_n^2[f'(b) - f'(a)] + O(h_n^4) \qquad (9\text{-}53)$$

$$= O(h_n^4)$$

Notice how the h_n^2 term serendipitously cancels out, leaving us with a truncation error that is of order h_n^4. The next (third) column of the Romberg table removes this term, and so forth.

Return to the electrostatics problem from the beginning of this section; we wanted to evaluate (9-34) numerically. We could quickly construct a program to compute the coefficients B_n using our existing Romberg function. However, remember how the Legendre polynomials are computed using the recursion relation. To get $P_n(x)$ we also compute $P_{n-1}(x), \ldots, P_0(x)$. It would be wasteful not to make use of these values. In fact, it's not difficult to construct the program so that it computes all the coefficients, B_n, simultaneously. I'll leave that as an exercise.

EXERCISES

12. Use Romberg integration to numerically evaluate the integrals below:

    (a) $\int_0^1 e^x \, dx$

    (b) $\int_0^{2\pi} \sin^4(8x) \, dx$

    (c) $\int_0^1 \sqrt{x} \, dx$

    (d) $\int_0^1 \sqrt{1 - x^2} \, dx$

    In each case, evaluate the integral analytically and graph the absolute error for the main diagonal of the Romberg table [i.e. $R(n, n)$]. How does the error vary with n? Does the error ever increase with increasing n (e.g., round-off effects)?

13. A popular integration scheme is *Simpson's rule*,

$$\int_a^b f(x) \, dx \approx \frac{h}{3} [f(a) + 4f(a + h) + 2f(a + 2h) + \ldots$$

$$+ 2f(b - 2h) + 4f(b - h) + f(b)]$$

    Show that this rule is equivalent to the second column of Romberg integration, that is, $R(n, 2)$.

14. Debye theory tells us that the heat capacity of a solid is

$$C_V(T) = 9kN \frac{T^3}{\theta_D^3} \int_0^{\theta_D/T} dx \, \frac{x^4 e^x}{(e^x - 1)^2}$$

    where θ_D is the Debye temperature, N is the number of atoms, and k is Boltzmann's constant. Produce a graph of the specific heat of copper ($\theta_D = 309$ K) which ranges from $T = 0$ K to 1083 K (melting point).

15. Write a program to compute and graph, $K(x)$, the complete elliptic integral of the first kind (see Section 2.2).

16. In Fresnel diffraction you meet the Fresnel integrals,

$$C(w) = \int_0^w \cos(\tfrac{1}{2}\pi x^2)\, dx \qquad S(w) = \int_0^w \sin(\tfrac{1}{2}\pi x^2)\, dx$$

Write functions that compute $C(w)$ and $S(w)$ and produce a graph of $S(w)$ versus $C(w)$ for $w = 0.0, 0.1, 0.2, \ldots, 5.0$. Your plot will be a Cornu spiral.

17. Write a program to compute the coefficients, B_n, for equation (9-34). Don't do it by brute force using rombf. Instead, write a new Romberg routine that uses legndr. Your program should compute all the coefficients simultaneously. Check your program with the split-sphere potential.

18. An azimuthally symmetric potential has a dipole moment of B_1, [see (9-34)]. Modify your program from the previous exercise to evaluate the potentials:

   (a) $V(\theta) = \cos(\theta)$ (b) $V(\theta) = \cos^3(\theta)$

$$\text{(c)} \ \ V(\theta) = \begin{cases} 1 & \text{if} \quad \theta < \pi/4 \\ -1 & \text{if} \quad \theta > 3\pi/4 \\ 0 & \text{otherwise} \end{cases}$$

   Using the coefficients, produce a contour plot of $\Phi(r, \theta)$. Compare with the potential for a point dipole with the same dipole moment.

19. Consider the following electrostatics problem: A hollow cylinder has radius R and height L. The potential on the top disk is an arbitrary function $V(\rho)$. All other sides are held at zero potential. Show that the separation of variables solution is

$$\Phi(\rho, z) = \sum_{s=1}^{\infty} A_s J_0\left(\zeta_{0,s}\frac{\rho}{R}\right) \sinh\left(\zeta_{0,s}\frac{z}{R}\right)$$

   where $\zeta_{0,s}$ is the sth zero of $J_0(x)$. Obtain an explicit expression for the coefficients A_s in terms of $V(\rho)$.

20. Write a program that numerically evaluates the electrostatic potential in the previous problem by numerically evaluating the integrals and summing the series. Produce a contour plot of $\Phi(\rho, z)$ for the potential (a) $V(\rho) = 1$; (b) $V(\rho) = 1 - \rho/R$; (c) $V(\rho) = e^{-\rho/R}$.

9.3 GAUSSIAN QUADRATURE*

Basic Idea

Our original formulation of the trapezoidal rule allows arbitrary values for the grid points x_i. For simplicity we used evenly spaced points, but can anything be gained if we use uneven intervals? Is there some optimal choice for the location of these grid points? If so, should some panels carry more weight than others? In other words, for an integration formula of the form

$$\int_a^b f(x)\,dx \cong w_1 f(x_1) + \ldots + w_N f(x_N) \qquad (9\text{-}54)$$

is there an optimal choice for the grid points (or nodes) x_i and the weights w_i?

The questions above lead us to formulate a new class of integration formulas, known collectively as *Gaussian quadrature*. We will use only the most common formula, namely Gauss–Legendre quadrature. There are many other kinds of Gaussian quadrature that treat specific types of integrands. For example, Gauss–Laguerre is optimal for integrals of the form $\int_0^\infty e^{-x} f(x)\,dx$. The derivation of the other Gaussian formulas is similar to our analysis of Gauss–Legendre quadrature.

The theory of Gaussian integration is based on the following theorem. Let $q(x)$ be a polynomial of degree N such that

$$\int_a^b q(x)\rho(x)x^k\,dx = 0 \qquad (9\text{-}55)$$

where $k = 1, 2, \ldots, N - 1$ and $\rho(x)$ is a specified weight function. Call $x_1, x_2, \ldots, x_N$ the roots of the polynomial $q(x)$. Using these roots as grid points plus a set of weights $w_1, w_2, \ldots, w_N$ we construct an integration formula of the form

$$\int_a^b f(x)\rho(x)\,dx \cong w_1 f(x_1) + w_2 f(x_2) + \ldots + w_N f(x_N) \qquad (9\text{-}56)$$

There exists a set of w's for which the integral formula will be *exact* if $f(x)$ is a polynomial of degree $< 2N$.

Think about this for a moment. In general, if we have N data values, we can fit an $N - 1$ order polynomial to the points. This gives us an integration formula that is exact for polynomials of degree $< N$. However, using Gaussian grid points (along with their weights) we have a formula that is exact for polynomials of degree $< 2N$. If our integrand is well approximated by a high-order polynomial, then our integral approximation should be very accurate.

Three-Point Gaussian–Legendre Rule

To demonstrate the construction of a Gaussian quadrature rule we'll work out the formula for three grid points in the interval $[-1, 1]$ with $\rho(x) = 1$. This gives us a Gaussian-Legendre formula. For integrals in the interval $[a, b]$, it is easy to transform them as

$$\int_a^b f(x)\,dx = \frac{b-a}{2}\int_{-1}^1 f(z)\,dz \qquad (9\text{-}57)$$

using the change of variable $x = \frac{1}{2}[b + a + (b - a)z]$.

The first step is to find the polynomial $q(x)$. We want a three-point rule so that $q(x)$ is a cubic

$$q(x) = c_0 + c_1 x + c_2 x^2 + c_3 x^3 \tag{9-58}$$

From the theorem, (9-55), we know that

$$\int_{-1}^{1} q(x)\, dx = \int_{-1}^{1} x q(x)\, dx = \int_{-1}^{1} x^2 q(x)\, dx = 0 \tag{9-59}$$

Plugging in and doing each integral we get the equations

$$2c_0 + \tfrac{2}{3}c_2 = \tfrac{2}{3}c_1 + \tfrac{2}{5}c_3 = \tfrac{2}{3}c_0 + \tfrac{2}{5}c_2 = 0 \tag{9-60}$$

A solution of the above equations gives us the polynomial[11]

$$q(x) = \tfrac{5}{2}x^3 - \tfrac{3}{2}x \tag{9-61}$$

Notice that this is just the Legendre polynomial $P_3(x)$, a result we might have anticipated given the orthogonality property of these polynomials.

Next we need to find the roots of $q(x) = P_3(x)$. This cubic is easy to factor. The roots are $x_1 = -\sqrt{3/5}$, $x_2 = 0$, $x_3 = \sqrt{3/5}$. Using these grid points in equation (9-56) gives us

$$\int_{-1}^{1} f(x)\, dx \cong w_1 f(\sqrt{\tfrac{3}{5}}) + w_2 f(0) + w_3 f(-\sqrt{\tfrac{3}{5}}) \tag{9-62}$$

Finally, to find the weights, we know that the above formula must be exact for $f(x) = 1, x, \ldots, x^5$. We can use this to work out the values of w_1, w_2, and w_3. It turns out to be sufficient to consider just $f(x) = 1, x,$ and x^2. For these three cases, after some computation, we arrive at the three equations

$$2 = w_1 + w_2 + w_3$$
$$0 = -\sqrt{\tfrac{3}{5}}w_1 + \sqrt{\tfrac{3}{5}}w_3 \tag{9-63}$$
$$\tfrac{2}{3} = \tfrac{3}{5}w_1 + \tfrac{3}{5}w_3$$

This linear system of equations is easy to solve; the solution is $w_1 = \tfrac{5}{9}$, $w_2 = \tfrac{8}{9}$, $w_3 = \tfrac{5}{9}$.

An alternative way of finding the weights is to use the identity

$$w_i = \frac{2}{(1 - x_i^2)\{(d/dx)P_N(x_i)\}^2} \tag{9-64}$$

---

[11] The general solution is $c_0 = 0$, $c_1 = -a$, $c_2 = 0$, $c_3 = 5a/3$ where a is some constant. This arbitrary constant cancels out in the second step.

where $N = 3$ in our example. This formula may be derived from the recurrence relation for Legendre polynomials.

After working out the grid points and weights in the above example, I must confess that one usually looks up these values in tables. Grid points and weights for various values of N are given in Table 9.1; for more extensive tables see Abramowitz and Stegun[12] or Stroud and Secrest.[13] If for some reason you need to compute these, your principal challenge will be to locate all the roots of the polynomial $q(x)$.

TABLE 9.1 Grid points and weights for Gauss–Legendre integration

$\pm x_i$	w_i	$\pm x_i$	w_i
$N = 2$		$N = 8$	
0.5773502692	1.0000000000	0.1834346425	0.3626837834
$N = 3$		0.5255324099	0.3137066459
0.0000000000	0.8888888889	0.7966664774	0.2223810345
0.7745966692	0.5555555556	0.9602898565	0.1012285363
$N = 4$		$N = 12$	
0.3399810436	0.6521451549	0.1252334085	0.2491470458
0.8611363116	0.3478548451	0.3678314990	0.2334925365
$N = 5$		0.5873179543	0.2031674267
0.0000000000	0.5688888889	0.7699026742	0.1600783285
0.5384693101	0.4786286705	0.9041172564	0.1069393260
0.9061798459	0.2369268850	0.9815606342	0.0471753364

There are various advantages and disadvantages in using Gaussian integration: The main benefit is that a very high-order accuracy is obtained for just a few points; often the method yields excellent results using fewer than 10 points. This is especially useful if $f(x)$ is expensive to compute. There are two main disadvantages. (1) The node points and weights must be computed or obtained from tables. This step is nontrivial if you want to use many node points. Using more than $N = 20$ points is rarely worth it since badly behaved functions will spoil the results in any case. (2) Unlike Romberg integration, the method does not lend itself to iteration nor is it easy to estimate the error.

Quantum Perturbation Theory

As a final example of the use of quadrature we consider quantum perturbation theory. The basic idea of perturbation theory is to start with a problem that is easy to solve, for example, the hydrogen atom. Next we change (perturb) the problem

---

[12] M. Abramowitz and I. Stegun, *Handbook of Mathematical Functions* (New York: Dover, 1972), chapter 25.

[13] A.H. Stroud and D. Secrest, *Gaussian Quadrature Formulas* (Englewood Cliffs, N.J.: Prentice Hall, 1966).

slightly, for example, apply a weak external field. The new problem is often significantly more difficult to solve even though the solution changes only slightly. Perturbation theory approximates the correction to the solution by making use of the change being small. The short discussion in this section is only an introduction to the theory; for more details see any of the standard quantum mechanics texts.[14]

We start with the time-independent Schrödinger equation for a particle of mass m_a in a potential $V(\mathbf{r})$

$$H\psi_n = E_n\psi_n(\mathbf{r}) \tag{9-65}$$

where H is the Hamiltonian

$$H = -\frac{\hbar^2}{2m_a}\nabla^2 + V(\mathbf{r}) \tag{9-66}$$

The energy levels and their corresponding wave functions are E_n and ψ_n, respectively. For simplicity, we assume these states are nondegenerate.

Suppose that we know the solution to (9-65) for a given Hamiltonian, H°, and want to compute an approximate solution for a slightly different Hamiltonian. We'll write our Hamiltonian as

$$H = H^\circ + H' \tag{9-67}$$

where H' is the perturbation. For H°, the energies and wave functions are

$$H^\circ\psi_n^\circ = E_n^\circ\psi_n^\circ \tag{9-68}$$

The wave functions are assumed to be orthogonal, so $\int d\mathbf{r}\psi_n^{\circ*}\,\psi_m^\circ = \langle\psi_n^\circ|\psi_m^\circ\rangle = 0$ if $n \neq m$.

First-order perturbation theory approximates E_n and ψ_n as

$$E_n \cong E_n^\circ + E_n' \tag{9-69}$$

$$\psi_n \cong \psi_n^\circ + \sum_m a_{mn}'\psi_m^\circ \tag{9-70}$$

Using (9-69) and (9-70) in (9-65) we get

$$(H^\circ + H')\left(\psi_n^\circ + \sum_m a_{mn}'\psi_m^\circ\right) = (E_n^\circ + E_n')\left(\psi_n^\circ + \sum_m a_{mn}'\psi_m\right) \tag{9-71}$$

---

[14] For example, D. Saxon, *Elementary Quantum Mechanics* (San Francisco: Holden-Day, 1968), chapter 7, section 3; L. Schiff, *Quantum Mechanics* (New York: McGraw-Hill, 1968), chapter 8, section 31.

or

$$H^\circ\psi_n^\circ + \underline{H'\psi_n^\circ} + \underline{H^\circ \sum a'_{mn}\psi_m^\circ} + \underline{\underline{H' \sum a'_{mn}\psi_m^\circ}}$$

$$= E_n^\circ\psi_n^\circ + \underline{E_n'\psi_n^\circ} + \underline{E_n^\circ \sum a'_{mn}\psi_m^\circ} + \underline{\underline{E_n' \sum a'_{mn}\psi_m^\circ}} \qquad (9\text{-}72)$$

where the first-order terms are underlined and the second-order terms are double underlined.

Using (9-68), the zeroth-order terms drop out. Retaining only first-order terms,

$$H'\psi_n^\circ + \sum a'_{mn}E_m^\circ\psi_m^\circ = E_n'\psi_n^\circ + E_n^\circ \sum a'_{mn}\psi_m^\circ \qquad (9\text{-}73)$$

Applying $\int d\mathbf{r}\psi_n^{\circ*}$ to both sides and knowing that the wave functions are orthogonal,

$$E_n' = \frac{\int d\mathbf{r}\psi_n^{\circ*} H'\psi_n^\circ}{\int d\mathbf{r}\psi_n^{\circ*}\psi_n^\circ} = \frac{\langle\psi_n^\circ| H' |\psi_n^\circ\rangle}{\langle\psi_n^\circ|\psi_n^\circ\rangle} \qquad (9\text{-}74)$$

The first-order energy shift due to the perturbation is thus equal to the expectation value of H'.

Particle in a Can

Let's work through an example using perturbation theory. Consider the particle in a box problem in a cylindrical geometry (i.e., the particle in a can). The container has radius R and height L. Because the particle is confined inside the can, the boundary conditions are

$$\psi(\rho = R, \phi, z) = \psi(\rho, \phi, z = 0) = \psi(\rho, \phi, z = L) = 0 \qquad (9\text{-}75)$$

that is, the wave function goes to zero at the interior surface of the can. This boundary condition is equivalent to having a potential that is zero inside the can and infinite at the boundary.

Since the potential inside the can is zero, the Schrödinger equation (9-65) is the Helmholtz equation (9-17), with $k^2 = 2m_aE/\hbar^2$. Separation of variables tells us the solution must be of the form [see equations (9-19), (9-20), and (9-22)],

$$\psi^\circ(\rho, \phi, z) = J_m(\sqrt{k^2 - \alpha^2}\rho)[c_1 \cos(m\phi) + c_2 \sin(m\phi)]$$

$$\times [c_3 \cos(\alpha z) + c_4 \sin(\alpha z)] \qquad (9\text{-}76)$$

The boundary condition at $z = 0$ requires that $c_3 = 0$; the condition at $z = L$ tells us that $\alpha = l\pi/L$ where $l = 1, 2, \ldots$. Finally, the boundary condition at $\rho = R$ requires that

$$\sqrt{k^2 - \alpha^2}R = \zeta_{m,n} \tag{9-77}$$

where $\zeta_{m,n}$ is the nth zero of $J_m(x)$ (see Section 9.1). Thus

$$E^0_{lmn} = \frac{\hbar^2}{2m_a}\left[\left(\frac{\zeta_{m,n}}{R}\right)^2 + \left(\frac{l\pi}{L}\right)^2\right] \tag{9-78}$$

are the energy levels for the particle in a can. You can easily check that the lowest energy level occurs when $l = 1$, $m = 0$, $n = 1$. The (unnormalized) ground state wave function is

$$\psi^0_{101} = J_0\left(\zeta_{0,1}\frac{\rho}{R}\right)\sin\left(\pi\frac{z}{L}\right) \tag{9-79}$$

If we take the radius and height of the can to be one Bohr radii, the ground state energy for an electron is $E^0_{101} = 214$ eV.

Now let's apply the perturbation

$$V'(\mathbf{r}) = V'_c\left(\frac{\rho}{R}\right)^{\gamma} \tag{9-80}$$

where V'_c is a constant. Using (9-74) and (9-79), we find the perturbation of the ground state to be

$$E'_{101} = \frac{\int_0^R d\rho\rho \int_0^{2\pi} d\phi \int_0^L dz (V'_c \rho^{\gamma}/R^{\gamma})J_0^2(\zeta_{0,1}\rho/R)\sin^2(\pi z/L)}{\int_0^R d\rho\rho \int_0^{2\pi} d\phi \int_0^L dz\, J_0^2(\zeta_{0,1}\rho/R)\sin^2(\pi z/L)} \tag{9-81}$$

or

$$E'_{101} = \frac{V'_c\dfrac{\pi L}{R^{\gamma}}\displaystyle\int_0^R d\rho\rho^{1+\gamma}J_0^2(\zeta_{0,1}\rho/R)}{\dfrac{\pi}{2}LR^2J_1^2(\zeta_{0,1})} \tag{9-82}$$

The problem is now reduced to quadrature. I'll leave it for you to finish up (see Exercise 9.25).

EXERCISES

21. (a) Estimate the integrals below using Gaussian quadrature:

(1) $I = \int_0^5 dx \, e^{-x}$

(2) $I = \int_0^1 dx \, \sqrt{1 - x^2}$

(3) $I = \int_0^1 dx \, \frac{1}{x}$

(4) $I = \int_0^1 dx \, \frac{\ln x}{x + 1}$

(5) $I = \int_{-1}^1 dx \, P_{10}(x)$

(6) $I = \int_{-1}^1 dx \, P_{10}^2(x)$

using the Gaussian quadrature formula with $N = 4, 8$, and 12 nodes. Compare the error in each case with the exact value of the integral. Note that some of the integrals have singularities.

(b) Compute each integral in part (a) using Romberg integration. Compare with your answers from part (a) and comment on your results.

22. Compute the weights and grid points for the two-point Gauss–Laguerre formula,

$$\int_0^\infty e^{-x} f(x) \, dx \cong w_1 f(x_1) + w_2 f(x_2)$$

Test your formula by estimating the integral $\int_0^\infty dx \, e^{-x} \cos x = \frac{1}{2}$.

23. (a) Show that the period of oscillation for a particle of mass m moving in a potential $V(x)$, which is symmetric about the origin, may be found by solving the integral (see Section 2.2)

$$T = 4 \sqrt{\frac{m}{2}} \int_0^{x_m} \frac{dx}{\sqrt{V(x_m) - V(x)}}$$

where $x_m = \max(x)$.

(b) Write a program to evaluate this integral using Gaussian quadrature for the potential $V(x) = |x|^\beta$. Produce a contour plot of the integral for $0 < x < 4$ and $0 < \beta < 4$.

24. Find the 10 lowest energy states of the particle in a can (take $R = L$). You may want to write a program to compute and sort through all the possibilities.

25. (a) Write a program to find the shift in the ground state energy for the perturbed particle in a can [see equation (9-82)]. Use Gaussian quadrature to evaluate the integral for $\gamma = \frac{1}{2}, 1, 2, 3$, and 4.

(b) Repeat part (a) using the perturbation $V'(\mathbf{r}) = V_c'|x/R|^\gamma$ for $\gamma = 1, 2, 3$, and 4.

26. (a) Consider a particle in a rectangular box of dimensions $L_x \times L_y \times L_z$. Use separation of variables to solve the Schrödinger equation for the free particle ($V(\mathbf{r}) = 0$ inside the box). Find the eigenfunctions and energy levels. Set up your coordinate system so the origin is in a corner of the box.

(b) Using your results from part (a), find the first order energy shift for the ground state due to the perturbation potential $V(x, y, z) = V_c'x/L_x$, where V_c' is a constant.

(c) Using your results from part (a), write a program which computes the energy shift, E', for an arbitrary separable potential, i.e., for potentials of the form $V(x, y, z) = V_x(x)V_y(y)V_z(z)$. Set up your program to find the energy shift for any state, not just the ground state.

(d) Using your program from part (c), find the energy shift in the first excited state for the potential $V(\mathbf{r}) = V_c' \exp(-r^2)$ with $L_x = 1$, $L_y = 2$ and $L_z = 3$.

27. Consider the two-dimensional integral

$$I = \int_a^b dx \int_c^d dy\, f(x, y)$$

Write a program that numerically estimates this integral by dividing it into a pair of integrals,

$$I = \int_a^b dx\, F(x) \qquad \text{where} \qquad F(x) = \int_c^d dy\, f(x, y)$$

Test your program by evaluating $\int_0^\pi \int_0^\pi \cos(x + y)\,dx\,dy$ and $\int_0^1 \int_0^1 (x + y + 1)^{-1}\,dx\,dy$ using Gaussian quadrature.

28. Using your program from the previous exercise, evaluate the rotational inertia of a thin wedge (see Figure 9.5). The axis of rotation is perpendicular to the plane of the wedge and through (a) the geometric center of the wedge and (b) the center of mass of the wedge.

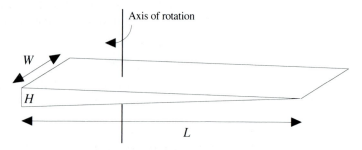

Figure 9.5 Wedge for Exercise 9.28.

BEYOND THIS CHAPTER

There are many more special functions besides the two discussed in this chapter; there is even a rich family of Bessel functions of which $J_m(x)$ is just one member. Baker presents algorithms for evaluating many special functions and their auxiliaries (e.g., zeros of functions).[15] Recently, special functions have been used increasingly for solving PDEs in nonrectangular coordinate systems by spectral methods.[16]

[15] L. Baker, *C Mathematical Function Handbook* (New York: McGraw-Hill, 1992).

[16] C. Canuto, M. Y. Hussaini, A. Quarteroni, and T. A. Zang, *Spectral Methods in Fluid Dynamics* (Berlin: Springer-Verlag, 1988), chapter 2.

Recall that our original formulation of the trapezoidal rule, (9-39), allowed us to place grid points at arbitrary locations. Consider the function in Figure 9.6. To evaluate $\int_a^b f(x)\,dx$ accurately we should use a fine grid spacing near the center of the interval. On the sides, the function is almost constant; even using a handful of grid points would give us an accurate answer. Integrals like this are suitable for *adaptive integration* schemes.[17] The idea is to start at one end and lay down grid points as we move across the interval. As we go we test if the grid size is adequate, increasing or decreasing it as needed. The most commonly used scheme is adaptive Simpson's rule.

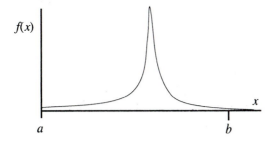

Figure 9.6 Function suitable for adaptive quadrature.

The standard quadrature methods may be extended to multidimensional integrals (sometimes called cubature).[18] As long as the number of dimensions is small (six or fewer), their efficiency is competitive. Multidimensional integration can be computation intensive, so high-accuracy techniques, such as Gaussian quadrature, are especially useful. For even higher-dimensional problems it turns out to be more efficient to essentially use the general trapezoidal rule and select the location of the grid points at random. This technique is known as *Monte Carlo integration*.[19]

The truncation error for Monte Carlo integration goes as $O(N^{-1/2})$ where N is the number of grid points. For one-dimensional integrals this is very poor as compared with trapezoidal rule with evenly spaced grid points. The good news is that Monte Carlo integration's truncation error is always $O(N^{-1/2})$, while the truncation error for deterministic rules deteriorates with dimension. For example, the truncation error for trapezoidal rule with evenly spaced grid points is $O(N^{-2/d})$ for a d-dimensional integral.[20]

---

[17] P. J. Davis and P. Rabinowitz, *Numerical Integration* (Waltham: Blaisdell, 1967), section 6.2.

[18] A. H. Stroud, *Approximate Calculation of Multiple Integrals* (Englewood Cliffs, N.J.: Prentice Hall, 1971).

[19] M. H. Kalos and P. A. Whitlock, *Monte Carlo Methods* (New York: Wiley, 1986), chapter 4; H. Gould and J. Tobochnik, *An Introduction to Computer Simulation Methods*, part 2 (Reading, Mass.: Addison-Wesley, 1988), chapter 10.

[20] F. James,"Monte Carlo Theory and Practice," *Rep. Prog. Phys.*, **43**, 1147–89 (1980).

APPENDIX 9A: FORTRAN LISTINGS

LISTING 9A.1   Subroutine `legndr`. Computes Legendre polynomials.

```
      subroutine legndr(n,x,p)
! Legendre polynomials
      real p(*)

      p(1)=1.        ! P(x)=0 for n=0
      p(2)=x         ! P(x)=x for n=1
      do i=3,(n+1)
        p(i) = ((2*i-3.)*x*p(i-1) - (i-2.)*p(i-2))/(i-1.)
      end do
      return
      end
```

LISTING 9A.2   Subroutine `bess`. Returns values of Bessel functions $[J_m(x) \ldots J_0(x)]$.

```
      subroutine bess(m_max,x,jj)
! Bessel functions
      parameter(max_max = 500, eps=1e-16)
      real*8 jj(*),j(max_max),norm,x

      m_top = max0(m_max,int(x)) + 15  ! Top value of m for
                                       ! recursion
      m_top = 2*((m_top-1)/2+1)        ! Round up to an even
                                       ! integer
      j(m_top+1)=0.
      j(m_top)=1.
      do m=m_top-2,0,-1        ! Downward recursion
        j(m+1) = 2*(m+1)/(x+eps)*j(m+2) - j(m+3)
      end do

      norm = j(1)              ! Note: Be careful, m=0,1,... but
      do m=2,m_top,2           ! vector goes j(1),j(2),...
        norm = norm + 2*j(m+1)
      end do
      do m=0,m_max             ! Send back only the values for
        jj(m+1) = j(m+1)/norm  ! m=0,...,m_max and discard
      end do                   ! m=m_max+1,...,m_top
      return
      end
```

LISTING 9A.3   Subroutine `zeroj`. Returns the *s*th zero of $J_m(x)$.

```
      subroutine zeroj(m_order,n_zero,z)
! Zeros of Bessel functions
```

```fortran
      parameter(max_max = 500, eps=1e-16)
      real*8 jj(max_max),beta,mu,z,deriv

      beta = (n_zero + 0.5*m_order - 0.25)*3.141592654
      mu = 4.*m_order**2
      z = beta - (mu-1.)/(8.*beta)
     &           - 4*(mu-1.)*(7.*mu-31)/(3*(8.*beta)**3)
      write (*,*) 'Initial guess is ',z
      do i=1,5
        call bess(m_order+1,z,jj)     ! Use recursion relation
        deriv = -jj(m_order+2)            ! to evaluate derivative
     &          + m_order/z*jj(m_order+1)
        z = z - jj(m_order+1)/deriv   ! Newton's root finding
      end do
      return
      end
```

LISTING 9A.4 Subroutine `rombf`. Computes integrals using the Romberg algorithm.

```fortran
      subroutine rombf(a,b,N,func,param,R,NR)
! Routine to compute integrals by Romberg algorithm
      real R(NR,NR),param(*)
      external func

      h=b-a          ! This is the coarsest panel
      R(1,1) = h/2 * (func(a,param) + func(b,param))
      np = 1         ! np is the number of panels
      do i=2,N
        h = h/2.
        np = 2*np
        sum = 0.
        do k=1,(np-1),2     ! This loop goes k=1,3,5,...,np-1
          sum = sum + func(a+k*h,param)
        end do

        ! Compute the first column entry R(i,1)
        R(i,1) = 0.5*R(i-1,1) + h*sum
        m=1
        do j=2,i       ! Compute the other columns
          m = 4*m
          R(i,j) = R(i,j-1) + (R(i,j-1)-R(i-1,j-1))/(m-1)
        end do
      end do
      return
      end
```

LISTING 9A.5   Function `my_f`. Defines the integrand function for `rombf`.

```
      real function my_f (x, param)
! Error function integrand
      real param(*)   ! Unused in this function
      my_f = exp (-x**2)
      return
      end
```

Chapter 10
Stochastic Methods

Many methods in computational physics involve a stochastic or random element. Not surprisingly, most applications of stochastic methods are in statistical mechanics. These algorithms are sometimes called Monte Carlo methods in honor of the famous casino in that European city-state. Central to any stochastic method is the generation of random numbers. In this chapter we discuss some basic stochastic techniques and apply them to problems in the kinetic theory of gases.

10.1 KINETIC THEORY

Molecular Dynamics

One of the first applications of probability theory in physics was in the kinetic theory of dilute gases.[1] Consider the following model for a monatomic gas: a system of volume V contains N particles. These particles interact, but since the gas is dilute the interactions are always two-body collisions. The criterion for a gas to be dilute is that the distance between the particles[2] is large compared to d,

---

[1] If you're a little rusty on probability theory you may want to review one of the standard texts, for example, W. Feller, *An Introduction to Probability Theory and its Applications* (New York: Wiley, 1971).

[2] Actually we should say the *average* distance between the particles since the particles are not uniformly spaced. We already see the probabilistic formulation creeping in.

the effective diameter of the particles. This effective diameter may be measured, for example, by scattering experiments. Our criterion for a gas to be considered dilute may be written as

$$d << \sqrt[3]{V/N} \tag{10-1}$$

An alternative view of this criterion is to say that a gas is dilute if the volume occupied by the particles is a small fraction of the total volume.

The interactions between particles in a dilute gas may be accurately modeled using classical mechanics. Assume the particles interact by a pairwise force that only depends on the relative separation

$$\mathbf{F}_{ij} = \mathbf{F}(\mathbf{r}_i - \mathbf{r}_j) = - \mathbf{F}_{ji} \tag{10-2}$$

where $\mathbf{F}_{ij}$ is the force on particle i due to particle j; the positions of the particles are $\mathbf{r}_i$ and $\mathbf{r}_j$, respectively. The explicit form for $\mathbf{F}$ may either be approximated from experimental data or computed theoretically using quantum mechanics.

Once we fix the interparticle force, the dynamics is given by the equation of motion

$$\frac{d^2}{dt^2} \mathbf{r}_i = \frac{1}{m} \sum_{\substack{j=1 \\ j \neq i}}^{N} \mathbf{F}_{ij} \tag{10-3}$$

where m is the mass of a particle. From the initial conditions, in principle, the future state can be computed by evaluating this system of ODEs. This numerical approach is called *molecular dynamics* and it has been very successful in computing microscopic properties of fluids.[3]

In Boltzmann's time there was no hope of evaluating (10-3) numerically, and even today molecular dynamics is limited to very small systems. To understand the scale of the problem, consider that in a dilute gas at standard temperature and pressure, the number of particles in a cubic centimeter, Loschmidt's number, is 2.687×10^{19}. A molecular dynamics simulation of a dilute gas containing a million particles represents a volume of 0.037 cubic microns. Even on a supercomputer, an hour of computer time will evolve the system for only a few nanoseconds of physical time.

Maxwell–Boltzmann Distribution

Instead of being overwhelmed by the huge numbers, we can use them to our advantage. The basic idea of statistical mechanics is to abandon any attempt to predict the instantaneous state of a single particle. Instead we obtain probabilities and compute average quantities, for example, the average speed of a particle. The

---

[3] M. Allen and D. Tildesley, *Computer Simulation of Liquids* (Oxford: Clarendon Press, 1987).

large numbers of particles now work in our favor because even in a very small volume we are averaging over a very large sample.

For a dilute gas we usually take the gas to be *ideal*, that is, we assume that a particle's energy is all kinetic energy,

$$E(\mathbf{r}, \mathbf{v}) = \tfrac{1}{2} m|\mathbf{v}|^2 \tag{10-4}$$

In a dilute gas this is a good approximation since the interparticle forces are short-ranged.

Our starting point is the fundamental axiom of the *canonical ensemble*:[4] Consider a system at thermodynamic equilibrium with temperature T. The probability that a particle in this system is at a position between $\mathbf{r}$ and $\mathbf{r} + d\mathbf{r}$ with a velocity between $\mathbf{v}$ and $\mathbf{v} + d\mathbf{v}$ is

$$P(\mathbf{r}, \mathbf{v}) \, d\mathbf{r} \, d\mathbf{v} = A \, \exp(-E(\mathbf{r}, \mathbf{v})/kT) \, d\mathbf{r} \, d\mathbf{v} \tag{10-5}$$

where $E(\mathbf{r}, \mathbf{v})$ is the energy of the particle and $k = 1.38 \times 10^{-23}$ J/K is Boltzmann's constant. The constant A is a normalization that is fixed by the condition that the integral of the probability over all possible states must equal unity. The differential elements, $d\mathbf{r} \, d\mathbf{v}$, on each side of the equation serve to remind us that $P(\mathbf{r}, \mathbf{v})$ is a *probability density*.

Since a particle's energy is independent of $\mathbf{r}$, the probability density may be written as

$$P(\mathbf{r}, \mathbf{v}) \, d\mathbf{r} \, d\mathbf{v} = [P(\mathbf{r}) \, d\mathbf{r}][P(\mathbf{v}) \, d\mathbf{v}]$$
$$= \left[\frac{1}{V} \, d\mathbf{r}\right][P(\mathbf{v}) \, d\mathbf{v}] \tag{10-6}$$

The particle is equally likely to be anywhere inside the volume V. For example, suppose that we demark a subregion α inside of our system. The probability that the particle is inside α is

$$P_\alpha = \iiint_\alpha d\mathbf{r} \, P(\mathbf{r}) = \frac{1}{V} \iiint_\alpha d\mathbf{r} = \frac{V_\alpha}{V} \tag{10-7}$$

where V_α is the volume of subregion α.

We may further simplify our expression for the probability by making use of the isotropy of the distribution. In spherical coordinates, the probability that a particle has a velocity between $\mathbf{v}$ and $\mathbf{v} + d\mathbf{v}$ is

$$P(\mathbf{v}) \, d\mathbf{v} = P(v, \theta, \phi)v^2 \cos\theta \, dv \, d\theta \, d\phi$$
$$= A e^{-\frac{1}{2}mv^2/kT} \, v^2 \, dv \, \cos\theta \, d\theta \, d\phi \tag{10-8}$$

---

[4] F. Reif, *Fundamentals of Statistical and Thermal Physics* (New York: McGraw-Hill, 1965).

Since the distribution of velocities is isotropic, the angular parts can be integrated to give

$$P(v) \, dv = 4\pi \left(\frac{m}{2\pi kT} \right)^{3/2} v^2 e^{-\frac{1}{2}mv^2/kT} \, dv \qquad (10\text{-}9)$$

where $P(v) \, dv$ is the probability that a particle's speed is between v and $v + dv$. Notice that we finally fixed the normalization constant A by imposing the condition that $\int_0^\infty P(v) \, dv = 1$. This velocity distribution is known as the *Maxwell–Boltzmann distribution* (see Figure 10.1).

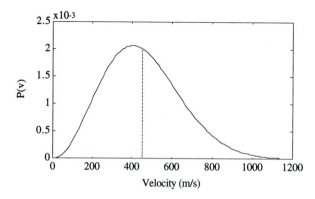

Figure 10.1   Maxwell–Boltzmann distribution of particle velocity for nitrogen at $T = 273$ K. The dashed line marks the average speed, $\langle v \rangle$.

Using (10-9) it is not difficult to compute various average quantities. For example, the average particle speed

$$\langle v \rangle = \int_0^\infty v P(v) \, dv = \frac{2\sqrt{2}}{\sqrt{\pi}} \sqrt{\frac{kT}{m}} \qquad (10\text{-}10)$$

and the root mean square (r.m.s.) particle speed

$$\sqrt{\langle v^2 \rangle} = \sqrt{\int_0^\infty v^2 P(v) \, dv} = \sqrt{3} \sqrt{\frac{kT}{m}} \qquad (10\text{-}11)$$

From this, the average kinetic energy of a particle is

$$\langle K \rangle = \langle \tfrac{1}{2}mv^2 \rangle = \tfrac{3}{2}kT \qquad (10\text{-}12)$$

in agreement with the equipartition theorem.

Finally, the most probable speed, v_{mp}, is not an average but rather is the speed at which $P(v)$ has a maximum,

$$\frac{d}{dv} P(v)\Big|_{v=v_{mp}} = 0 \quad \Rightarrow \quad v_{mp} = \sqrt{2}\sqrt{\frac{kT}{m}} \tag{10-13}$$

Notice that $v_{mp} < \langle v \rangle < \sqrt{\langle v^2 \rangle}$, but they are all of comparable magnitude and approximately equal to the speed of sound

$$v_s = \sqrt{\gamma}\sqrt{\frac{kT}{m}} \tag{10-14}$$

where $\gamma = c_p/c_v$ is the ratio of specific heats. In a monatomic gas, $\gamma = \frac{5}{3}$.

Collision Frequency and Mean Free Path

In general, the particles in a dilute gas interact when their separation is of the order of their effective diameter. While the interaction between particles is continuous, it is also short-ranged, so it is useful to think of the particles as colliding. In the *hard sphere model* we picture the dilute gas as a cloud of tiny billiard balls of diameter d. Particles collide elastically when their separation equals this diameter. We use this model throughout this chapter because it does a surprisingly good job of representing a real gas.

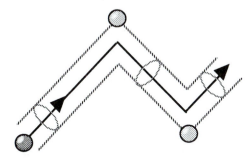

Figure 10.2 Test particle moving amid a field of stationary particles.

The average number of collisions per particle per unit time is called the *collision frequency*, f. A related quantity is the *mean free path*, λ, which is average distance traveled by a particle between collisions. To compute these two quantities for a hard sphere gas consider the following picture. A test particle moves with speed v_r amid a sea of stationary particles (see Figure 10.2). The particle travels a zigzag path much like the ball in a pinball machine.[5] In a time increment t_o, the test particle travels a distance $l = v_r t_o$.

---

[5] A primitive mechanical entertainment device in common use before the advent of video games.

Imagine a cylindrical tube of radius d centered on the path of the test particle. This tube has "elbows" at the locations of collisions, but we approximate its volume as $\pi d^2 l$. The number of stationary particles contained within this cylinder equals the number of collisions experienced by the test particle. On the other hand, since the gas is homogeneous, the number of particles in the tube is

$$M_{\text{tube}} = (\text{number of particles per unit volume}) \times (\text{volume of tube})$$
$$= (N/V)(\pi d^2 v_r t_o) \tag{10-15}$$

which is also the number of collisions in a time t_o.

The collision frequency is

$$f = \frac{\langle M_{\text{tube}} \rangle}{t_o} = \frac{N}{V} \pi d^2 \langle v_r \rangle \tag{10-16}$$

There remains one problem: We don't really have just one moving particle; all particles move. This issue is resolved by identifying $\langle v_r \rangle$ as the average *relative* speed between particles

$$\langle v_r \rangle = \langle |\mathbf{v}_1 - \mathbf{v}_2| \rangle = \int\!\!\int d\mathbf{v}_1 \, d\mathbf{v}_2 |\mathbf{v}_1 - \mathbf{v}_2| P(\mathbf{v}_1) P(\mathbf{v}_2)$$
$$= \frac{4}{\sqrt{\pi}} \sqrt{\frac{kT}{m}} \tag{10-17}$$

where $P(\mathbf{v})$ is given by (10-8). From the collision frequency, the mean free path is obtained as

$$\lambda = (\text{average particle speed}) \times (\text{average time between collisions}) \tag{10-18}$$
$$= \langle v \rangle \frac{1}{f} = \frac{V}{N\pi d^2} \frac{\langle v \rangle}{\langle v_r \rangle}$$

Using (10-10) and (10-17),

$$\lambda = \frac{V}{\sqrt{2} N \pi d^2} \tag{10-19}$$

Notice that the mean free path depends only on density and particle diameter; it is independent of temperature.

Using kinetic theory, we can design a numerical simulation of a dilute gas. Instead of solving the deterministic equations of motion, we will build a stochastic model. The probability arguments discussed in this section give us the framework for the numerical scheme. Section 10.2 discusses how to generate random numbers, the foundation of any stochastic method. In Section 10.3, we bring it all together to formulate the Monte Carlo simulation of a dilute gas.

EXERCISES

1. Plot $\sqrt[3]{V/N}$ as a function of temperature for a dilute gas at one atmosphere of pressure (use ideal gas law). Use a temperature range of 0 K to 500 K; mark the liquification temperatures of nitrogen and oxygen. The effective diameters of N_2 and O_2 are $d = 3.78$ Å and 3.64 Å, respectively. Show that inequality (10-1) is satisfied by air, but not by a wide margin.

2. Using the identities

$$\int_0^\infty e^{-x^2/a^2} dx = \frac{\sqrt{\pi}a}{2}; \qquad \int_0^\infty x e^{-x^2/a^2} dx = \frac{a^2}{2}$$

   (a) Confirm that the Maxwell–Boltzmann distribution is correctly normalized.
   (b) Derive the average speed, (10-10).
   (c) Derive the r.m.s. speed, (10-11).

3. Derive our expression for the average relative speed, (10-17), by explicitly solving the integrals. You will probably want to make the change of variable to center-of-mass coordinates.

4. Write a program to numerically compute the median velocity, $v_{1/2}$, which is defined as $\int_0^{v_{1/2}} P(v)\, dv = \frac{1}{2}$. Plot $v_{1/2}$ as a function of temperature for nitrogen.

5. (a) Plot the mean free path in nitrogen as a function of pressure from 10^{-6} atmospheres to 10 atmospheres. The effective diameter of N_2 is $d = 3.78$ Å; assume $T = 300$ K.
   (b) Plot the number of particles in a cubic mean free path as a function of pressure. Explain why the number of particles *decreases* with increasing pressure.

6. An important model in kinetic theory and one that is often used in molecular dynamics simulations is the *hard disk gas*. This is the two-dimensional analog of the hard sphere gas; the particles are elastic disks moving in a plane. For a dilute hard disk gas, find (a) the most probable speed; (b) the r.m.s. speed; (c) the average relative speed; and (d) the mean free path.

10.2 RANDOM NUMBER GENERATORS

Uniform Deviates

To write a computer program that implements a stochastic algorithm, we first need to know how to generate random numbers. Most languages include a random number generator as one of the functions in their math library.[6] In MATLAB, this function is called `rand`; an example of its use is

```
>> R = rand(n)
```

---

[6] Because these generators use a deterministic algorithm, they are sometimes refered to as psuedorandom number generators. Since the definition of random is problematic, I prefer to avoid these randomer-than-thou arguments.

The variable R is an n×n matrix with each element set to an independent random value; rand(m, n) returns an m×n matrix. As with most built-in generators, the rand function generates *uniform deviates*, which means it returns random numbers in the interval [0, 1). The distribution is uniform so all values in the interval are equally probable. I will use the variable $\mathscr{R}$ to refer to a uniform deviate.

Uniform deviate generators are almost always implemented using the linear congruential method.[7] Given an initial integer "seed" value, I_1, a sequence of integers, I_n, is generated using the mapping

$$I_{n+1} = (aI_n + c)\bmod M \qquad (10\text{-}20)$$

where a, c, and M are integer constants. In MATLAB, this would be

```
I(n+1) = floor( rem( a*I(n) + c, M) )
```

Since MATLAB has no integer arithmetic, the rem (remainder) function is used in place of modulo. In FORTRAN, this would be

```
I(n+1) = mod( a*I(n) + c, M )
```

Using I_n our uniform deviate is computed as $\mathscr{R} = I_n/M$.

The quality of this generator is highly dependent on the choice of a, c, and M. There is no single, ideal choice because there are many tests for validating generators. MATLAB uses $a = 7^5$, $c = 0$, and $M = 2^{31} - 1$; this choice is justified at length by Park and Miller.[8] If you are writing your own uniform deviate generator in another language, I strongly recommend that you look at their article. Notice that if $I_1 \neq 0$, then $I_n \neq 0$ since M is a (Mersenne) prime.

There is considerable superstition as to how one should set the initial seed. There are three schools of thought. (1) Use a simple initial seed such as $I_1 = 1$. (2) Use a large integer. I sometimes use my social security number (which happens to be prime!). (3) Select the seed in a "blind" fashion, for example, by reading the computer's internal clock. I don't feel you can make a convincing argument why any of these is superior. However, no matter how you select the initial seed, you should record its value. For debugging purposes we often want to run a program using identical conditions, that is, with the same set of random numbers.

A few tips and warnings about using uniform deviate generators:

1. Beware of generators whose seed is shorter than 4 bytes. They are unsuitable for scientific work because they have a short period. For example, the

---

[7] D. Knuth, *Seminumerical Algorithms*, vol. 2 of *The Art of Computer Programming* (Reading, Mass.: Addison-Wesley, 1981), section 3.2.

[8] S. K. Park and K. W. Miller, "Random Number Generators: Good Ones Are Hard To Find," *Comm. A.C.M.*, **32**, 1192–1201 (1988).

standard ANSI C uniform deviate generator repeats the same sequence of numbers after about 33,000 calls.[9]

2. The lowest-order bits (least significant digits) in the numbers generated by the linear congruential method are not very random.

3. Resist the temptation to "improve" a generator by building a Rube Goldberg machine; for a cautionary tale see Knuth.[10] One allowed exception is to generate a set of numbers and then shuffle them.[11]

Invertible Distributions

The uniform deviate is the basic building block in the construction of most random number generators. Let's start with some simple examples of how it can be transformed to deliver a variety of distributions. In the following examples, we use the MATLAB rand function to obtain a random number uniformly distributed in [0, 1).

As a simple example, a single random number, x, which is uniformly distributed in the interval $[a, b)$, may be obtained as

$$x = a + (b - a)\mathcal{R} \tag{10-21}$$

The random variable, y, defined as

$$y = \mathcal{R}_1 + \mathcal{R}_2 - 1 \tag{10-22}$$

is triangle distributed in $[-1, 1)$; $\mathcal{R}_1$, $\mathcal{R}_2$ are independent uniform deviates. Its distribution is

$$P(y) = \begin{cases} 1 + y & -1 \le y \le 0 \\ 1 - y & 0 \le y < 1 \end{cases} \tag{10-23}$$

This result may be extended to construct a Gaussian distributed random variable (see the exercises).

Uniform deviates may be easily transformed to generate random numbers from *invertible distributions*. A simple example is the exponential distribution

$$P(u) = \frac{1}{\lambda} e^{-u/\lambda} \tag{10-24}$$

---

[9] S. Harbison and G. Steele Jr., *C, A Reference Manual* (Englewood Cliffs, N.J.: Prentice Hall, 1984), p. 354.

[10] D. Knuth, *Seminumerical Algorithms*, vol. 2 of *The Art of Computer Programming* (Reading Mass.: Addison-Wesley, 1981), p. 4.

[11] W. Press, B. P. Flannery, S. A. Teukolsky, and W. T. Vetterling, *Numerical Recipes in FORTRAN*, 2d ed. (Cambridge: Cambridge University Press, 1992), section 7.1.

where $u \in [0, \infty)$. This distribution has mean value $\langle u \rangle = \lambda$ and variance $\langle (u - \langle u \rangle)^2 \rangle = \lambda^2$. Consider a new random variable, R, defined as

$$u = -\lambda \ln(1 - R) \qquad (10\text{-}25)$$

so

$$R = 1 - e^{-u/\lambda}; \qquad dR = \frac{1}{\lambda} e^{-u/\lambda} \, du \qquad (10\text{-}26)$$

Performing this change of variable just as we normally do in integral calculus,

$$P(u) \, du = 1 \cdot dR = P(R) \, dR \qquad (10\text{-}27)$$

Thus the variable R is uniformly distributed in $[0, 1)$, that is, it is a uniform deviate and $R = \mathcal{R}$. Using (10-25), an exponentially distributed random value u may be computed as

```
u = - lambda * log( 1 - rand(1) );
```

in MATLAB.

A more common distribution is the *Gaussian* (or normal) distribution,

$$P(x) \, dx = \frac{1}{\sqrt{2\pi}\sigma} e^{-(x - \mu)^2/2\sigma^2} \, dx \qquad (10\text{-}28)$$

It has mean value, $\langle x \rangle = \mu$, and a variance, $\langle (x - \langle x \rangle)^2 \rangle = \sigma^2$. You are probably familiar with the polar transformation trick used for integrating a Gaussian[12]; to make the Gaussian an invertible distribution we play the same game. Consider the product of two similar Gaussian distributions,

$$P(x)P(y)dxdy = \frac{1}{2\pi\sigma^2} e^{-(x-\mu)^2/2\sigma^2} e^{-(y-\mu)^2/2\sigma^2} \, dx \, dy$$

$$= \frac{1}{2\pi\sigma^2} e^{-[(x-\mu)^2+(y-\mu)^2]/2\sigma^2} \, dx \, dy \qquad (10\text{-}29)$$

Introducing the polar coordinates, $x = \rho \cos \theta + \mu$, $y = \rho \sin \theta + \mu$, we have

$$P(\rho)P(\theta) \, \rho \, d\rho \, d\theta = \frac{1}{2\pi\sigma^2} e^{-\rho^2/2\sigma^2} \rho \, d\rho \, d\theta \qquad (10\text{-}30)$$

If we define a supplementary change of variable,

---

[12] F. Reif, *Fundamentals of Statistical and Thermal Physics* (New York: McGraw-Hill, 1965), appendix A2.

$$u = \rho^2/2\sigma^2; \qquad a = \frac{1}{2\pi}\,\theta \qquad (10\text{-}31)$$

then

$$P(u)P(a)\ du\ da = (e^{-u}\ du)(1\ da) \qquad (10\text{-}32)$$

The variable u is exponentially distributed, while the variable a is uniformly distributed in $[0, 1)$. This procedure for obtaining Gaussian distributed random numbers is known as the *Box–Muller transformation*.

In MATLAB, we may obtain a pair of independent, Gaussian distributed random numbers, x and y, using

```
R  =  rand(2,1);          % R(1) and R(2) are uniform deviates
u  =  -log( 1 - R(1) );   % u is exponentially distributed
a  =  R(2);
rho  =  sigma * sqrt(2*u);
theta  =  2*pi*a;
x  =  rho*cos(theta) + mu;   % x and y are independent,
                             % Gaussian
y  =  rho*sin(theta) + mu;   % distrib. numbers, with
                             % mean = mu, variance = sigma
```

The code listed above can be significantly condensed and streamlined. Finally, note that MATLAB's rand function can be set to produce Gaussian distributed random numbers automatically; see the reference manual for details.

To understand why these transformations work, we introduce the cumulative distribution function,

$$F(x) = \int_a^x P(x)\ dx \qquad (10\text{-}33)$$

Note that $F(a) = 0$ and $F(b) = 1$. Figure 10.3 shows a schematic of a typical distribution $P(x)$ and its corresponding $F(x)$. Using the cumulative distribution function, we map the interval $[0, 1)$ into the interval $[a, b)$. When the slope of F is small, the interval dR gets mapped into a large interval dx. In this way the transformation correctly maps the uniform deviate $\mathscr{R}$ into the random variable x. Although the method works in general, if the function $F(x)$ is difficult to invert, that is, if $F^{-1}(\mathscr{R})$ does not have a simple form, other techniques may be more efficient.

Discrete Distributions

So far we have considered continuous probability distributions where the random variable is a real number. However, some random processes have only a discrete set of outcomes. Say that the random variable i takes on integer values in the

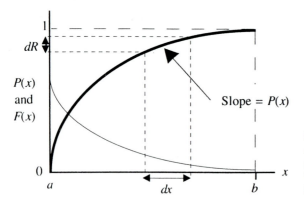

Figure 10.3   Probability distribution $P(x)$ (light line) and its cumulative distribution function $F(x)$ (heavy line). The interval dR is mapped into the interval dx.

interval $[a, b]$ and is distributed with probability P_i. For example, with a die roll, the interval is $[1, 6]$ and $P_i = \frac{1}{6}$ for $i = 1, \ldots, 6$. As a more complicated example, the probability of getting k ''heads'' when flipping N coins is given by the *binomial distribution*,

$$P_k = \frac{N!}{(N - k)!k!}(\tfrac{1}{2})^N \tag{10-34}$$

where $k = 0, \ldots, N$.

For the die roll example, the random variable i may be generated as $i = \lceil 6\mathfrak{R} \rceil$ where $\mathfrak{R}$ is a uniform deviate in $[0, 1)$. The ceiling of x, $\lceil x \rceil$, is the smallest integer greater than x (i.e., round-up x to an integer). In MATLAB, the random variable i may be generated as

```
i = ceil(6*rand(1));
```

For the coin toss example, call j the outcome of a single toss. We may select it as $j = \lfloor 2\mathfrak{R} \rfloor$; the floor of x, $\lfloor x \rfloor$, is the largest integer less than x. In MATLAB, we may generate j using

```
j = floor(2*rand(1));
```

If k is the number of ''heads'' out of N tosses, it may be selected as

```
k=0;
for i=1:N
   k = k + floor(2*rand(1));
end
```

or more compactly, k = sum(floor(2*rand(N, 1))).

Another algorithm for evaluating a single coin toss is

```
if ( rand(1) < 1/2 )
  j = 0;   % Tails
else
  j = 1;   % Heads
end
```

This second approach may seem cumbersome but it has the advantage that it may be easily generalized to handle any discrete probability distribution.

A simple algorithm for selecting a random variable i with distribution P_i is

```
R = rand(1);   % Uniform deviate in [0,1)
sum = 0;
for i=a:b
sum = sum + P(i);   % Cumulative sum of P(i)
  if ( R < sum )
    break;   % Jump out of the for loop
  end
end
```

We usually need many random values drawn from a distribution. In that case, this algorithm is more efficient if we use the discrete cumulative distribution

$$F_i = \sum_{j=a}^{i} P_j \tag{10-35}$$

We compute F_i once and use it as follows:

```
R = rand(1);   % Uniform deviate in [0,1)
for i=a:b
  if ( R < F(i) )
    break;   % Jump out of the for loop
  end
end
```

The idea is to select the random variable i by the condition

$$F_{i-1} < \Re < F_i \tag{10-36}$$

(see Figure 10.4). The simple search scheme above requires, on average, $O(N)$ operations where N is the number of possible values for i (i.e., $N = b - a + 1$). A more efficient search algorithm can typically reduce this to $O(\ln N)$ operations.[13]

---

[13] W. Press, B. P. Flannery, S. A. Teukolsky, and W. T. Vetterling, *Numerical Recipes in FORTRAN*, 2d ed. (Cambridge: Cambridge University Press, 1992), chapter 8.

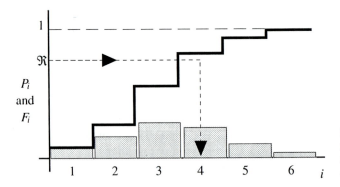

Figure 10.4 Selection of a discrete random number using the generating function. Shaded histogram is $P(i)$ and dark staircase is $F(i)$.

Acceptance–Rejection

Acceptance–rejection is a useful technique for building generators of arbitrary distributions. It works equally well for both continuous and discrete random variables. The general idea is analogous to throwing darts at a dartboard. There is a low probability of hitting the bull's-eye because its area is a small fraction of the area of the board.

Consider a continuous random variable x distributed in $[a, b)$ with a probability distribution $P(x)$. We select a value P_{max} with the condition

$$P_{max} \geq P(x) \tag{10-37}$$

for all x. The scheme for selecting a random value for x is: (1) pick a trial value $x_{try} = a + (b - a)\mathcal{R}_1$; (2) compute $P(x_{try})$; (3) accept the value as the generated random number if

$$\frac{P(x_{try})}{P_{max}} \geq \mathcal{R}_2 \tag{10-38}$$

where $\mathcal{R}_1$ and $\mathcal{R}_2$ are independent uniform deviates; and (4) if this condition is not satisfied, x_{try} is rejected and we return to step (1) and try again.

Figure 10.5 shows a geometric interpretation of this scheme. Picture a rectangle bounded by $x = [a, b)$ and $y = [0, P_{max})$. We select a random point (x_{try}, y_{try}) inside this rectangle, where $y_{try} = P_{max}\mathcal{R}_2$. If this point lands in the shaded region [i.e. below the curve $P(x)$], then it is accepted. The larger the value of $P(x_{try})$, the more likely it is that we'll accept x_{try}.

The acceptance–rejection scheme is exact if P_{max} satisfies (10-37). However, it is most efficient when $P_{max} = \max(P(x))$ since this minimizes the number of rejections. Finally, the scheme is easy to adapt to discrete random variables; replace the curve in Figure 10.5 with the histogram of the distribution.

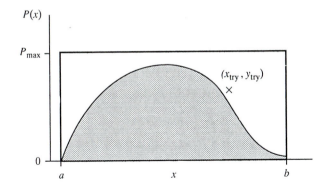

$P(x)$

P_{max}

(x_{try}, y_{try})
$\times$

0

a x b

Figure 10.5 Schematic illustrating acceptance–rejection method.

EXERCISES

7. (a) Show that the values returned by the rand function are uniformly distributed by plotting a histogram of the distribution. Produce histograms for 100, 1000, and 10^4 random numbers.

(b) Produce similar histograms for exponentially distributed random values.

(c) Produce similar histograms for Gaussian distributed random values.

8. The moments of a distribution are defined as

$$\langle x^i \rangle = \int_a^b x^i P(x)\, dx$$

where the random variable, x, takes values in the interval $[a, b]$. Obtain the moments for the following distributions:

(a) Uniform deviate;

(b) Exponential distribution, (10-24);

(c) Gaussian distribution, (10-28) for $\mu = 0$.

(d) Show that the variance may be written in terms of the first and second moments, $\langle (x - \langle x \rangle)^2 \rangle = \langle x^2 \rangle - \langle x \rangle^2$. Compute the variance of each of the distributions from parts (a) to (c).

9. The moments of a distribution may be estimated given a sample of random numbers from that distribution. Given the numbers x_j, this estimate may be computed as $\langle x^i \rangle \cong \frac{1}{N} \sum_{j=1}^N x_j^i$. Write a program that computes this estimate and compare your numerical estimates with the analytic results from the previous exercise.

10. An extension of the exponential distribution is the *gamma distribution*,

$$P(x) = \frac{e^{-x} x^{m-1}}{(m-1)!}$$

where $x \in [0, \infty)$. Show that x may be generated using

$$x = - \sum_{i=1}^{m} \ln(1 - \mathcal{R}_i)$$

where $\mathcal{R}_i$ is a uniform deviate.

11. Consider a random variable $x \in [0, b)$ that is distributed as

$$P(x) \, dx = \frac{(n + 1)x^n}{b^{n+1}} \, dx$$

Find the transformation that maps the uniform deviate $\mathcal{R}$ into x.

12. Consider a random variable $x \in [0, \infty)$ that is distributed as

$$P(x) \, dx = \frac{2}{a^2} \, x e^{-x^2/a^2} \, dx$$

Find the transformation that maps the uniform deviate $\mathcal{R}$ into x.

13. One way to obtain a Gaussian distributed random number, z, is to make use of the central limit theorem and compute

$$z = \sum_{i=1}^{N} (\mathcal{R}_i - \tfrac{1}{2})$$

where the $\mathcal{R}_i$ is a uniform deviate. Write a function that implements this algorithm and show that the resulting values of z are indeed Gaussian distributed if N is sufficiently large (how large?). What is the variance of z and how does it depend on the choice of N?

14. The Poisson distribution is a discrete probability distribution common in statistical mechanics. It is defined as

$$P_i = \frac{e^{-\lambda}\lambda^i}{i!}$$

where $i \in [0, \infty)$.

(a) Write a function that uses the cumulative distribution to generate Poisson distributed random integers. Assume that λ is never very large so that a simple search algorithm is adequate. Test your function by showing that the mean value $\langle i \rangle = \lambda$.

(b) Modify your function to use a more sophisticated search algorithm for large values of λ. For what values of λ does the advanced method pay off?

15. (a) Write a function that uses acceptance–rejection to generate the distribution $P(x) = \tfrac{3}{4}(1 - x^2)$ where $x \in (-1, 1)$. Test your routine by producing histograms and computing the first few moments (see Exercise 10.9).

(b) Repeat part (a) for the binomial distribution [equation (10-34)].

(c) Repeat part (a) for Poisson distributed random numbers (see Exercise 10.14).

16. Consider the dice game known as craps. On the first throw if you roll 7 or 11 you win, otherwise the roll establishes your "mark." You continue throwing until either you roll your mark (and win) or roll a 7 or an 11 (and lose). Write a program that simulates the continuous playing of craps. Determine (a) the probability of winning; (b) the average number of dice rolls in a game; (c) the probability of rolling 10 times without hitting your mark.

10.3 DIRECT SIMULATION MONTE CARLO

General Algorithm

We now turn to the problem of constructing a numerical simulation for a dilute gas. Again, we don't want to compute the trajectory of every particle. Instead, we'll use kinetic theory to build a stochastic model. The scheme is loosely based on the Boltzmann equation; it was popularized as a practical numerical algorithm by G. A. Bird.[14] He named it direct simulation Monte Carlo (DSMC), and it has been called "the dominant predictive tool in rarefied gas dynamics for the past decade."[15]

The DSMC algorithm is like molecular dynamics in that the state of the system is given by the positions and velocities of the particles, $\{\mathbf{r}_i, \mathbf{v}_i\}$, for $i = 1, \ldots$. , N. In this section we study homogeneous problems but formulate the DSMC algorithm for inhomogeneous systems, in anticipation of the next section.

The evolution of the system is integrated in time steps, τ, which are typically on the order of the mean collision time for a particle. At each time step, the particles are first moved as if they did not interact with each other. Every particle's position is reset as $\mathbf{r}_i^{\text{new}} = \mathbf{r}_i + \mathbf{v}_i\tau$. After the particles move, some are selected to collide. The rules for this random selection process are obtained from kinetic theory. After the velocities of all colliding particles have been reset, the process is repeated for the next time step.

Intuitively, we would want to select only particles that were near each other as collision partners. In other words, particles on opposite sides of the system should not be allowed to interact. To implement this condition, the particles are sorted into spatial cells and only particles in the same cell are allowed to collide. We could invent more complicated schemes, but this one works well as long as the dimension of a cell is no larger than a mean free path.

Collisions

In each cell, a set of representative collisions is processed at each time step. All pairs of particles in a cell are considered to be candidate collision partners, *regardless* of their positions within the cell. In the hard sphere model, the collision probability for the pair of particles, i and j, is proportional to their relative speed,

$$P_{\text{coll}}(i, j) = \frac{|\mathbf{v}_i - \mathbf{v}_j|}{\displaystyle\sum_{m=1}^{N_c} \sum_{n=1}^{m-1} |\mathbf{v}_m - \mathbf{v}_n|} \tag{10-39}$$

where N_c is the number of particles in the cell [see equation (10-16)]. Notice that the denominator serves to normalize this discrete probability distribution.

---

[14] G.A. Bird, *Molecular Gas Dynamics* (Oxford: Clarendon Press, 1976).
[15] E.P. Muntz, "Rarefied Gas Dynamics," *Ann. Rev. Fluid Mech.*, **21**, 387–417 (1989).

It would be computationally expensive to use (10-39) directly because of the double sum in the denominator. Instead, the following acceptance–rejection scheme is used to select collision pairs.

1. A pair of candidate particles, i and j, is chosen at random.
2. Their relative speed is computed.
3. The pair is accepted as collision partners if

$$\frac{|\mathbf{v}_i - \mathbf{v}_j|}{v_r^{max}} > \mathcal{R} \tag{10-40}$$

where v_r^{max} is the maximum relative velocity in the cell and $\mathcal{R}$ is a uniform deviate in [0, 1).

4. If the pair is accepted, the collision is processed and the velocities of the particles are reset.
5. After the collision is processed or if the pair is rejected, return to step 1.

This acceptance–rejection procedure exactly selects collision pairs according to (10-39). The method is also exact if we overestimate the value of v_r^{max}, although it is less efficient in the sense that more candidates are rejected. On the whole, it is computationally cheaper to make an intelligent guess that overestimates v_r^{max} rather than recompute it at each time step.

After the collision pair is chosen, their postcollision velocities, $\mathbf{v}_i^*$ and $\mathbf{v}_j^*$, need to be evaluated. Conservation of linear momentum tells us that the center of mass velocity remains unchanged by the collision,

$$\mathbf{v}_{cm} = \tfrac{1}{2}(\mathbf{v}_i + \mathbf{v}_j) = \tfrac{1}{2}(\mathbf{v}_i^* + \mathbf{v}_j^*) = \mathbf{v}_{cm}^* \tag{10-41}$$

From conservation of energy, the magnitude of the relative velocity is also unchanged by the collision,

$$v_r = |\mathbf{v}_i - \mathbf{v}_j| = |\mathbf{v}_i^* - \mathbf{v}_j^*| = v_r^* \tag{10-42}$$

Equations (10-41) and (10-42) give us four constraints for the six unknowns in $\mathbf{v}_i^*$ and $\mathbf{v}_j^*$.

The two remaining unknowns are fixed by the angles, θ and ϕ, for the relative velocity

$$\mathbf{v}_r^* = v_r[(\sin\theta\,\cos\phi)\hat{\mathbf{x}} + (\sin\theta\,\sin\phi)\hat{\mathbf{y}} + (\cos\theta)\hat{\mathbf{z}}] \tag{10-43}$$

For the hard sphere model, these angles are uniformly distributed over the unit sphere. The azimuthal angle is uniformly distributed between 0 and 2π, so it is selected as $\phi = 2\pi\mathcal{R}$. The θ angle is distributed according to the probability density,

$$P(\theta)\, d\theta = \cos\theta\, d\theta \tag{10-44}$$

Using the change of variable $q = \sin\theta$, we have $P(q)\, dq = (1)\, dq$, so q is uniformly distributed in the interval $[-1, 1]$. We don't really need to find θ; instead we compute

$$q = 2\mathcal{R} - 1$$
$$\sin\theta = q \tag{10-45}$$
$$\cos\theta = \sqrt{1 - q^2}$$

where $\mathcal{R}$ is a uniform deviate in $[0, 1)$. We can then use $\sin\theta$ and $\cos\theta$ in (10-43). The postcollision velocities are set as

$$\mathbf{v}_i^* = \mathbf{v}_{cm}^* + \tfrac{1}{2}\mathbf{v}_r^*$$
$$\mathbf{v}_j^* = \mathbf{v}_{cm}^* - \tfrac{1}{2}\mathbf{v}_r^* \tag{10-46}$$

and we go on to select the next collision pair.

Finally we ask, "How many total collisions should take place in a cell during a time step?" From the collision frequency (10-16), the total number of collisions in a cell during a time τ is

$$M_{coll} = \tfrac{1}{2}N_c f\tau = \frac{N_c^2 \pi d^2 \langle v_r \rangle \tau}{2V_c} \tag{10-47}$$

where V_c is the volume of the cell. However, we don't really want to compute $\langle v_r \rangle$ since that involves doing a sum over all $\tfrac{1}{2}N_c^2$ pairs of particles in the cell.

Recall that collision candidates go through an acceptance–rejection procedure. The ratio of total accepted to total candidates is

$$\frac{M_{coll}}{M_{cand}} \approx \frac{\langle v_r \rangle}{v_r^{max}} \tag{10-48}$$

since the probability of accepting a pair is proportional to their relative velocity. Using (10-47) and (10-48),

$$M_{cand} = \frac{N_c^2 \pi d^2 v_r^{max} \tau}{2V_c} \tag{10-49}$$

which tells us how many candidates we should select over a time step τ. Notice that if we set v_r^{max} too high, we still process the same number of collisions on average, but the program is inefficient since many candidates are rejected.

DSMC Program

The program dsmceq (Listing 10.1) uses the DSMC algorithm to compute the relaxation to equilibrium of a gas. The system is assumed to be homogeneous in the y- and z-directions, so only the x-component of position is recorded. The system is divided into cells along the x-direction and the particles are sorted into these cells by the function sorter (Listing 10.2). While the problem we study in this section is also homogeneous in the x-direction, it is better to construct the more general algorithm for use in the next section. The boundaries at $x = 0$ and $x = L$ are periodic. If a particle crosses the right boundary, it reappears on the left and vice versa (see Figure 10.6).

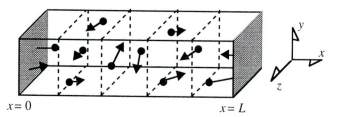

Figure 10.6   Schematic of the dilute gas system simulated by dsmceq. Notice that the boundaries at $x = 0$, L are periodic.

A useful concept in these types of simulations is that of *representative particles*. The volume of the system is 1 cubic micron. If each particle in the simulation were to represent one argon atom, we would need about 27 million particles. Instead, each particle represents effnum real atoms (see line 13). In other words, each particle in the simulation is considered to represent a large number of real molecules that are roughly at the same position with roughly the same velocity. The totally democratic dynamics of the real system is represented in the simulation by a parliamentary subset. This concept allows us to rescale length and time to model larger systems. Of course the simulation will not be accurate if the number of particles is too small. Surprisingly, using 20 or more particles per cell is usually enough to give good results.

In the dsmceq program, all the particles (argon atoms) have the same initial speed (see lines 18–19). The y- and z-components of velocity are initially zero, while the x-component is set to $\pm v_{init}$ where the sign is selected by a coin flip. We wouldn't want to set all the particle velocities equally because then the relative velocities would all be zero and collisions would never take place (remember that the boundaries are periodic).

The width of a cell, L_c, should be set to a fraction of a mean free path. This ensures that a cell is locally homogeneous. The time step is selected as $\tau = a L_c / \langle v \rangle$ where the value of a is set to a fraction less than one ($a = 0.2$ in dsmceq). Thus a particle, on average, will spend several time steps in a cell.

LISTING 10.1   Program `dsmceq`. Simulates relaxation to equilibrium of a dilute gas using the DSMC algorithm. Uses `sorter` (Listing 10.2) and `colide` (Listing 10.3).

```
1    % dsmceq - Dilute gas simulation using DSMC algorithm
2    % This version illustrates the approach to equilibrium
3    clear;  help dsmceq;    % Clear memory and print header
4    % Initialize variables  (MKS units)
5    rand('seed',1);            % Initialize rand
6    npart = input('Enter number of simulation particles - ');
7    boltz = 1.3806e-23;      % Boltzmann's constant
8    mass = 6.63e-26;         % Mass of argon atom
9    diam = 3.66e-10;         % Effective diameter of argon atom
10   T = 273;                 % Temperature  (K)
11   density = 1.78;          % Mass density of argon at STP
12   L = 1e-6;                % System size is one micron
13   eff_num = density/mass*L^3/npart;
14   fprintf('Each particle represents %g atoms\n',eff_num);
15   v_init = sqrt(3*boltz*T/mass);     % Initial velocities
16   x = L*rand(npart,1);     % Assign random positions
17   % Assign x velocities as +/- v_init
18   v = zeros(npart,3);
19   v(:,1) = v_init * (1 - 2*floor(2*rand(npart,1)));
20   ncell = 15;               % Number of cells
21   tau = 0.2*(L/ncell)/v_init;     % Set timestep tau
22   vrmax = 3*v_init*ones(ncell,1); % Estimated max rel. speed
23   selxtra = zeros(ncell,1);          % Used by routine "colide"
24   coeff = 0.5*eff_num*pi*diam^2*tau/(L^3/ncell);
25   vmag = sqrt(v(:,1).^2 + v(:,2).^2 + v(:,3).^2);
26   vbin = 50:100:1050;   % Bins for histogram
27   hist(vmag,vbin);  title('Initial speed distribution');
28   xlabel('Speed');  ylabel('Number');
29   pause;
30   coltot = 0;     % Count total collisions
31   nstep = input('Enter total number of timesteps - ');
32   for istep = 1:nstep
33     x(:) = x(:) + v(:,1)*tau; % Update x position of particle
34     x = rem(x+L,L);           % Periodic boundary conditions
35
36     [cell_n, index, Xref] = sorter(x,npart,ncell,L);
37     [v, vrmax, selxtra, col] = colide(v,cell_n, ...
38                   index,Xref,ncell,vrmax,tau,selxtra,coeff);
39     coltot = coltot + col;
40     vmag = sqrt(v(:,1).^2 + v(:,2).^2 + v(:,3).^2);
41     hist(vmag,vbin);
42     string = sprintf('Done %g of %g steps; %g collisions',...
43                                    istep,nstep,coltot);
44     title(string);  xlabel('Speed');  ylabel('Number');
45     pause(1);
```

```
46  end
47  vmag = sqrt(v(:,1).^2 + v(:,2).^2 + v(:,3).^2);
48  string = sprintf('Final distribution, time = %g s',
                                          nstep*tau);
49  hist(vmag,vbin);  title(string);
50  xlabel('Speed');  ylabel('Number');
```

LISTING 10.2 Function `sorter`. Produces sorted lists used by `colide` to select random particles from a cell.

```
1   function [cell_n, index, Xref] = sorter(x,npart,ncell,L)
2   % sorter - Function to sort particles into cells
3   % Inputs
4   %    x     - Positions of particles
5   %    npart - Number of particles in the system
6   %    ncell - Number of cells in the system
7   %    L     - System size
8   % Output
9   %    cell_n - Number of particles in a cell
10  %    index  - Index into Xref array
11  %    Xref   - Cross-reference array
12  [xsort, Xref] = sort(x);
13  jx = ceil(ncell*x/L);      % Particle i is in cell jx(i)
14  cell_n = zeros(ncell,1);
15  for ipart=1:npart
16    jcell = jx(ipart);
17    cell_n(jcell) = cell_n(jcell) + 1;
18  end
19  index = cumsum(cell_n);
20  index(2:ncell) = index(1:ncell-1) + 1;
21  index(1) = 1;
22  return;
```

LISTING 10.3 Function `colide`. Called by `dsmceq` and `dsmcne` to evaluate collisions using the DSMC algorithm.

```
1   function [v,crmax,selxtra,col] = colide(v,cell_n,...
2                  index,Xref,ncell,crmax,tau,selxtra,coeff)
3   % colide - Function to process collisions in cells
4   % Inputs
5   %    v      - Velocities of the particles
6   %    cell_n - Number of particles in a cell
7   %    index  - Index for drawing from Xref list
8   %    Xref   - Cross-reference list for drawing particles
9   %    ncell  - Number of cells in the system
```

```
10  %      crmax   - Estimated maximum relative speed in a cell
11  %      tau     - Time step
12  %      selxtra - Extra selections carried over from last
                       timestep
13  %      coeff   - Coefficient in computing number of selected
                       pairs
14  % Outputs
15  %      v       - Updated velocities of the particles
16  %      crmax   - Updated maximum relative speed
17  %      selxtra - Extra selections carried over to next timestep
18  %      col     - Total number of collisions processed
19  %
20  col = 0;            % Count number of collisions
21  for jcell=1:ncell
22   number = cell_n(jcell);
23   if ( number > 1 )  % If more than one particle is in the cell
24    %% Determine number of candidate collision pairs
25    %% are to be selected in this cell
26    select = coeff*number^2*crmax(jcell) + selxtra(jcell);
27    nsel = floor(select);         % Number of pairs to be
                                        selected
28    selxtra(jcell) = select-nsel; % Carry over left-over
                                        fraction
29    crm = crmax(jcell);           % Current maximum relative
                                        speed
30    for isel=1:nsel
31     % Pick two particles at random out of this cell
32     k = floor(rand(1)*number);
33     kk = rem(ceil(k+rand(1)*(number-1)),number);
34     ip1 = Xref(k+index(jcell));    % First particle
35     ip2 = Xref(kk+index(jcell));   % Second particle
36     cr = norm( v(ip1,:)-v(ip2,:) );  % Their relative speed
37     if( cr > crm )               % If relative speed larger than
                                        crm
38       crm = cr;                  % Reset crm to larger value
39     end
40     if( cr/crmax(jcell) > rand(1) ) % Accept or reject
41       col = col+1;                    % Collision counter
42       vcm = 0.5*(v(ip1,:) + v(ip2,:)); % Center of mass
                                            velocity
43       cos_th = 1 - 2*rand(1);       % Cosine and sine of
44       sin_th = sqrt(1 - cos_th^2);  % collision angle
                                          theta
45       phi = 2*pi*rand(1);          % Collision angle phi
46       vrel(1) = cr*cos_th;         % Compute
                                          post-collision
47       vrel(2) = cr*sin_th*cos(phi); % relative velocity
48       vrel(3) = cr*sin_th*sin(phi);
```

```
49          v(ip1,:) = vcm + 0.5*vrel;          % Update
50          v(ip2,:) = vcm - 0.5*vrel;          % velocities
51        end
52      end
53      crmax(jcell) = crm; % Update max relative speed
54    end
55  end
56  return;
```

The sorting function `sorter` (Listing 10.2) creates three lists that are used to select particles at random from cells. Figure 10.7 illustrates how the three lists are built. The first, `Xref`, is just a list of particle names sorted by their x-coordinate. The number of particles in a cell is given by `cell_n`. Finally, `index` is just the cumulative sum of `cell_n`. When drawing particles at random from a given cell, `index` and `cell_n` tell us where in the list `Xref` to look. In the example illustrated in Figure 10.7, suppose that we wanted a random particle from cell 4. We should choose one from `Xref` between 6 (=`index(4)`) and 8 (=`index(4) + cell_n(4) -1`). The three candidates are `Xref(6)`=99, `Xref(7)`=2, and `Xref(8)`=86.

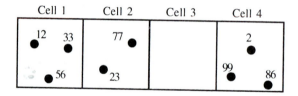

```
Xref = (12, 33, 56, 23, 77, 2, 86, 99, ...)
cell_n = (3, 2, 0, 3, ...)
index = (1, 4, 6, 6, ...)
```

Figure 10.7   Illustration of particle sorting as done by `sorter` function.

The function `colide` (Listing 10.3) processes collisions using the DSMC algorithm. For each cell, we first determine the number of collision candidates to be selected (lines 26–27). Each candidate pair is drawn at random from the particles in the cell (lines 32–35). Given their relative speed, the pair is accepted or rejected (lines 36–40); if the pair is accepted, the particles are said to collide. The postcollision velocities of the particles are computed using (10-43), (10-45), and (10-46) (see lines 42–50).

After each time step, the `dsmceq` program displays a histogram of the speed distribution of the particles. The initial speed distribution is a delta function centered in the 400–500 m/s bin. Figure 10.8 shows the distribution after 10 steps (and 300 collisions) for a system of 300 particles. At this point the distribution has already significantly relaxed toward equilibrium despite the extremely improbable

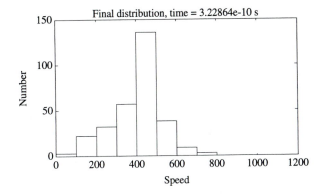

Figure 10.8 Speed distribution as obtained from dsmceq for $N = 300$ particles. After 10 time steps, there have been 300 collisions.

initial condition and that each particle has only been in two collisions. Figure 10.9 shows the distribution after 50 steps (and 1556 collisions). This latter histogram shows that the system has almost completely relaxed to equilibrium in about a nanosecond.

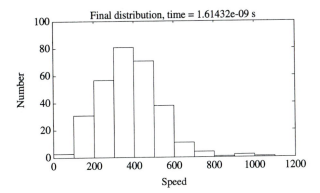

Figure 10.9 Speed distribution as obtained from dsmceq for $N = 300$ particles. After 50 time steps, there have been 1556 collisions.

EXERCISES

17. Show that in a collision, the magnitude of the relative velocity remains unchanged. Is this result modified if the particles have dissimilar masses?

18. Modify dsmceq to compute the expected equilibrium speed distribution histogram using the Maxwell–Boltzmann distribution. Plot this distribution along with the instantaneous speed distribution histogram and demonstrate that the program correctly approaches equilibrium.

19. Modify dsmceq to compute

$$H(t) = \sum_{bins} N_h(v) \ln N_h(v) \, \Delta v$$

where $N_h(v)$ is the number of particles that have velocities in the range $[v, v + \Delta v]$.

Show that for a system initially out of equilibrium, H decreases with time until the system equilibrates.[16]

20. Modify dsmceq to use specular walls instead of periodic boundaries at $x = 0, L$. At a specular wall, a particle is reflected elastically. The pressure at a wall is defined as the time-averaged change in momentum of particles that strike the wall per unit area. Initialize the particles with a Maxwell–Boltzmann distribution and measure the pressure at the walls. Compare with the expected value as given by the ideal gas law.

21. Modify your program from the previous exercise to measure the x-velocity distribution of particles that strike a wall. Plot this distribution as a histogram. Show that particles arriving at a wall are distributed according to the biased Maxwell–Boltzmann distribution,

$$P(v_x) = \pm \frac{m}{kT} v_x e^{-mv_x^2/2kT}$$

with the sign being plus for the left wall and minus for the right wall.

22. While the total number of particles in the dsmceq program remains constant, the number of particles in a given cell fluctuates. Sample and compute the correlation in number density between cells. Compare with the theoretical result,

$$\langle \Delta N_i \Delta N_j \rangle = \langle N_i \rangle \delta_{i,j} - \frac{\langle N_i \rangle \langle N_j \rangle}{\sum_k \langle N_k \rangle}$$

where $\langle N_i \rangle$ is the average number of particles in cell i and ΔN_i is the fluctuation in the number of particles (i.e., $\Delta N_i = N_i - \langle N_i \rangle$).

10.4 NONEQUILIBRIUM STATISTICAL MECHANICS*

Steady States

The statistical mechanics of equilibrium systems is well developed, resting on the firm foundation of ensemble theory. Unfortunately, we have no similar general theory for nonequilibrium systems. As such it would be overly ambitious to start off trying to tackle a complex problem such as turbulence. In this section we consider simple systems that are out of equilibrium but at a *steady state*. In other words, quantities such as density and temperature may vary in space but are stationary in time.

As a first example of a nonequilibrium steady state, consider a dilute gas in a box of length L and cross section A. Suppose that the particles are tinted either black or white. Particles that reflect off the left wall are turned black, while those contacting the right wall are turned white. A particle's pigmentation is unaffected by reflections off the other walls or by collisions. After a time we reach a steady state, as illustrated in Figure 10.10.

---

[16] A. Bellemans and J. Orban, *Phys. Lett.*, **24A**, 620 (1967).

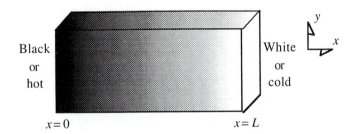

Figure 10.10 Schematic illustrating a nonequilibrium steady state with a constant pigment (or temperature) gradient.

In this system, black particles diffuse to the right and white particles diffuse to the left. This means there is a net flux of pigment in the system. If we assume that the flux, F_ρ, is proportional to the gradient of pigment, then

$$F_\rho = -D \frac{\partial}{\partial x} \rho(x, t) \tag{10-50}$$

where $\rho(x, t)$ is the density of pigment and the constant of proportionality, D, is the *coefficient of diffusion*. Notice the negative sign on the right-hand side of this equation; if pigment increases from right to left (negative gradient), then the flux is from left to right (positive flux).

The time evolution is given by the equation of continuity,

$$\frac{\partial}{\partial t} \rho(x, t) = -\frac{\partial}{\partial x} F_\rho \tag{10-51}$$

At the steady state, the flux must be constant across the system. Thus,

$$\rho(x) = 1 - x/L \tag{10-52}$$

given the boundary conditions $\rho(0) = 1$ and $\rho(L) = 0$.

Consider Figure 10.10 again but instead of having the walls of different pigment, set them at different temperatures. Particles leaving the hot, left wall have, on average, a large kinetic energy as compared with those leaving the cold, right wall. The scenario is slightly more complicated because, unlike pigment, kinetic energy is exchanged in collisions. The general picture, however, remains the same.

We define the number density, $n(x, t)$, as the number of particles per unit volume and the energy density, $e(x, t)$, as the kinetic energy per unit volume. At the particle level, they are defined as

$$n(x, t) = \frac{1}{A dx} \sum_{i=1}^{N} \delta[x \le x_i < x + dx] \tag{10-53}$$

$$e(x,\ t) = \frac{1}{Adx} \sum_{i=1}^{N} \tfrac{1}{2}mv_i^2 \delta[x \le x_i < x + dx] \tag{10-54}$$

The sums are over all N particles, but with the Kronecker delta functions only particles located between x and $x + dx$ are counted. These are strictly mechanical variables, so there is no problem with their definition.

Next, we use the equipartition theorem, (10-12), to define a local, instantaneous temperature as

$$T(x,\ t) \equiv \frac{2}{3k} \frac{e(x,\ t)}{n(x,\ t)} \tag{10-55}$$

Aside from a conversion factor involving Boltzmann's constant k, this temperature is the average kinetic energy per particle. If you've had a rigorous training in equilibrium statistical mechanics, you should instinctively cringe at (10-55). The proper thermodynamic definition of temperature is based on entropy and involves an average over an ensemble of states. All the same, it is useful to extend definitions of thermodynamic quantities, such as temperature, to nonequilibrium systems by using equilibrium identities, such as the equipartition theorem.[17]

Returning to the problem at hand, the energy flux through the system is

$$F_e = -\alpha \frac{\partial}{\partial x} T(x,\ t) \tag{10-56}$$

where α is the thermal conductivity. A related quantity is the thermal diffusion coefficient, κ (for a dilute, monatomic gas, $\kappa = 2\alpha/3kn$). If the thermal conductivity is a constant, there is a linear temperature gradient across the system.

Viscosity

Consider a dilute gas contained between two walls moving in opposite directions with velocities $\pm u_w$ in the y-direction (see Figure 10.11). You can picture the walls as infinite planes or take the boundaries perpendicular to the walls to be periodic. This simple flow problem is called planar *Couette flow*.

The momentum density per unit volume is defined as

$$\mathbf{p}(x,\ t) = \frac{1}{Adx} \sum_{i=1}^{N} m\mathbf{v}_i \delta[x \le x_i < x + dx] \tag{10-57}$$

---

[17] For a more rigorous discussion, see J. McLennan, *Introduction to Non-Equilibrium Statistical Mechanics* (Englewood Cliffs, N.J.: Prentice Hall, 1989), chapter 1.

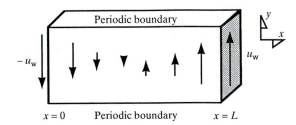

Figure 10.11 Planar Couette flow. Left and right walls move at constant velocities in opposite directions. The steady state velocity profile in the fluid is linear [equation (10-58)].

where $\mathbf{v}_i$ is the velocity of particle i. The fluid velocity is defined as $\mathbf{u}(x,\,t) = \mathbf{p}(x,\,t)/mn(x,\,t)$. At the steady state the velocity of the fluid is

$$\mathbf{u}(x) = u_w\left(\frac{2x}{L} - 1\right)\hat{\mathbf{y}} \qquad (10\text{-}58)$$

that is, we have a linear velocity profile across the system. Particles leave the left (right) wall with a net downward (upward) momentum and this y-momentum diffuses across the system. Assuming the net flux of y-momentum varies linearly with the velocity gradient, we may write it as

$$F_{p_y} = -\eta\,\frac{du_y}{dx} \qquad (10\text{-}59)$$

where u_y is the y-component of the fluid velocity and η is the *viscosity* of the fluid.

Before continuing, let's reconcile this definition with your intuitive notions about viscosity. Picture a highly viscous fluid, say syrup, in a cup. If we quickly stir the syrup then let it relax, the motion quickly comes to a halt. The reason is that the syrup quickly transports its momentum to the sides of the cup. The faster the rate of transport is, the more viscous the fluid.

We can obtain an approximate expression for the viscosity of a dilute gas using a heuristic argument first proposed by Maxwell. Consider a vertical plane located at $x = x^*$ (see Figure 10.12). From purely dimensional arguments, we know that the total flux of particles crossing this plane from right to left is $a\langle v\rangle n$ where n is the number density and a is a dimensionless constant of order unity.

Particles crossing from right to left had their last contact with other particles at a distance of about one mean free path from the plane (see Figure 10.12). Since

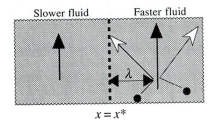

Figure 10.12 Schematic for Maxwell's back-of-the-envelope estimate of viscosity.

the y-velocity of the fluid is $u_y(x)$, these particles, on average, carry a y-momentum of $mu_y(x^* + \lambda)$. Similarly, particles crossing from left to right, on average, carry a y-momentum of $mu_y(x^* - \lambda)$.

Assembling our results, the y-momentum flux is approximately

$$F_{p_y} \approx [a\langle v \rangle nmu_y(x^* - \lambda)] - [a\langle v \rangle nmu_y(x^* + \lambda)]$$

$$\approx -2a\langle v \rangle nm\lambda \, \frac{u_y(x^* + \lambda) - u_y(x^* - \lambda)}{2\lambda} \tag{10-60}$$

$$\approx -2a\langle v \rangle nm\lambda \, \left. \frac{du_y}{dx} \right|_{x=x^*}$$

From (10-59), the viscosity is

$$\eta \approx 2anm\langle v \rangle \lambda \tag{10-61}$$

Using slightly more sophisticated derivations we find that a lies between $\frac{1}{6}$ and $\frac{1}{4}$. Using a significantly more rigorous approach (Chapman–Enskog theory) we find

$$\eta = \frac{5\pi}{32} \, nm\langle v \rangle \lambda \tag{10-62}$$

for a hard sphere gas.[18]

DSMC Nonequilibrium Program

The program dsmcne (Listing 10.4) simulates planar Couette flow in a dilute gas. The particles (argon atoms) are initialized near the steady state with temperature T; their thermal velocities are set as

$$v_x, v_y, v_z = \sqrt{\frac{2kT}{m}} \, \sqrt{-\ln(1 - \mathcal{R}_1)} \, \sin(2\pi\mathcal{R}_2) \tag{10-63}$$

where $\mathcal{R}_1$ and $\mathcal{R}_2$ are independent uniform deviates [see (10-28) to (10-32)]. Furthermore, a linear y-velocity profile is set up across the system (line 28). The wall velocities are entered in terms of Mach number (Ma $\equiv u_w/v_s$ where v_s is the sound speed). We want use a high wall speed (e.g., Ma $= 0.3$) to make the velocity profile noticeable above the random fluctuations in the system (remember that we use only a few hundred particles).

---

[18] J. McLennan, *Introduction to Non-Equilibrium Statistical Mechanics* (Englewood Cliffs, N.J.: Prentice Hall, 1989), chapter 4.

LISTING 10.4 Program `dsmcne`. Measures viscosity in a dilute gas using the DSMC algorithm. Uses `sorter` (Listing 10.2), `colide` (Listing 10.3), `mover` (Listing 10.5), and `sampler` (Listing 10.6).

```
1    % dsmcne - Program to simulate a dilute gas using DSMC
              algorithm
2    % This version simulates planar Couette flow
3    clear;  help dsmcne;    % Clear memory and print header
4    % Initialize variables   (MKS units)
5    rand('seed',1);         % Initialize rand
6    npart = input('Enter number of simulation particles - ');
7    boltz = 1.3806e-23;     % Boltzmann's constant
8    mass = 6.63e-26;        % Mass of argon atom
9    diam = 3.66e-10;        % Effective diameter of argon atom
10   T = 273;                % Initial temperature
11   density = 1.78;         % Mass density of argon at STP
12   L = 1e-6;               % System size is one micron
13   xwall = [0 L];          % Positions of walls
14   Volume = L^3;           % Volume of the system
15   eff_num = (density/mass)*Volume/npart;
16   fprintf('One simulation particle = %g atoms\n',eff_num);
17   mfp = 1/(sqrt(2)*pi*diam^2*density/mass);
18   fprintf('System width is %g mean free paths \n',L/mfp);
19   mpv = sqrt(2*boltz*T/mass);  % Most probable initial velocity
20   vwall_m = input('Enter wall velocity as Mach number - ');
21   vwall = vwall_m * sqrt(5/3 * boltz*T/mass) * [-1 1];
22   fprintf('Wall velocities = %g and %g m/s\n',
                           vwall(1),vwall(2));
23   x = L*rand(npart,1);    % Assign random positions
24   % Assign thermal velocities using Gaussian random numbers
25   v = mpv * sqrt(-log(ones(npart,3)-rand(npart,3))) ...
26                           .* sin(2*pi*rand(npart,3));
27   % Add velocity gradient
28   v(:,2) = v(:,2) + vwall(1)+(vwall(2)-vwall(1))*(x(:)/L);
29   ncell = 20;             % Number of cells
30   tau = 0.2*(L/ncell)/mpv;   % Set timestep tau
31   vrmax = 3*mpv*ones(ncell,1); % Estimated max rel. speed in
                              cells
32   selxtra = zeros(ncell,1);  % Used by collision routine
                              "colide"
33   coeff = 0.5*eff_num*pi*diam^2*tau/(Volume/ncell);
34   ave_n = zeros(ncell,1);    % Average cell number density
35   ave_u = zeros(ncell,3);    % Average cell velocity
36   ave_T = zeros(ncell,1);    % Average cell temperature
37   dvtot = zeros(1,2);        % Total momentum change at a wall
38   dverr = zeros(1,2);        % Used to find error in dvtot
39   nsamp = 0;                 % Total number of samples
40   tsamp = 0;                 % Sample time
```

```
41   nstep = input('Enter total number of timesteps - ');
42   for istep = 1:nstep
43     [x, v, strikes, delv] =
                         mover(x,v,npart,xwall,mpv,vwall,tau);
44     [cell_n, index, Xref] = sorter(x,npart,ncell,L);
45     [v, vrmax, selxtra, col] = colide(v,cell_n,index,Xref, ...
46                         ncell,vrmax,tau,selxtra,coeff);
47     fprintf('Finished %g of %g steps, Collisions = %g\n', ...
48                         istep,nstep,col);
49     fprintf('Strikes on left wall = %g, right wall = %g \n', ...
50                         strikes(1),strikes(2));
51     if(istep > nstep/10) % Start taking samples
52       [ave_n,ave_u,ave_T,nsamp]=sampler(x,v,npart,ncell,L, ...
53                         ave_n,ave_u,ave_T,nsamp);
54       dvtot = dvtot + delv;
55       dverr = dverr + delv.^2;
56       tsamp = tsamp + tau;
57     end
58   end
59   % Normalize statistics
60   ave_n = ave_n/nsamp;
61   for i=1:3
62     ave_u(:,i) = ave_u(:,i)/nsamp;
63   end
64   ave_T = mass/(3*boltz) * (ave_T/nsamp);
65   dverr = dverr/(nsamp-1) - (dvtot/nsamp).^2; % Normalize
66   dverr = sqrt(dverr*nsamp);
67   xcell(:) = ((1:ncell)-0.5)/ncell * L;
68   plot(xcell,ave_n); xlabel('position'); ylabel('Density');
69   pause;
70   plot(xcell,ave_u); xlabel('position'); ylabel('Velocities');
71   pause;
72   plot(xcell,ave_T); xlabel('position'); ylabel('Temperature');
73   force = (eff_num*dvtot*mass)/tsamp /L^2;
74   ferr = (eff_num*dverr*mass)/tsamp /L^2;
75   fprintf('Force per unit area is \n');
76   fprintf('Left wall:    %g +/- %g \n',force(1),ferr(1));
77   fprintf('Right wall:   %g +/- %g \n',force(2),ferr(2));
78   vgrad = (vwall(2)-vwall(1))/L;  % Velocity gradient
79   visc = 1/2*(-force(1)+force(2))/vgrad;  % Average viscosity
80   viscerr = 1/2*(ferr(1)+ferr(2))/vgrad;  % Error
81   fprintf('Viscosity = %g +/- %g\n',visc,viscerr);
82   eta = 5*pi/32*density*(2/sqrt(pi)*mpv)*mfp;
83   fprintf('Theoretical value of viscosity is %g \n',eta);
```

As in the program from Section 10.3, the routine `sorter` is used to sort the particles into cells and the routine `colide` evaluates collisions in those cells. The function mover (Listing 10.5) moves the particles and evaluates reflections off the thermal walls at $x = 0, L$. When a particle reaches a wall, its velocity is reset according to the biased Maxwellian distribution,

$$P(v_x) = \pm \frac{m}{kT_w} v_x e^{-mv_x^2/2kT_w} \tag{10-64}$$

$$P(v_y) = \sqrt{\frac{m}{2\pi kT_w}} e^{-m(v_y - u_w)^2/2kT_w} \tag{10-65}$$

$$P(v_z) = \sqrt{\frac{m}{2\pi kT_w}} e^{-mv_z^2/2kT_w} \tag{10-66}$$

where u_w and T_w are the y-velocity and temperature of the wall, respectively. The sign on the x-velocity is positive for the left wall and negative for the right wall. After a particle's velocity is thermalized, it is allowed to move away from the wall for whatever fraction of a time step remains (lines 40–43).

To understand why the v_x distribution is biased, consider Figure 10.13. On the left we have a fluid at equilibrium with a thermal wall. The picture on the right is similar except the wall has been replaced with a reservoir of fluid held at the same temperature as the wall. Particles reflected off a thermal wall should have the same distribution as particles entering the system from a thermal reservoir at temperature T_w. Clearly, the distribution is biased toward the faster moving particles (they are more likely to cross the boundary). If it is still not clear how this works, do Exercise 10.21.

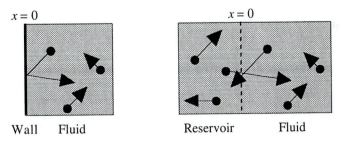

Figure 10.13 Equivalence of a thermal wall and a thermal reservoir.

LISTING 10.5 Function mover. Used by the dsmcne program to update particle positions. Mover also evaluates collisions between the particles and the thermal walls.

```
1    function [x,v,strikes,delv] = mover(x,v,npart, . . .
2                              xwall,mpv,vwall,tau);
```

```
3    % mover - Function to move particles by free flight
4    %          Also handles collisions with walls
5    % Inputs
6    % x     - Positions of the particles
7    % v     - Velocities of the particles
8    % npart - Number of particles in the system
9    % xwall - Positions of walls
10   % mpv   - Most probable velocity off the wall
11   % vwall - Wall velocities
12   % tau   - Time step
13   % Output
14   % x,v     - Updated positions and velocities
15   % strikes - Number of particles striking each wall
16   % delv    - Change of y-velocity at each wall
17   %
18   strikes = [0 0];  delv = [0 0];
19   direction = [1 -1];   % Direction of particle leaving wall
20   x_old = x;            % Remember original position
21   % Update x position of particle
22   x(:) = x_old(:) + v(:,1)*tau;
23   for i=1:npart
24     if( x(i) <= xwall(1) )
25       flag=1; % Particle strikes left wall
26     elseif( x(i) >= xwall(2) )
27       flag=2; % Particle strikes right wall
28     else
29      flag=0; % Particle strikes neither wall
30     end
31     if( flag > 0 )
32       strikes(flag) = strikes(flag) + 1;
33       delv(flag) = delv(flag) - v(i,2);
34       % Reset velocities as biased Maxwellian
35       v(i,1) = direction(flag)*sqrt(-log(1-rand(1))) * mpv;
36       phi = 2*pi*rand(1);
37       v_parll = sqrt(-log(1-rand(1))) * mpv;
38       v(i,2) = v_parll*sin(phi) + vwall(flag);
39       v(i,3) = v_parll*cos(phi);
40       % Time of flight after leaving wall
41       dtr = tau*(x(i)-xwall(flag))/(x(i)-x_old(i));
42       % Position after leaving wall
43       x(i) = xwall(flag) + v(i,1)*dtr;
44       delv(flag) = delv(flag) + v(i,2);
45     end
46   end
```

LISTING 10.6 Function `sampler`. Used by the `dsmcne` program to sample the number density, fluid velocity, and temperature in the cells.

```
1    function [ave_n,ave_u,ave_T,nsamp]= . . .
2      sampler(x,v,npart,ncell,L,ave_n,ave_u,ave_T,nsamp)
3    % sampler - function to sample density and velocities
4    % Inputs
5    % x      - Particle positions
6    % v      - Particle velocities
7    % npart - Number of particles
8    % ncell - Number of cells
9    % L      - System size
10   % ave_n - Accumulated ave. number of particles in a cell
11   % ave_u - Accumulated average fluid velocity in a cell
12   % ave_T - Accumulated average temperature in a cell
13   % nsamp - Number of samples taken
14   % Outputs
15   % ave_n, ave_u, ave_T, nsamp - Updated values
16   %
17   jx=ceil(ncell*x/L);
18   sum_n = zeros(ncell,1);
19   sum_v = zeros(ncell,3);
20   sum_v2 = zeros(ncell,1);
21   for ipart=1:npart
22     jcell = jx(ipart);    % Particle ipart is in cell jcell
23     sum_n(jcell) = sum_n(jcell)+1;
24     sum_v(jcell,:) = sum_v(jcell,:) + v(ipart,:);
25     sum_v2(jcell) = sum_v2(jcell) + . . .
26                  v(ipart,1)^2 + v(ipart,2)^2 + v(ipart,3)^2;
27   end
28   ave_n = ave_n + sum_n;
29   for i=1:3
30     sum_v(:,i) = sum_v(:,i)./sum_n;
31   end
32   ave_u = ave_u + sum_v;
33   ave_T = ave_T + sum_v2./sum_n - . . .
34                (sum_v(:,1).^2 + sum_v(:,2).^2 + sum_v(:,3).^2);
35   nsamp = nsamp + 1;
36   return;
```

After an initial relaxation period, the `dsmcne` program calls the routine `sampler` (Listing 10.6) to sample the cells. This routine measures the instantaneous number density, fluid velocity, and temperature in each cell. These are added to the running sums `ave_n`, `ave_u` and `ave_T`. In `dsmcne`, these averages are normalized and plotted. Notice that the instantaneous thermal kinetic energy in cell j is defined as

$$e_j(t) = \sum_{i \in j} \frac{1}{2} m |\mathbf{v}_i - \mathbf{u}_j|^2 \tag{10-67}$$

where the sum runs over all the particles in the cell and $\mathbf{u}_j$ is the instantaneous fluid velocity in cell j. Since the fluid is moving, we have to remove the center of mass kinetic energy when computing the thermal kinetic energy.

The total y-momentum flux, F_{p_y}, may be measured from the change in momentum of particles that reflect off the walls. From the momentum–impulse theorem, this flux is related to the time-average drag force on a wall as

$$\langle f_{\mathrm{drag}} \rangle = -\frac{1}{t} \sum m \, \Delta v_y = -\langle F_{p_y} \rangle \tag{10-68}$$

where Δv_y is the change in y-velocity for a particle striking the wall. The sum is over all particle collision with the wall over a time t. The change in velocity is measured by the routine mover. The average viscosity is then

$$\eta = \frac{\langle f_{\mathrm{drag}} \rangle}{du_y/dx} \tag{10-69}$$

Finally, the dsmcne program computes a viscosity at each wall and averages the two values.

Chapman–Enskog theory predicts a viscosity for argon gas of $\eta = 2.08 \times 10^{-5}$ N · s/m². For a short run (1000 steps) using 300 particles and a wall speed of $u_w = 0.3$ Ma, the dsmcne program obtains the estimate of $\eta = 1.81 \pm 1.09 \times 10^{-5}$ N · s/m²; the answer is reasonable but the error bar is unacceptably large. However, doing a longer run (10,000 steps) with a larger system (2000 particles), we get the more satisfying value of $\eta = 2.09 \pm 0.14 \times 10^{-5}$ N · s/m². You should come to the conclusion that Monte Carlo simulations, by their statistical nature, often require long runs to accumulate enough samples.

EXERCISES

23. In a dilute hard sphere gas, the thermal diffusivity varies with temperature as $\kappa(T) = \kappa_0 \sqrt{T}$ where κ_0 is a constant. Solve the diffusion equation,

$$\frac{\partial T}{\partial t} = \frac{\partial}{\partial x} \kappa \frac{\partial}{\partial x} T(x,t)$$

and find the steady state temperature profile in a one-dimensional system with boundary conditions $T(x = 0) = T_a$ and $T(x = L) = T_b$. Plot your solution for $T_a = 300K$, $T_b = 400K$ and compare with the linear profile obtained when we assume κ is a constant.

24. Modify dsmcne so that particles are labeled "black" or "white." Particles that reflect off a wall are turned black with probability q and white with probability $(1-q)$. Set the walls at equal temperature and make them stationary; give them different values of q to set up a pigmentation gradient across the system (see Figure 10.10). Measure the average pigment flux and compute the self-diffusion coefficient using (10-50). Compare your results with

$$ D = \frac{6\pi}{32} \langle v \rangle \lambda $$

the value given by Chapman–Enskog theory. Be sure to run your system for at least 10,000 time steps.

25. Modify dsmcne to measure energy flux, F_e, in a system with a temperature gradient. Set up your simulation with stationary walls at different temperatures. Compute thermal conductivity, α, and compare your results with, $\alpha = 15k/4\, m\, \eta$, the value given by Chapman-Enskog theory. Be sure to run your system for at least 10,000 time steps.

26. For the small systems simulated by the dsmcne program, our definition of temperature does not exactly reproduce the correct thermodynamic temperature.
    (a) Do several equilibrium runs and show that the time average of the instantaneous temperature is

    $$ \langle T \rangle = \frac{\langle N_c \rangle}{\langle N_c \rangle + 1} T_w $$

    where $\langle N_c \rangle$ is the average number of particles in a cell.
    (b) When $\langle N_c \rangle$ is very small ($\langle N_c \rangle < 10$), another problem arises. Explain what causes it and how to avoid it.

27. In Couette flow, the velocity of a fluid near a wall does not exactly equal the velocity of the wall; this phenomenon is known as *slip*. Do a variety of runs using dsmcne to show that

    $$ u_y^o = u_y(x)|_{wall} \approx u_w \pm \lambda \frac{du_y}{dx} $$

    where u is the velocity of the fluid and u_w is the velocity of the wall. Be sure your system is at least 10 mean free paths wide.

28. Consider the following simple one-dimensional flow problem: A constant acceleration, g, is applied to the particles in the y-direction. The walls are fixed at constant temperature and are stationary. This is called planar *Poiseuille flow*. Modify dsmcne to simulate planar Poiseuille flow and confirm that the velocity profile is

    $$ u_y(x) = \frac{\rho g}{2\eta} \left[\frac{L^2}{4} - (x - L/2)^2 \right] + u_y^o $$

    where u_y^o is the velocity of the fluid at the wall (see Exercise 10.27). By measuring the curvature of this parabolic profile determine the viscosity.

29. In Couette flow, the velocity gradient in the fluid produces viscous heating. The temperature profile is parabolic and given by

$$T(x) = \frac{\eta \gamma^2}{2\alpha} ((L/2)^2 - x^2) + T_w$$

where $\gamma = du_y/dx$ is the velocity gradient and T_w is the temperature of the fluid at the walls. Modify dsmcne and fit the temperature profile to the above quadratic. Compute the ratio of thermal conductivity to viscosity and compare your result with $\alpha/\eta = 15k/4m$, the value given by Chapman–Enskog theory.

BEYOND THIS CHAPTER

I picture my readers' eyebrows rising to the tops of their heads as they discover that this chapter does *not* cover such topics as the Ising model or quantum Monte Carlo. My excuse is that a proper coverage of stochastic methods really requires a full-length book. Many of the topics I have omitted are discussed by Gould and Tobochnik.[19] The Metropolis algorithm and its application to the Ising model is covered in depth by Binder[20] and by Heermann.[21] Several introductory articles on the various flavors of quantum Monte Carlo have recently appeared in *Computers in Physics*.[22] In the more general field of stochastic processes, Gardiner's book is an excellent introduction.[23]

We've seen two different ways to model a fluid: using partial differential equations and, on a more microscopic level, using particles. In general, the latter is computationally much more expensive. However, particle simulations thrive in certain "ecological niches." For example, the PDE description sometimes breaks down. Define the Knudsen number

$$\text{Kn} \equiv \frac{\lambda}{L} = \frac{\text{(mean free path)}}{\text{(characteristic length)}} \tag{10-70}$$

[19] H. Gould and J. Tobochnik, *An Introduction to Computer Simulation Methods*, part 2 (Reading, Mass.: Addison Wesley, 1988).

[20] *Monte Carlo Methods in Statistical Physics*, Topics Current Physics, vol. 7, ed. K. Binder (Berlin: Springer, 1979); *Applications of the Monte Carlo Method in Statistical Physics*, ed. K. Binder (Berlin: Springer, 1984).

[21] D. Heermann, *Computer Simulation Methods in Theoretical Physics* (Berlin: Springer, 1986).

[22] J. Tobochnik, H. Gould, and K. Mulder, "An Introduction to Quantum Monte Carlo," *Comput. Phys.*, **4**, 431–35 (1990); P. Reynolds, J. Tobochnik, and H. Gould, "Diffusion Monte Carlo," *Comput. Phys.*, **4**, 662–68 (1990); M. Lee and K. Schmidt, "Green's Function Monte Carlo," *Comput. Phys.*, **6**, 192-97 (1992); J. Tobochnik, G. Batrouni, and H. Gould, "Quantum Monte Carlo on a Lattice," *Comput. Phys.*, **6**, 673-80 (1992).

[23] C.W. Gardiner, *Handbook of Stochastic Methods for Physics, Chemistry and the Natural Sciences* (Berlin: Springer-Verlag, 1985).

The continuum description of a fluid begins to break down when $Kn > \frac{1}{10}$. Three important cases where this occur are (1) flow in narrow channels, such as the flow under the write head of a disk drive; (2) sharp fronts, such as shock waves; and (3) rarefied gas flows, such as high-altitude flight. The DSMC algorithm is ideally suited for these scenarios.

This chapter presents only very basic DSMC algorithms; there are many possible extensions and improvements. To model true gases more realistically, one can use a more sophisticated potential than hard spheres. One successful model is the variable hard sphere potential[24] for which the effective cross section of the particles is a function of their relative speed. The scattering angles, θ and ϕ, are still selected according to the hard sphere distribution [see (10-43)–(10-45)]. The DSMC method can also simulate chemistry by including an extra selection process at each collision. Particles react chemically when their relative kinetic energy surpasses the activation energy of the reaction.[25]

More recently, advances in massively parallel computer architectures have renewed interest in *lattice gases*. A lattice gas simulation is similar to molecular dynamics except that the particles are constrained to move on a lattice with only a small, discrete set of allowed velocities. For a recent review, see Doolen et al.[26] An even more promising variant is the lattice Boltzmann model.[27] In this model, the binary logic used in the lattice gas (i.e., a site either has a particle or it doesn't) is replaced with a continuous probability distribution (i.e., a floating point number).

APPENDIX 10A: FORTRAN LISTINGS

LISTING 10A.1 Program `dsmceq`. Simulates relaxation to equilibrium in a dilute gas using the DSMC algorithm. Uses `sorter` (Listing 10A.2), `colide` (Listing 10A.3), and `rand` (Listing 10A.7).

```
      program dsmceq
! Program to simulate a dilute gas using DSMC algorithm
! This version illustrates the approach to equilibrium
      parameter( npmax = 5000, ncmax = 200, nbmax = 11 )
      real x(npmax), v(npmax,3), vmag(npmax), vbin(nbmax)
```

[24] G.A. Bird, "Monte-Carlo Simulation in an Engineering Context," in *Rarefied Gas Dynamics: Technical Papers from the* 12th *International Symposium*, (New York: Prog. in Astro. and Aero. AIAA, 1980), p. 239.

[25] B. L. Haas and J.D. McDonald, "Validation of Chemistry Models Employed in a Particle Simulation Method," *J. Therm. and Heat Transfer*, **7**, 42–48 (1993).

[26] *Lattice Gas Methods for Partial Differential Equations*, ed. G.D. Doolen, U. Frisch, B. Hasslacher, S. Orszag, and S. Wolfram (Redwood City, Calif.: Addison-Wesley, 1990).

[27] G. McNamara and G. Zanetti, "Using the Boltzmann Equation to Simulate Lattice Gas Automata," *Phys. Rev. Lett.*, **61**, 2332–35 (1988).

```
real selxtra(ncmax), vrmax(ncmax)
integer Xref(npmax), index(ncmax), cell_n(ncmax)
integer col, coltot
real mass, L
integer*4 iseed
common /dsmc/ x,v,npart,ncell,Xref,index,cell_n

pi = 4.*atan(1.)        ! = 3.14159. . .
iseed = 1
write (*,*) 'Enter number of simulation particles - '
read (*,*) npart
boltz = 1.3806e-23      ! Boltzmann's constant
mass = 6.63e-26         ! Mass of argon atom
diam = 3.66e-10         ! Effective diameter of argon atom
T = 273.                ! Temperature (K)
density = 1.78          ! Mass density of argon at STP
L = 1.e-6               ! System size is one micron
eff_num = density/mass*L**3/npart
write (*,*) 'Each simulation particle represents ',
&                                  eff_num,' atoms'
 v_init = sqrt(3*boltz*T/mass)     ! Initial speed of
                                   ! particles

do i=1,npart
  x(i) = L*rand(iseed)   ! Assign random position
  ! Assign x velocity as +/- v_init
  v(i,1) = v_init * int(1 - 2*int(2*rand(iseed)))
  v(i,2) = 0.    ! Initial y and z velocities
  v(i,3) = 0.    ! are zero
end do

ncell = 15              ! Number of cells
tau = 0.2*(L/ncell)/v_init   ! Set time step tau
do i=1,ncell
  vrmax(i) = 3*v_init ! Used by collision routine
  selxtra(i) = 0       ! Used by collision routine
end do
coeff = 0.5*eff_num*pi*diam**2*tau/(L**3/ncell)
do i=1,nbmax
  vbin(i) = 50. * float(2*i-1)   ! Bins for velocity
end do
coltot = 0.             ! Count total collisions

write (*,*) 'Enter total number of time steps'
read (*,*) nstep
do istep=1,nstep
  do i=1,npart
```

```
            x(i) = x(i) + v(i,1)*tau      ! Update x position of
                                          ! particle
            x(i) = amod(x(i)+L,L)         ! Periodic boundary
                                          ! conditions
         end do
         call sorter(L)                   ! Sort particles into cells
         call colide(vrmax,tau,selxtra,coeff,iseed,col)
         coltot = coltot + col     ! Count total collisions
         write (*,*) 'Done ',istep,' of ',nstep,' steps; ',
     &                       coltot,' collisions'
       end do
       do i=1,npart  ! Compute magnitude of particle velocities
          vmag(i) = sqrt(v(i,1)**2 + v(i,2)**2 + v(i,3)**2)
       end do
! Print out the plotting variables -
!    vmag,vbin
!
       open(11,file='vmag.dat')
       open(12,file='vbin.dat')
       do i=1,npart
         write (11,1001) vmag(i)
       end do
       do i=1,nbmax
         write (12,1001) vbin(i)
       end do
       stop
1001   format(e12.6)
       end
```

LISTING 10A.2 Subroutine `sorter`. Produces sorted lists used by `colide` to select random particles from a cell.

```
       subroutine sorter(L)
! Routine to sort particles into cells
       parameter( npmax = 5000, ncmax = 200, nbmax = 11 )
       real x(npmax), v(npmax,3)
       integer Xref(npmax), index(ncmax), cell_n(ncmax)
       real L
       integer jx(npmax), temp(ncmax)
       common /dsmc/ x,v,npart,ncell,Xref,index,cell_n

       do jcell=1,ncell
         cell_n(jcell) = 0
         temp(jcell) = 0
       end do
```

```
do ipart=1,npart
  jx(ipart) = min0(int(ncell*x(ipart)/L)+1,ncell)
  cell_n(jx(ipart)) = cell_n(jx(ipart)) + 1
end do
m=1
do jcell=1,ncell
  index(jcell) = m
  m = m + cell_n(jcell)
end do
do ipart=1,npart
  jcell = jx(ipart)
  temp(jcell)=temp(jcell)+1
  k = index(jcell) + temp(jcell) - 1
  Xref(k) = ipart
end do
return
end
```

LISTING 10A.3   Subroutine colide. Called by dsmceq and dsmcne to evaluate collisions using the DSMC algorithm. Uses rand (Listing 10A.7).

```
      subroutine colide(vrmax,tau,selxtra,coeff,iseed,col)
! Program to simulate a dilute gas using DSMC algorithm
! This version illustrates the approach to equilibrium
      parameter( npmax = 5000, ncmax = 200, nbmax = 11 )
      real x(npmax), v(npmax,3)
      real selxtra(ncmax), vrmax(ncmax)
      integer Xref(npmax), index(ncmax), cell_n(ncmax)
      integer col
      real vcm(3),vrel(3)
      common /dsmc/ x,v,npart,ncell,Xref,index,cell_n
      data pi/3.141592654/

      col = 0  ! Count the total number of collisions
      do jcell=1,ncell
        number = cell_n(jcell)
        if( number .gt. 1 ) then
! Determine how many candidate collision pairs to be
! selected
          select = coeff*number**2*vrmax(jcell) + selxtra(jcell)
          nsel = int(select)        ! Number of pairs to be
                                    ! selected
          selxtra(jcell) = select-nsel ! Carry over fraction
          crm = vrmax(jcell)
          do isel=1,nsel
! Pick two particles at random out of this cell
```

```
          k = int(rand(iseed)*number)
          kk = mod( int(k + rand(iseed)*(number-1)) + 1,
    &          number)
          ip1 = Xref(k+index(jcell))
          ip2 = Xref(kk+index(jcell))
          cr = 0.
          do k=1,3
            cr = cr + (v(ip1,k)-v(ip2,k))**2
          end do
          cr = sqrt(cr)              ! Relative velocity
          if( cr .gt. crm ) then
            crm = cr                 ! Max. relative velocity
          end if
          if( cr/vrmax(jcell) .gt. rand(iseed) ) then
            ! Collision occurs
            col = col + 1
            do k=1,3
              vcm(k) = 0.5*(v(ip1,k) + v(ip2,k)) !C.o.m. vel.
            end do
            cos_th = 1. - 2.*rand(iseed)       ! Cosine(theta)
            sin_th = sqrt(1. - cos_th**2)      ! Sine(theta)
            phi = 2*pi*rand(iseed)
            vrel(1) = cr*cos_th                ! Relative velocity
            vrel(2) = cr*sin_th*cos(phi)       ! after collision
            vrel(3) = cr*sin_th*sin(phi)
            do k=1,3
              v(ip1,k) = vcm(k) + 0.5*vrel(k)  ! Post-collision
              v(ip2,k) = vcm(k) - 0.5*vrel(k)  ! velocities
            end do
          end if
        end do
        vrmax(jcell) = crm  ! Reset maximum relative velocity
      end if
    end do
    return
    end
```

LISTING 10A.4 Program dsmcne. Measures viscosity in a dilute gas using the DSMC algorithm. Uses sorter (Listing 10A.2), colide (Listing 10A.3), mover (Listing 10A.5), sampler (Listing 10A.6), and rand (Listing 10A.7).

```
      program dsmcne
! Program to simulate a dilute gas using DSMC algorithm
! This version measures the viscosity in a dilute gas
      parameter( npmax = 5000, ncmax = 200, nbmax = 11 )
      real x(npmax), v(npmax,3)
```

```
   real selxtra(ncmax), vrmax(ncmax)
   integer Xref(npmax), index(ncmax), cell_n(ncmax), col
   real mass, L, mpv, mfp, xwall(2), vwall(2)
   real ave_n(ncmax), ave_u(ncmax,3), ave_T(ncmax),
&xcell(ncmax)
   real strikes(2), delv(2)
   real dvtot(2), dverr(2), force(2), ferr(2)
   integer*4 iseed
   common /dsmc/ x,v,npart,ncell,Xref,index,cell_n

   pi = 4.*atan(1.)  ! = 3.14159. . .
   iseed = 1
   write (*,*) 'Enter number of simulation particles -
   read (*,*) npart
   boltz = 1.3806e-23    ! Boltzmann's constant
   mass = 6.63e-26       ! Mass of argon atom
   diam = 3.66e-10       ! Effective diameter of argon atom
   T = 273.              ! Initial temperature (K)
   density = 1.78        ! Mass density of argon at STP
   L = 1.e-6             ! System size is one micron
   Volume = L**3         ! Volume of the system
   xwall(1) = 0.         ! Position of left wall
   xwall(2) = L          ! Position of right wall
   eff_num = (density/mass)*Volume/npart
   write (*,*) 'Each simulation particle represents ',
&                                  eff_num,' atoms'
   mfp = 1./(sqrt(2.)*pi*diam**2*density/mass) ! Mean free
                                               ! path
   write (*,*) 'System width is ',L/mfp,' mean free paths'
   mpv = sqrt(2*boltz*T/mass)
   write (*,*) 'Enter wall velocity as Mach number'
   read (*,*) vwall_m
   vwall(2) = vwall_m * sqrt(5./3. * boltz*T/mass)
   vwall(1) = -vwall(2)
   write (*,*) 'Wall velocity is ',vwall(2),' m/s'
   do i=1,npart
     x(i) = L*rand(iseed)   ! Assign random position
       do j=1,3
         temp = sqrt(-alog(1.-rand(iseed))) *
&                                  cos(2*pi*rand(iseed))
         v(i,j) = mpv*temp
       end do
     v(i,2) = v(i,2) + vwall(1)+(vwall(2)-vwall(1))*x(i)/L
   end do
   ncell = 20                ! Number of cells
   tau = 0.2*(L/ncell)/mpv   ! Set time step tau
   do i=1,ncell
```

```fortran
        vrmax(i) = 3*mpv     ! Used by collision routine
        selxtra(i) = 0        ! Used by collision routine
      end do
      coeff = 0.5*eff_num*pi*diam**2*tau/(Volume/ncell)

      do jcell=1,ncell
        ave_n(jcell) = 0.        ! Average number density
        do k=1,3
          ave_u(jcell,k) = 0.   ! Average velocity in a cell
        end do
        ave_T(jcell) = 0.        ! Average velocity square
      end do
      do j=1,2
        dvtot(j) = 0.    ! Total momentum change at wall
        dverr(j) = 0.    ! Used to estimate error in dvtot
      end do
      nsamp = 0.              ! Total number of samples
      tsamp = 0.              ! Sample time

      write (*,*) 'Enter total number of time steps'
      read (*,*) nstep
      do istep=1,nstep
        call mover(xwall,mpv,vwall,tau,strikes,delv,iseed)
        call sorter(L)                   ! Sort particles into cells
        call colide(vrmax,tau,selxtra,coeff,iseed,col)
        if( mod(istep,nstep/20) .lt. 1 ) then
          write (*,*) 'Done ',istep,' of ',nstep,' steps; ',
     &                      col,' collisions'
        end if
          write (*,*) 'Strikes on left wall = ',strikes(1),
     &                ', right wall = ',strikes(2)
        if( istep .gt. nstep/10 ) then   ! Start taking samples
          call sampler(L,ave_n,ave_u,ave_T,nsamp)
          do j=1,2
            dvtot(j) = dvtot(j) + delv(j)
            dverr(j) = dverr(j) + delv(j)**2
          end do
          tsamp = tsamp + tau
        end if
      end do

! Normalize statistics
      do i=1,ncell
        ave_n(i) = ave_n(i)/nsamp
        do j=1,3
          ave_u(i,j) = ave_u(i,j)/nsamp
        end do
```

```
          ave_T(i) = mass/(3*boltz) * ave_T(i)/nsamp
          xcell(i) = (i-0.5)*L/ncell
       end do
       do j=1,2
          dverr(j) = dverr(j)/(nsamp-1) - (dvtot(j)/nsamp)**2
          dverr(j) = sqrt(dverr(j)*float(nsamp))
          force(j) = (eff_num*dvtot(j)*mass)/tsamp /L**2
          ferr(j) = (eff_num*dverr(j)*mass)/tsamp /L**2
       end do
       write (*,*) 'Force per unit area at each wall is '
       write (*,*) 'Left wall:  ',force(1),' +/- ',ferr(1)
       write (*,*) 'Right wall: ',force(2),' +/- ',ferr(2)
       vgrad = (vwall(2)-vwall(1))/L
       visc = 1/2.*( -force(1)+force(2)  )/vgrad
       viscerr = 1/2.*( ferr(1)+ferr(2)  )/vgrad
       write (*,*) 'Viscosity = ',visc,' +/- ',viscerr
       write (*,*) 'Theoretical value of viscosity is'
       write (*,*) 5.*pi/32.*density*(2./sqrt(pi)*mpv)*mfp
! Print out the plotting variables -
!   xcell,ave_n,ave_u,ave_T
!
       open(11,file='ave_n.dat')
       open(12,file='ave_u.dat')
       open(13,file='ave_T.dat')
       open(14,file='xcell.dat')
       do i=1,ncell
          write (11,1001) ave_n(i)
          write (12,1002) ave_u(i,1),ave_u(i,2),ave_u(i,3)
          write (13,1001) ave_T(i)
          write (14,1001) xcell(i)
       end do
       stop
1001   format(e12.6)
1002   format(3(e12.6,2x))
       end
```

LISTING 10A.5   Function mover. Used by the dsmcne program to update particle positions. It also evaluates collisions between the particles and the thermal walls. Uses rand (Listing 10A.7).

```
       subroutine mover(xwall,mpv,vwall,tau,strikes,delv,iseed)
       ! Subroutine to move particles for DSMC code
       parameter( npmax = 5000, ncmax = 200, nbmax = 11 )
       real x(npmax), v(npmax,3), xwall(2), vwall(2)
       real mpv, delv(2), strikes(2), direction(2)
       integer Xref(npmax), index(ncmax), cell_n(ncmax), flag
```

```fortran
      integer*4 iseed
      static twopi,direction
      common /dsmc/ x,v,npart,ncell,Xref,index,cell_n
      data twopi,direction/6.283185307,1.,-1./

      do i=1,2
        strikes(i) = 0.    ! Strikes counts # of impacts for each
                           ! wall
        delv(i) = 0.       ! Delv is change in velocity after
                           ! impact
      end do
      do i=1,npart
        x_old = x(i)                    ! Remember original position
        x(i) = x_old + v(i,1)*tau       ! Update x position of
                                        ! particle
        if( x(i) .le. xwall(1) ) then
          flag = 1    ! Particle strikes left wall
        else if( x(i) .ge. xwall(2) ) then
          flag = 2    ! Particle strikes right wall
        else
          flag = 0    ! Particle doesn't strike either wall
        end if
        if( flag .gt. 0 ) then
          strikes(flag) = strikes(flag) + 1
          delv(flag) = delv(flag) - v(i,2) !Record change in
                                           !velocity
          ! Reset velocities as biased Maxwellian
          v(i,1)=direction(flag)*sqrt(-alog(1-rand(iseed)))*mpv
          phi = twopi*rand(iseed)
          v_parll = sqrt(-alog(1.-rand(iseed))) * mpv
          v(i,2) = v_parll*sin(phi) + vwall(flag)
          v(i,3) = v_parll*cos(phi)
          ! Dtr = time of flight after leaving wall
          dtr = tau*(x(i)-xwall(flag))/(x(i)-x_old)
          x(i) = xwall(flag) + v(i,1)*dtr !Position after hitting
                                          !wall
          delv(flag) = delv(flag) + v(i,2) !Record change in
                                           !velocity
        end if
        if( x(i) .lt. 0 .or. x(i) .gt. xwall(2) ) then
          write (*,*) 'Error - particle out of system'
        end if
      end do
      return
      end
```

LISTING 10A.6 Function `sampler`. Used by the dsmcne program to sample the number density, fluid velocity, and temperature in the cells.

```fortran
      subroutine sampler(L,ave_n,ave_u,ave_T,nsamp)
! sampler - function to sample density and velocities
      parameter( npmax = 5000, ncmax = 200, nbmax = 11 )
      real x(npmax), v(npmax,3)
      real L
      real ave_n(ncmax), ave_u(ncmax,3), ave_T(ncmax)
      real sum_n(ncmax), sum_v(ncmax,3), sum_v2(ncmax)
      integer Xref(npmax), index(ncmax), cell_n(ncmax)
      common /dsmc/ x,v,npart,ncell,Xref,index,cell_n

      do i=1,ncell                ! Zero the sums used to accumulate
        sum_n(i) = 0.             ! statistics on density, velocity,
        do j=1,3                  ! and energy
          sum_v(i,j) = 0.
        end do
        sum_v2(i) = 0.
      end do
      do ipart=1,npart
        ! Particle 'ipart' is located in cell 'jcell'
        jcell=min0( int(ncell*x(ipart)/L)+1, ncell)
        sum_n(jcell) = sum_n(jcell)+1
        do k=1,3
          sum_v(jcell,k) = sum_v(jcell,k) + v(ipart,k)
        end do
        sum_v2(jcell) = sum_v2(jcell) +
     &          v(ipart,1)**2 + v(ipart,2)**2 + v(ipart,3)**2
      end do
      ! Accumulate running sums of density, u and T for cells
      do i=1,ncell
        ave_n(i) = ave_n(i) + sum_n(i)
        do j=1,3
          ave_u(i,j) = ave_u(i,j) + sum_v(i,j)/sum_n(i)
        end do
        ave_T(i) = ave_T(i) + sum_v2(i)/sum_n(i)
     &     - (sum_v(i,1)**2 + sum_v(i,2)**2
     &     +sum_v(i,3)**2)/sum_n(i)**2
      end do
      nsamp = nsamp + 1  ! Increment number of samples
      return
      end
```

LISTING 10A.7 Function rand. Returns random numbers uniformly distributed in [0,1) (i.e., uniform deviates).

```
function rand(seed)
integer*4 seed
static a,m
real*8 a,m,temp
data a,m/16807.0, 2147483647.0/

temp = a*seed
seed = dmod(temp,m)
rand = seed/m

return
end
```

LISTING 10A.8 Function randn. Returns normal (Gaussian) distributed random numbers with zero mean and unit variance. Uses rand (Listing 10A.7).

```
      function randn(seed)
! Random number generator for normal (Gaussian) distribution
      integer*4 seed

      randn = sqrt(-2*alog(1.-rand(seed)))
     &              * cos(6.283185307 * rand(seed))
      return
      end
```

Selected Solutions

Chapter 1

11.

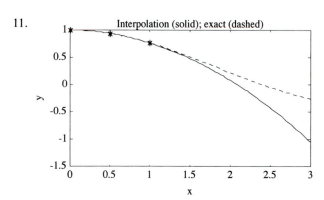

16.

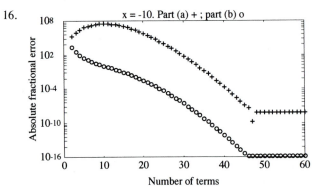

17. 1 s = 2.42×10^{14} h/eV.

23.

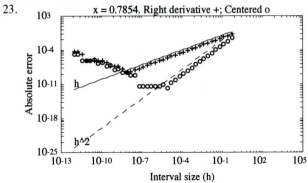

Chapter 2

3.

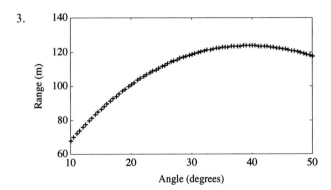

6.

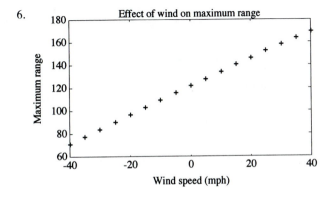

10.

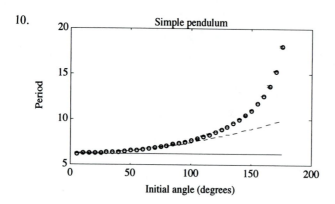

11.

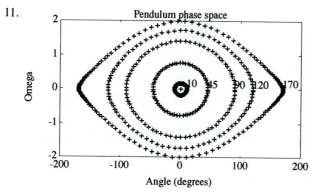

15. For $a_o = 100g$, $T_d = 0.2$, $g/L = 1$, and an initial angle of $\theta_o = 170°$,

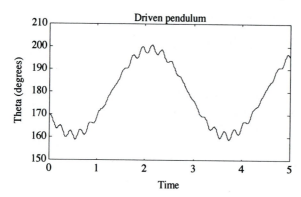

Chapter 3

7. For $\mathbf{v} = [0\ 1\ 0]$, $\mathbf{E} = [0\ 1\ 0]$, $\mathbf{B} = [0\ 0\ 1]$, and $\tau = 10^{-13}$,

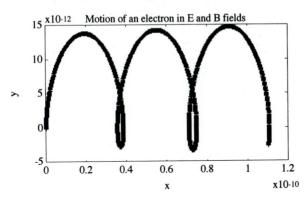

11. For $r_o = 1$ AU, $v_o = 2\pi$, $\tau = 0.02$ yr,

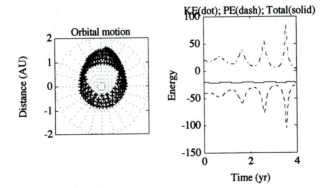

12. For $z = 0.1$ m, $\theta = 0°$, and $\tau = 0.1$ s,

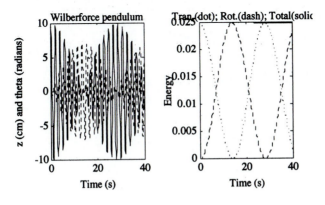

14. The drag force is $\mathbf{F}(\mathbf{v}) = D|\mathbf{v}|\mathbf{v}$. Using an initial radial distance of 1 AU, an initial velocity of 2π AU/yr, and a drag parameter of $D = 0.1$,

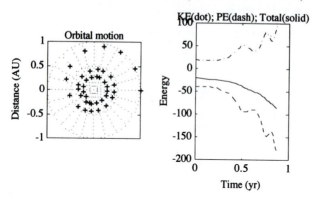

15.

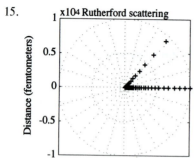

22.

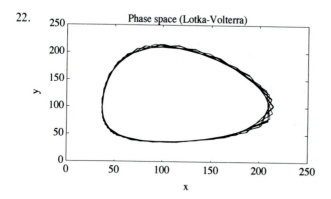

24.

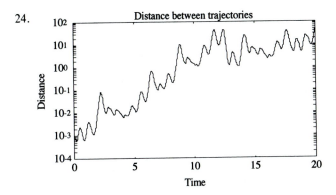

Chapter 4

4. For $[x \quad y] = [1 \quad 1]$ and $[A \quad B] = [1 \quad 3]$,

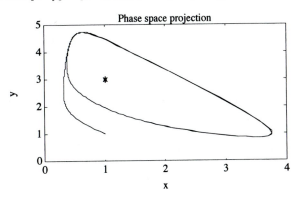

6.

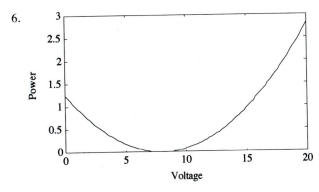

8.

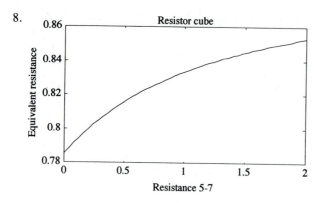

13.

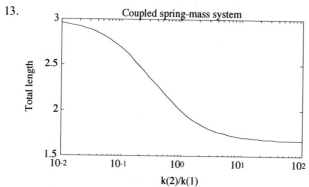

16.

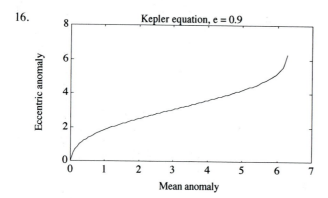

18. $\alpha = 0.2014$.

23. $[x \quad y] = [0.06245 \quad 0.1265]$.

Chapter 5

3. $\alpha = \dfrac{\sum xy}{\sum x^2}$ and $\sigma_\alpha^2 = \dfrac{1}{\sum x^2}$.

6. For a time step of $\tau = 0.1$, we obtain the results shown below. The fit is rather good; $\chi^2 \approx 4.91$ for 30 data points. The fit is given by $T(\theta) = 6.295 \pm 0.019 - 0.0896 \pm 0.0498\ \theta + 0.5020 \pm 0.0283\ \theta^2$.

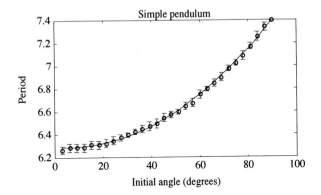

13.

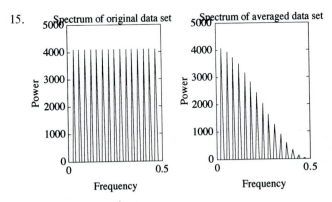

15.

18. (a) $\omega^2 = \frac{1}{2}(\omega_\theta^2 + \omega_z^2) \pm \frac{1}{2}\sqrt{(\omega_\theta^2 - \omega_z^2)^2 + \varepsilon^2/mI}$.

19. (a) $\omega = 0,\ \sqrt{2}\omega_o,\ 2\omega_o$ where $\omega_o = \sqrt{k/m}$.

21. For 1024 time steps with $\tau = 0.1$ and $\theta_o = 170°$,

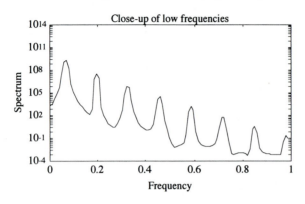

Close-up of low frequencies

Chapter 6

3.

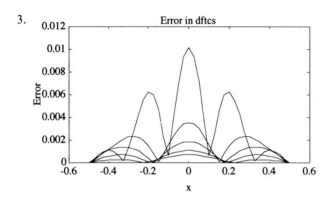

Error in dftcs

5. For $N = 41$ grid points and $\tau = 2 \times 10^{-4}$,

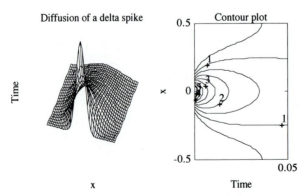

Diffusion of a delta spike

Contour plot

9. (b) For $\tau = 0.015$,

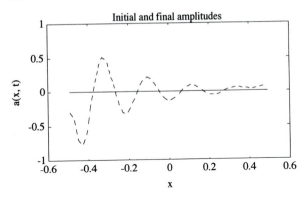

12. For $N = 50$, $\tau = 0.015$,

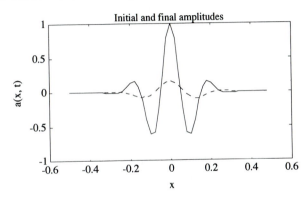

15.

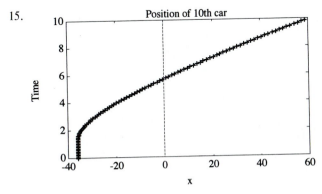

Chapter 7

3. $\Phi(x, y, z) = \displaystyle\sum_{n,m} c_{n,m} \sin\left(\frac{n\pi x}{L}\right) \sin\left(\frac{m\pi y}{L}\right) \cosh\left[\frac{\sqrt{n^2 + m^2}\,\pi(z - L/2)}{L}\right]$

where $c_{n,m} = \begin{cases} \dfrac{16\Phi_o}{nm\pi^2} \operatorname{sech}(\sqrt{n^2 + m^2}\,\pi/2) & m \text{ and } n \text{ odd} \\[2mm] 0 & \text{otherwise} \end{cases}$

8.

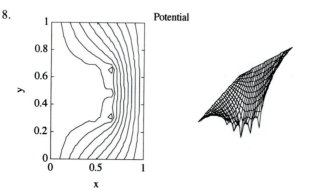

Chapter 8

2. $|\xi|^2 = 1 + \left\{-\left(\dfrac{c\tau}{h}\right)^2 + \left(\dfrac{c\tau}{h}\right)^4\right\}(\cos(kh) - 1)^2.$

9.

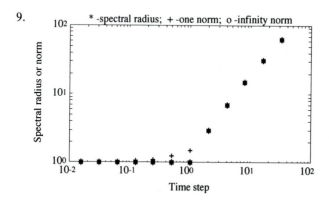

16.

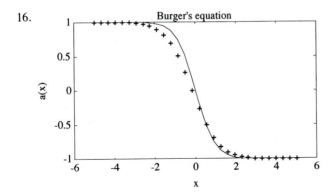

Chapter 9

10.

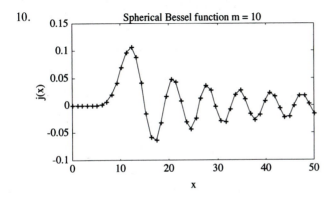

11.

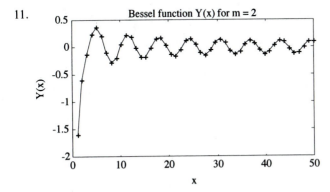

14.

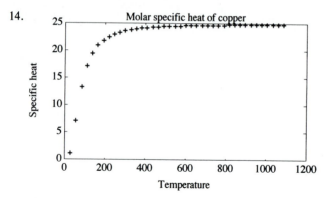

22. $\int_0^\infty f(x)e^{-x}\,dx \cong (\frac{1}{2} + \frac{1}{4}\sqrt{2})f(2 - \sqrt{2}) + (\frac{1}{2} - \frac{1}{4}\sqrt{2})f(2 + \sqrt{2}).$

28. (a) $I = \dfrac{M}{6}(L^2 + W^2).$

Chapter 10

4.

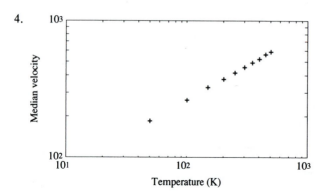

6. (a) $\langle v \rangle = \sqrt{\dfrac{\pi kT}{2m}}$;   (c) $\langle |\mathbf{v}_1 - \mathbf{v}_2| \rangle = \sqrt{\dfrac{\pi kT}{m}}.$

12. $a\sqrt{-\ln(1 - \mathcal{R})}.$

16. (c) After 10,000 trials,

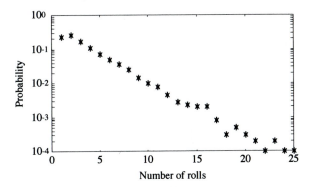

19. For 300 particles,

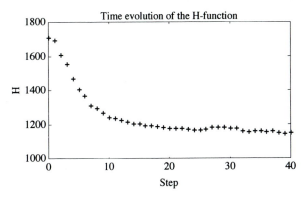

22. For 300 particles and 100 time steps,

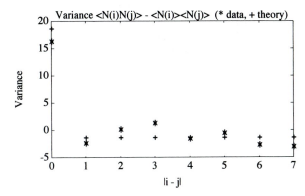

Index

Garcia

The MATH WORKS Inc.

BUSINESS REPLY MAIL

FIRST CLASS   PERMIT NO. 82   NATICK, MA

POSTAGE WILL BE PAID BY ADDRESSEE

THE MATHWORKS, INC.
24 Prime Park Way
Natick, MA 01760-9889

NO POSTAGE
NECESSARY IF
MAILED IN THE
UNITED STATES

DATE DUE

NOV 2 6			
DEC 5 2018			

DEMCO 38-297